Mathematics in India

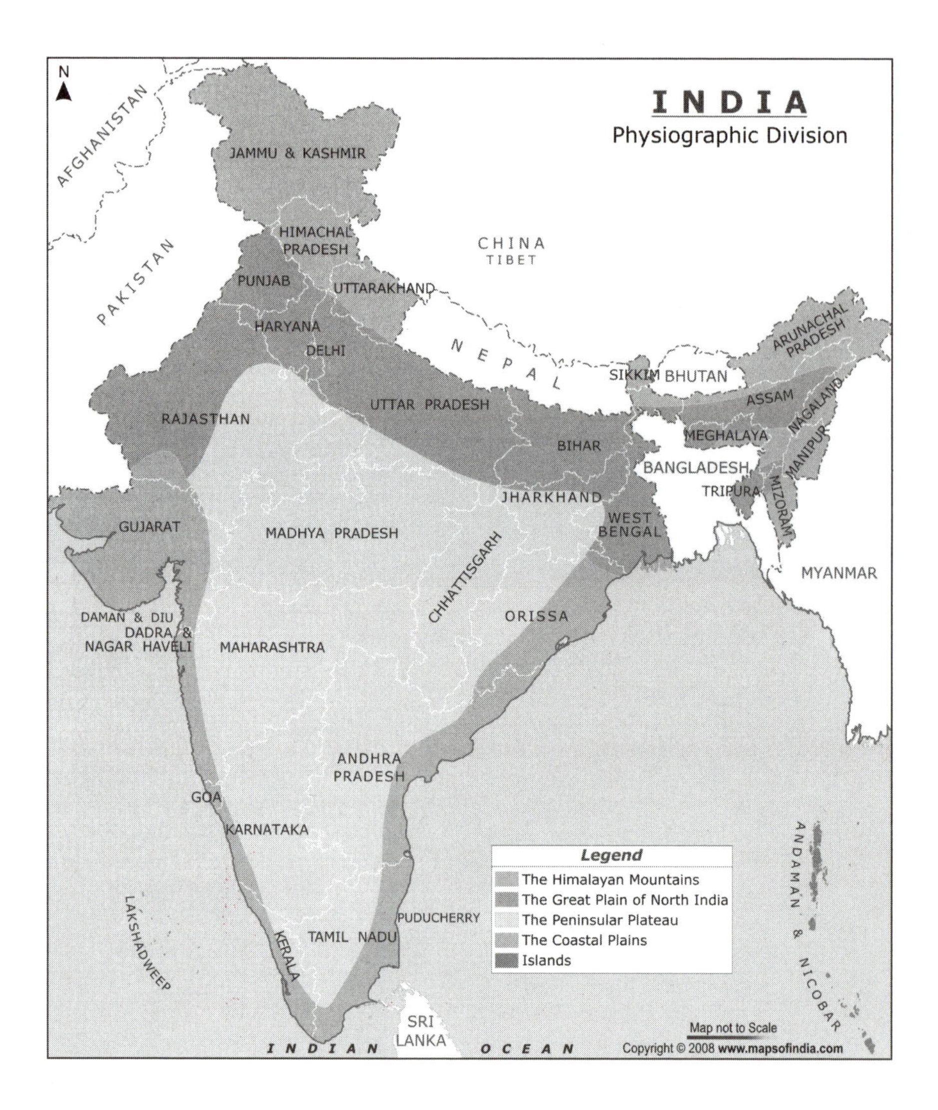

Geographical regions and modern states of India. *Source: mapsofindia.com*

Mathematics in India

Kim Plofker

PRINCETON UNIVERSITY PRESS
PRINCETON AND OXFORD

Published by Princeton University Press,
41 William Street, Princeton, New Jersey 08540

In the United Kingdom: Princeton University Press,
6 Oxford Street, Woodstock, Oxfordshire OX20 1TW

Library of Congress Cataloging-in-Publication Data

Plofker, Kim, 1963–
Mathematics in India / Kim Plofker.
p. cm.
Includes bibliographical references and index.
ISBN 978-0-691-12067-6 (hardcover : alk. paper)
1. Mathematics—India—History. 2. Mathematics—India—bibliography. I. Title.
QA27.I4P56 2009
510.954—dc22
2008028186

British Library Cataloging-in-Publication Data is available

The publisher would like to acknowledge the author of this volume
for providing the camera-ready copy from which this book was printed.

This book has been composed in LaTeX.

Printed on acid-free paper. ∞

press.princeton.edu

Printed in the United States of America

1 3 5 7 9 10 8 6 4 2

Contents

Preface

"Why is it so hard to find information about Indian math?" Many researchers in the history of Indian mathematics have heard (or asked) this plaintive question at one time or another. Usually it's posed by a frustrated non-Indologist colleague engaged in some attempt to integrate the Indian tradition into the history of mathematical sciences elsewhere in the world: for example, teaching a general history of math course or writing a general history of a mathematical topic.

There's no denying that the Indian tradition presents some unique challenges for anyone interested in the history of mathematics. It's not that information about the subject isn't available, but it's frequently difficult to separate reliable information from speculation or invention, or to extract from it a coherent and consistent overview of the historical development of Indian mathematical sciences. What other branch of history of math can show, for example, a pair of articles by two widely published researchers, appearing side by side in the same edited volume, whose estimates of their subject's approximate date of origin differ by as much as two thousand years?[1]

As I explain in more detail in the following chapters, these difficulties are due in large part to the uncertainty of early Indian chronology, the absence of historical or biographical data in many Indian technical works, and the ways that Sanskrit literature deals with authority, intertextuality, and tradition. There are many missing links in the chain of historical fact tracing out the development of Indian mathematical sciences; some of these links will someday be uncovered by new research, while others may remain forever conjectural.

This does not mean that we can't construct a reasonable narrative for the history of Indian mathematics based on the available data and plausible inferences. The narrative currently accepted by most mainstream historians as consistent with the textual record, linguistic and archaeological evidence, and the known history of other mathematical traditions goes more or less like this: The earliest urban Indian cultures, centered in the river valleys in the northwest of the South Asian subcontinent around the third millennium BCE, have left no clear record of their mathematical knowledge, although we can infer from the complexity of their infrastructure and global trade that this knowledge must have been substantial. From the second mil-

[1] Compare [Oha2000], p. 342, and [Kak2000a], p. 328; the former follows the mainstream opinion placing the emergence of quantitative astronomy in Vedic India in the second millennium BCE, while the latter suggests it goes back as far as the fourth millennium.

lennium BCE onward, the northwestern region (and eventually the entire subcontinent) was dominated by Indo-European cultures whose language was an early form of Sanskrit. Their earliest surviving texts mostly reflect basic mathematical knowledge supporting a simple ritual calendar and the economy of a pastoral society. In the first millennium BCE, Sanskrit texts began to show more sophisticated techniques in geometry for religious ritual and in the computations of mathematical astronomy; the latter subject may have been influenced by knowledge of Mesopotamian astronomy transmitted from the Achaemenid empire. Mathematical methods for commerce and other purposes continued to develop in India through the start of the current era, and a mature decimal place value arithmetic was established well before the middle of the first millennium CE. Spurred by interest in the astrological doctrines learned from Greeks settling in western India, Indian scholars of this period incorporated into their own astronomy some of their underlying models and techniques, such as Hellenistic spherical geometry, celestial coordinate systems, and trigonometry of chords.

Over the next thousand years or so, the Indian mathematical sciences flourished as one of the richest and most fascinating scientific traditions ever known. Using rules composed mostly in Sanskrit verse and detailed prose commentaries on them, and without the formal deductive proof structure that we now routinely associate with mathematics, Indian mathematicians brilliantly explored topics in arithmetic, algebra, geometry, trigonometry, numerical approximations, combinatorics, series (including infinite series and infinitesimal methods), and a host of other fields. Mathematical subjects were closely linked with the discipline of mathematical astronomy; the professional lives of its practitioners were generally organized around family traditions of scholarship, court patronage, and informal collegial networks rather than official institutions of learning and formal credentials. Before the end of the first millennium CE, Indian mathematics and astronomy had influenced scientific traditions in Islamic West Asia, much of Southeast Asia, and China. In the second millennium, Indian exchanges with Islamic sciences significantly increased and direct encounters with European sciences followed. The Indian mathematical tradition remained active until it was displaced in the nineteenth and twentieth centuries by the heliocentric astronomy and Western mathematics promoted by European colonial powers.

Almost every one of the above statements is disputed by some historian (although we all seem to agree at least that Indian mathematics is brilliant and fascinating). The various objections to this mainstream narrative range from mere numerological fantasy to serious scholarly critique. In almost all cases, though, the contested issues ultimately depend on some debatable point of interpretation or speculation, for which neither the mainstream nor the revisionist view can point to incontrovertible documentary evidence to settle the question beyond a doubt. Many historians of mathematics, confronted with this uncertainty, have fallen back on temporizing or compromising between two opposing views, splitting the difference between their widely divergent estimates of dates or periods, or evading the difficulty by skimming over the history of Indian mathematics as briefly as possible. So readers go on wondering why discussions of Indian mathematics often seem sketchy or confusing.

My objective in this book is to present a condensed form of the mainstream narrative summarized above, which remains the most generally accepted "best guess" about the development of Indian mathematics. I have tried to lay out its chief arguments with some samples of the sources on which it is based, while being honest about the areas where direct supporting evidence is lacking, and explaining specific points where it differs from various revisionist hypotheses. Except for the broad and brief overview in the introductory chapter, I include citations of published works on both sides of the issue in my descriptions of controversial points, so that readers can pursue the debates in more detail if they choose. There is a great deal of exciting research currently in progress concerning Indian history and Indian mathematics, some of which will certainly amplify and modify parts of today's standard narrative. The bibliography attempts to represent an adequate though necessarily partial subset of this new research, as well as standard editions and studies from earlier periods. The technical content of the mathematical material should be accessible to anyone with a solid precalculus background and a willingness to explore basic concepts in astronomy and Indology.

This book was written with the support of a research fellowship (project number 613.000.430) from the Exacte Wetenschappen division of the Nederlandse Organisatie voor Wetenschappelijk Onderzoek, at the Mathematical Institute of the University of Utrecht in 2004–2006. I am especially grateful to my NWO project supervisor and colleague, Prof. Jan P. Hogendijk, who not only provided invaluable advice and criticism about multiple drafts of the book but selflessly plowed through reams of administrative paperwork to support its creation. Heartfelt thanks are also due to other colleagues, staff, and students at the Mathematical Institute for their interest in this project, particularly Prof. Henk Bos, and the members of the Utrecht Studygroup for the History of Astronomy. During my time in the Netherlands I was also an Affiliated Fellow at the International Institute for Asian Studies in Leiden; I am greatly indebted to its then Director, Prof. Wim Stokhof, to Dr. Saraju Rath, the rest of the Fellows and staff of IIAS and the Kern Library, and other Indologists in Leiden and Amsterdam for a wonderful experience blending history of science with the study of India.

Some of the background research for this work was carried out during the course of research fellowships at the Dibner Institute for the History of Science and Technology (2000–2002) and with the American Institute of Indian Studies in Jaipur (2003–2004). While in Jaipur, I was fortunate to be able to work with manuscript collections at the Sri Ram Charan Museum of Indology and at the Jain Vidya Sansthan. I am greatly obliged to my AIIS supervisor, Prof. Basant Jaitly of the University of Rajasthan, to the late Dr. S.C. Sharma, to Dr. S.G. Sharma of the SRCMI, and to Dr. K.C. Sogani and all the staff of the Mahavir Digambar Jain Pandulipi Sanrakshan Kendra at the JVS for their welcome and their kindness. I am also indebted to Brown University and to Mrs. Isabelle Pingree for various opportunities to work as a Visiting Lecturer and Visiting Scholar at Brown, which has long been a world-renowned center for the study of the history of the exact sciences, particularly in the Indian tradition.

Parts of the material presented here are expanded or revised versions of work appearing in other publications. The overall structure of the book and several discussions of Sanskrit textual sources, especially in chapters 5 and 6, are based on my chapter "Mathematics in India" [Plo2007b] in Victor Katz's *Mathematics of Egypt, Mesopotamia, China, India, and Islam: A Sourcebook* [Kat2007]. These two works are designed partly to complement each other, as the *Sourcebook* chapter focuses on translations of substantial excerpts from about a dozen original sources in Indian mathematics, while the present book offers a broad survey with shorter source excerpts and more detailed historical background. Table 4.3 and some discussions in appendix A are based on material in [Plo2008b], section 7.3.2 is partly based on [Plo2005b], and section 6.2.5 is a revised version of [Plo1996b].

The photographs in figures 3.1 and 3.2 were generously made available for my use by Bill Casselman. The image shown in figure 5.7 is reproduced from [Hay1995], p. 549. I thank Takanori Kusuba for sharing the photocopy of MS Benares 104595 that appears in figure 6.3, and Benno van Dalen for the image in figure 8.3.

In addition to the obligations acknowledged above, I am grateful to many colleagues who contributed comments, criticism, and source materials that helped to shape and improve the book. Johannes Bronkhorst, Christopher Minkowski, and Bill Casselman were kind enough to provide careful readings and detailed critiques of large portions of the manuscript. Helpful feedback and references were also supplied by David Brown, Jamil Ragep, S.R. Sarma, Toke Knudsen, Frits Staal, P.P. Divakaran, M.D. Srinivas, Davide P. Cervone, Amy Shell-Gellasch and other members of the Euler Society and the History of Mathematics Special Interest Group of the MAA Book Club, S.M.R. Ansari, Dennis Duke, Takanori Kusuba, Shemsi Alhaddad, Benno van Dalen, R.C. Gupta, Saraju Rath, Todd Timmons, Dominik Wujastyk, David Mumford, Aldine Aaten, Joel Haack, Michio Yano, Steven Wepster, David Metcalf, Setsuro Ikeyama, Fernando Gouvêa, S. Balachandra Rao, R. Sridharan, K. Ramasubramanian, Takao Hayashi, Homer White, Peter Scharf, Silvia d'Intino, Victor Katz, Glen Van Brummelen, Clemency Montelle, Amrit Gomperts, C.S. Seshadri, Joseph Dauben, Agathe Keller, Karine Chemla, M.S. Sriram, Gary Tubb, Jeff Oaks, Kengo Harimoto, Dennis Duke, Jens Høyrup, Eleanor Robson, David King, Arlo Griffiths, Dhruv Raina, and Annette Imhausen, as well as by the anonymous reviewers at Princeton University Press. I thank them all for their generosity and their kindness in taking an interest in this work. I owe the initial inspiration for its creation to David Ireland, formerly at Princeton University Press. My present editors at Princeton, Vickie Kearn and Deborah Tegarden, and their superb staff have earned my deep gratitude for their support and unfailing assistance, as has my sister, Amy Plofker, for her timely help with the proofs. Whatever errors and failings remain in the final version of the book are nobody's fault but my own.

My greatest debt is to my teacher, the late David Pingree of the Department of the History of Mathematics at Brown University. His monumental erudition and wisdom nourished the study of the history of Indian exact sciences for nearly half a century, and his passing is an irreplaceable loss to the field. This book is dedicated to his memory.

śrī gurudevāya namaḥ

homage to the revered teacher

List of Abbreviations

ĀpSS *Āpastamba-śulba-sūtra*

BauSS *Baudhāyana-śulba-sūtra*

KāSS *Kātyāyana-śulba-sūtra*

MāSS *Mānava-śulba-sūtra*

RJV *Ṛg-Jyotiṣa-vedāṅga*

YJV *Yajur-Jyotiṣa-vedāṅga*

Chapter One

Introduction

1.1 BACKGROUND AND AIMS OF THIS BOOK

The mathematical heritage of the Indian subcontinent has long been recognized as extraordinarily rich. For well over 2500 years, Sanskrit texts have recorded the mathematical interests and achievements of Indian scholars, scientists, priests, and merchants. Hundreds of thousands of manuscripts in India and elsewhere attest to this tradition, and a few of its highlights—decimal place value numerals, the use of negative numbers, solutions to indeterminate equations, power series in the Kerala school—have become standard episodes in the story told by general histories of mathematics. Unfortunately, owing mostly to various difficulties in working with the sources, the broader history of Indian mathematics linking those episodes still remains inaccessible to most readers. This book attempts to address that lack.

The European scholars who encountered Indian mathematical texts in the eighteenth and nineteenth centuries were often completely at sea concerning the ages of the texts, their interrelationships, and even their identities. The sheer number of such works and the uncertainty surrounding even the most basic chronology of Sanskrit literature gave rise to great confusion, much of which survives to this day in discussions of Indian mathematics. This confusion was compounded by the fact that authors of different mathematical texts sometimes had the same name, and different texts themselves sometimes bore the same title. Even when the background and content of the best-known treatises were sorted out in the early nineteenth century, historians still had many vexing problems to contend with. Much mathematical material was embedded in the very unfamiliar context of medieval Indian astronomy and astrology. The style of its presentation, in highly compressed Sanskrit verse, was equally alien in appearance. Yet the material also bore many similarities, from its decimal numerals to its trigonometric formulas, to certain features of Western mathematics.

Into this new historiographic territory came the early authors of general histories of mathematics, foraging for grand narratives. Historians from Montucla to Moritz Cantor and Cajori incorporated into their overviews of world mathematics many of the newly gleaned facts about the Indian tradition. Their accounts established a standard if seriously incomplete picture of Indian mathematics that still serves as the basic framework for its treatment in most modern histories. Meanwhile, in India, researchers such as Bāpudeva Śāstrī, Sudhākara Dvivedī, and S. B. Dikshit unearthed vast amounts of additional information that, being published mostly in Sanskrit and Hindi, had little impact on the work of non-Indologists.

B. Datta's and A. N. Singh's *History of Hindu Mathematics*, published in the mid-1930s, rapidly became the standard text on the subject in English, with a far broader range of sources and a more careful treatment of original texts than most general histories could boast. Other surveys followed, including C. N. Srinivasiengar's *History of Ancient Indian Mathematics*, in 1967, and T. A. Sarasvati Amma's *Geometry in Ancient and Medieval India* and A. K. Bag's *Mathematics in Ancient and Medieval India*, both in 1979. Indian mathematics has also been featured in several more general studies of Indian science and of non-Western mathematics, such as S. N. Sen's 1966 *Bibliography of Sanskrit Works in Astronomy and Mathematics* and G. G. Joseph's 1991 *Crest of the Peacock*. In addition, a large body of specialist literature on Sanskrit exact sciences—astronomy, mathematics and the disciplines that historically accompanied them, such as astrology—has appeared in English over the last few decades. Examples of this literature include David Pingree's biobibliographical *Census of the Exact Sciences in Sanskrit* and his *Jyotiḥśāstra: Astral and Mathematical Literature*, in Jan Gonda's *History of Indian Literature* series, the articles of R. C. Gupta and others in the journal *Gaṇita Bhāratī*, and editions and translations of Sanskrit texts, such as Takao Hayashi's *Bakhshālī Manuscript* and Pushpa Jain's *Sūryaprakāśa*.

Why, then, is it still so difficult for the nonspecialist to find trustworthy information on many aspects of the Indian mathematical tradition? The inadequacy of the old "grand narratives" in this regard still plagues many modern historians of mathematics who have to rely on them. Early surveys of Indian sources tended to portray them as a record of "discoveries" or "contributions," classified according to modern mathematical categories and important in proportion to their "originality" or "priority." The context for understanding Indian mathematics in its own right, as a part of Indian literature, science, and culture, was generally neglected. Up-to-date specialist literature supplying that context is often difficult for nonspecialists to identify or obtain, and sometimes difficult to understand. Finally, much of the desired data is simply absent from India's historical record as presently known, and the resulting informational vacuum has attracted a swirling chaos of myths and controversies to bewilder the uninitiated.

Additionally, the historiography of science in India has long been co-opted for political purposes. Most notoriously, some nineteenth-century colonial officials disparaged local intellectual traditions, which they termed "native learning," in order to justify Westernized education for future colonial servants. Many nationalists responded in kind by promoting various separatist or Hindu nationalist historiographies, often including extravagant claims for the autonomy or antiquity of their scientific traditions. The influence of all these attitudes persists today in politicized debates about history, religion, and culture in Indian society.

The present work attempts to trace the overall course of Indian mathematical science from antiquity to the early colonial era. Its chief aim is to do justice to its subject as a coherent and largely continuous intellectual tradition, rather than a collection of achievements to be measured against the mathematics of other cultures. For that reason, the book is divided roughly chronologically, with emphasis on various historical perspectives, rather than according to mathematical topics, as in the classic surveys by Datta and Singh and Sarasvati Amma. Of course, this account remains greatly indebted to the labors of these and other earlier scholars, without whose groundbreaking achievements it would not have been possible.

The rest of this chapter discusses the historical setting and some of the chief historiographic difficulties surrounding Indian mathematics, as well as the role of mathematics in Sanskrit learning. Chapter 2 considers the evidence concerning mathematical concepts in the earliest extant Indian texts, while chapter 3 examines what we know from the (mostly fragmentary) sources in the first several centuries of the Classical Sanskrit period, starting in the late first millennium BCE. These reveal, among other things, the development of written number forms, particularly the now universal decimal place value numerals, and the circulation of mathematical ideas between India and neighboring cultures.

The middle of the first millennium CE saw the appearance of the first surviving complete Sanskrit texts in the medieval Indian tradition of mathematical astronomy. Chapter 4 explores these early texts and the snapshot they provide of mathematical sciences in their day. The establishment of mathematics as an independent textual genre—attested to in works dealing exclusively with the topics and techniques of calculation, rather than their application to astronomical problems—apparently followed soon afterward, as far as we know from the extant texts. The development, subject matter, and structure of this genre and its continuing relation to mathematical astronomy are discussed in chapter 5. Aspects of its social and intellectual context are treated in chapter 6: who were the people who were studying and writing about mathematics in medieval Indian society, what did they perceive its nature and significance to be, and how did this relate to the emergence in the early second millennium CE of important canonical mathematical texts? Chapter 7 continues this theme with a discussion of the best-known (and in many ways the most remarkable) of the pedagogical lineages in Indian mathematics, the famous Kerala school of Mādhava.

Chapter 8 explores the impact of the contacts between Indian and Islamic mathematics, which increased after Central and West Asian incursions into the subcontinent during the second millennium. The story closes in Chapter 9 with a survey of some of the early modern developments that gave place, during the British colonial period, to the cultural and intellectual transition from "Indian mathematics" to Indian participation in modern mathematics. This narrative is supplemented by two appendices at the end of the book. The first supplies some background on the relevant linguistic and literary features of Sanskrit. The second lists the biographical information available on some of the most historically significant Indian writers on mathematics and attempts to separate out the widespread legends concerning them from the (usually scanty) established facts.

This material includes more discussion of astronomy than is typical for works on Indian mathematics. But it is not really possible to understand the structure and context of mathematics in India without recognizing its close connections to astronomy. Most authors of major Sanskrit mathematical works also wrote on astronomy, often in the same work. Astronomical problems drove the development of many mathematical techniques and practices, from ancient times up through the early modern period.

Equally crucial for our understanding of this subject is an awareness of some of the historiographic controversies involving ancient Indian texts. The whole framework of the history of Sanskrit mathematical science ultimately hinges on the question of when and how these texts were composed, and it is a question that still has no universally accepted answer. The discussion in this book for the most part hews to the standard or conservative scholarly consensus about the basic chronology of Indian history and science. Many of the generally accepted conclusions in this consensus are nonetheless not definitively proved, and many revisionist or minority views have achieved a wide popular currency.

These issues profoundly affect the inferences that we can draw about mathematics in India, and most readers will probably be much less familiar with them than with the historical background of mathematics in other cultures, such as ancient Greece or seventeenth-century Europe. It therefore seems appropriate to devote some space in the relevant chapters to explaining a few of the most influential debates on these topics. The aim is to steer a middle course between unnecessarily perplexing the reader with far-fetched speculations and ignoring valid criticisms of established hypotheses. Therefore, formerly controversial or surprising claims are not emphasized here if they are now universally accepted or discarded. There should be no need nowadays to point out, for example, that Āryabhaṭa's decimal arithmetic is not associated with Greek sources or that Mādhava's power series for trigonometric functions predate by centuries Newton's and Leibniz's versions of them.

1.2 HISTORY AND SOUTH ASIA

Traditional Indian culture and literature are frequently said to have an ahistorical perspective, supposedly preoccupied with timeless spiritual knowledge rather than the recording of mundane events. This is a rather misleading oversimplification. It is true that chronicles of purely historical events (as opposed to the legends of the ancient Epics and Purāṇas, only distantly inspired by history) are rare in Sanskrit literature. The historian of India, particularly early India, can follow no chronological trail blazed by an ancient predecessor like Thucydides or Sima Qian. Studies of artifacts—archaeology, epigraphy, numismatics—and some literary references provide most of the known data about what happened and when in premodern South Asia. The current big picture of Indian history has been built up only slowly from these data, and has changed (and continues to change) significantly in the process.

The geographical locus of classical Indian culture is the South Asian subcontinent, encompassing most of the modern nations of India, Pakistan, Nepal, Bangladesh, and Sri Lanka. (Throughout this book the term "India" or "the subcontinent" will generally refer to this larger region rather than the territory bounded by the modern state of India.) Evidence concerning the historical roots of this culture is quite sparse. The earliest known texts in an Indian language are the collections of religious hymns and rituals called the Vedas, composed in an archaic form of Sanskrit known as Vedic Sanskrit, or Old Indo-Aryan. Their language and subject matter clearly reveal their kinship with the various cultures known as Indo-European. For example, the Vedic hymns refer to various Indo-European themes and motifs, such as fire sacrifices to the members of a divine pantheon with many counterparts among, for example, Greek and Norse deities, including a male thunder-god as leader; large herds of cattle; the two-wheeled, two-horse chariots used for battle and sport; and a sacred ritual drink (called *soma* in Vedic and *haoma* in Old Iranian). Moreover, Vedic Sanskrit is unmistakably descended, like the members of the Celtic, Germanic, Hellenic, Italic, Iranian, and other linguistic groups, from a closely related group of ancestral dialects reconstructed by linguists as Proto-Indo-European.

The origin and diffusion of the common ancestral Indo-European cultures are still quite problematic. The similarities and differences among the various reconstructed Proto-Indo-European dialects may provide some clues to their geographical distribution. For example, the Indo-Iranian ancestral dialect appears to have been farthest from the Germanic and Celtic, with ancestors of Greek and Armenian somewhere between them. Many linguists hypothesize that this reflects an Indo-European origin roughly in the middle of the regions over which these languages later spread: somewhere around the Black Sea or Caspian Sea, perhaps. The relative positions of the various dialect groups consequently were more or less maintained as the groups migrated outward into new territories, eventually becoming Celtic and Germanic languages in the northwest, Iranian and Indo-Aryan in the southeast, and so on.

When did this hypothesized diffusion occur? Most reconstructions place it somewhere in the fourth or third millennium BCE. Textual evidence provides some data points concerning later chronology. By the early second millennium BCE, the Anatolian Indo-European language called Hittite was spoken in Asia Minor; a few centuries afterward, an Indo-Aryan language (more archaic than Vedic Sanskrit) was in use in the Mitanni kingdom in what is now Iraq and Syria; an early form of Greek was written in the Linear B script in Crete and the Greek mainland in the thirteenth century BCE; and there are comparatively abundant records by the early first millennium of Indo-European languages and cultures in Iran, Greece, Asia Minor, northwestern

Europe, Central Asia, and elsewhere.

According to this scenario, speakers of Indo-Iranian (the immediate common ancestor of Indian and Iranian languages) were living in eastern Iran and western Afghanistan around the end of the third millennium BCE. Some of them spread westward into Iran, where the Iranian language subfamily then developed. Others moved eastward over the Afghan highlands into the Panjab, where some earlier populations had recently shifted to the east and south, probably due to environmental changes that dried up local rivers. The Indo-Iranian newcomers may have been taking advantage of the resulting increase in elbow room. There, perhaps in the late second millennium, they composed the earliest Vedic hymns in the Old Indo-Aryan tongue that had evolved from Indo-Iranian. (Alternatively, perhaps earlier Indo-Aryan speakers already settled in Iran were split by a wedge of Iranian speakers, which displaced some of them west into what became the Mitanni realm and the rest east into India.) Subsequently they assimilated the cultures, territories, and to a large extent populations of non-Indo-European groups in nearby parts of the subcontinent. By the middle of the first millennium BCE, Indo-Aryan culture was widespread in northern India, and dominant in its political centers. (Languages of the non-Indo-European family called Dravidian, such as Tamil and Telugu, retained their primacy in southern India, although they and their speakers were strongly influenced by Indo-Aryan language and culture.)

This, the standard account of the origin and growth of Vedic India, is sometimes referred to as the Aryan invasion theory (AIT). However, most modern Indologists prefer other terms such as "immigration" or "influx" to "invasion," which connotes earlier assumptions, now discarded, of large-scale military conquest in the Panjab. The word "Aryan" likewise has unfortunate racialist connotations, but it remains the standard linguistic designation for the Indian branch of the Indo-Iranian descendants of Proto-Indo-European. The AIT label itself, however, has become so loaded with ideological overtones that it seems best to avoid it. Here I rely instead on more general terms, such as "standard hypothesis" or "majority view," to refer to the historical narrative described in the preceding paragraphs.

There are numerous difficulties with most of the features of this hypothesis. In the first place, the archaeological record of Indo-European diffusion is not clearly established. Nor is it clear how relatively small Indo-European population groups might have established so great a cultural, political, and linguistic dominance over such a broad geographical extent between about 3000 and 1000 BCE. But if the Indo-European diffusion was primarily a linguistic and cultural evolution rather than a mass migration of foreign populations, we would expect to find a good deal of continuity in genetics and material culture within the regions of expansion rather than a record of sudden disruption by hordes of new arrivals.

Such continuity is very apparent in northern South Asia, where there is a long record of settled communities with domesticated animals and grain agriculture. The so-called Indus Valley culture, which flourished in and around Sind and the Panjab in the mid-third millennium, left archaeological traces similar to those found in nearby sites dating from as early as the seventh millennium BCE and as late as the first. The remains of these communities, including major Indus Valley urban centers such as Harappa and Mohenjo-Daro, show extensively developed agriculture, architecture, manufacture, and trade. They also preserve a collection of still undeciphered graphic symbols that may have been part of an ancient script, or perhaps just nonlinguistic signs. More recently discovered sites in Central Asia were probably linked to such centers, which also traded with Sumerian cities in Mesopotamia. Even after the previously mentioned ecological displacement of many inhabitants toward the east and south in the early second millennium and the decline of the major cities, the Indus Valley and related cultures apparently persisted throughout the Vedic period.

These facts have led some historians to suggest that this prehistoric urban-agrarian culture *was* Vedic culture. In this alternative reconstruction, there is no need to link the Vedas and their language to a presumed Indo-European expansion over the Afghan highlands; they can be accounted for as an autonomous development within the Indus Valley culture or one of its relatives (the so-called indigenous Aryan theory). However, this suggestion requires an explanation of the evident cultural and linguistic links between these alleged "autochthonous Aryans" and their counterparts in lands north and west of South Asia.

One proposed explanation is that the Indus Valley region was actually the original homeland of Indo-European culture: instead of a few Indo-Europeans trickling into the subcontinent through the mountain passes, most Indo-Europeans trickled out of it (whence the alternative name, Out of India, for this hypothesis). But this proposal creates at least as many problems as it solves. It is difficult to compare the evidence of Vedic Sanskrit culture with that of the Indus Valley and related cultures: the former is mostly textual while the latter is exclusively archaeological. But there do seem to be some significant differences between the two. For example, early Vedic hymns do not refer to cities or wheat, well known in the Indus culture. At the same time, Indus culture sites do not contain remains of characteristic Indo-European goods such as horses or chariots.

Linguistically, the Out of India hypothesis is seriously inadequate. Vedic Sanskrit exhibits some linguistic influences from non-Indo-European Indian languages that are not found in other Indo-European language families. How could this have happened if all Indo-European languages originated together in India? In addition, a number of plants and animals whose names occur in different Indo-European language families, allowing reconstruction of corresponding words in Proto-Indo-European, are found only in temperate climates north of the subcontinent, suggesting that Proto-Indo-European dialects were spoken outside India. Finally, as noted above, the reconstructed relationships among these dialects appear to correspond roughly to the relative spatial locations of the language families they ultimately evolved into. This correspondence is hard to explain if we assume that all the dialects diffused in the same direction, via the same narrow channel, from a place of origin near the southeastern edge of the Eurasian continent. And of course, the Out of India hypothesis still leaves us with all the abovementioned difficulties in accounting for Indo-European expansion in other regions.

Consequently, the standard historical narrative, in which Vedic culture is largely based on Indo-European influence from northwest of the subcontinent in the second millennium BCE, still appears the simplest and most consistent explanation. However, it must be stressed that there is little definite evidence concerning the ways in which this influence operated, the genetic makeup or geographic origin of the people involved, and the relationships between Vedic and other early Indian cultures.

Events in Indo-Aryan India began to connect to recorded history elsewhere only around the middle of the first millennium BCE. This period saw what is known as the "second urbanization" of the subcontinent, with new major urban centers, the first to emerge since the decline of the Indus civilization, arising mostly in the eastern valley of the Ganges. By the late sixth century, the Persian empire had expanded as far as the northwestern Gandhara region on the Indus River. Alexander seized control of Gandhara from the Persians in the 320s. Almost simultaneously, a large kingdom was consolidated in northern India under Candragupta Maurya, who may have participated in the battles to check Alexander's advance across the Panjab.

The birth of the religious-philosophical traditions of Jainism and Buddhism also occurred in the middle of the first millennium. Mahāvīra, the founder of Jainism, was born probably in the late sixth century, and the Buddha perhaps somewhat later. Their teachings, frequently linked to reformist movements within late Vedic thought, are possibly derived from non-Vedic religious beliefs in northeastern India, based on the concepts of karmic retribution and cycles of rebirth. Their influence in the late first millennium was considerable, even among the elite. Alexander's contemporary Candragupta Maurya is said to have embraced Jainism; his grandson, the emperor Aśoka, in the mid-third century BCE adopted Buddhist beliefs.

The inscribed stone monuments of Aśoka's reign contain the oldest securely dated writing in an Indian language (in this case, a Middle Indo-Aryan language related to Sanskrit). It may be that writing systems had been in use in India before Aśoka's monuments were carved, but we have no positive evidence for this. The Vedas are the only extant Indian texts known to be much older than Aśoka's time, and they were preserved by a sacred oral tradition rather than in written form. On the other hand, it may be that writing was a fairly recent innovation in Aśoka's India, possibly stimulated by contact with the Persian empire.

Aśoka's inscriptions also testify to a remarkable geographical range for the political influence, or at least the scattered political penetration, of the Mauryan empire: they occur as far north as Gandhara and as far south as modern Karnataka, and on both the western and the eastern coasts. Moreover, they record the launch of Buddhist missionary expeditions to Greek kingdoms in the west and to Sri Lanka. The teachings of Buddha and Mahāvīra diffused rapidly throughout the subcontinent, although in the early Common Era these movements lost ground to an emerging complex of beliefs and practices that we now call Hinduism, namely, the worship of a modified pantheon combining Vedic and pre-Vedic deities and dominated by the gods Viṣṇu and Śiva.

For several centuries during and after Aśoka's reign, Indian contacts with neighboring cultures were frequent and often turbulent. In the northwest, successors to Alexander (the so-called Indo-Greeks) blended Greek and Indian cultures in their dominions. They in turn were followed by the Śakas or "Indo-Scythians" and "Indo-Parthians" arriving from central Asia, starting around the first century BCE. Some Śaka groups subsequently expanded southward into western India, under pressure from incursions by the Yuezhi of Mongolia, founders of the Kuṣāṇa empire. The Kuṣāṇas were strongly established in northern and western India by the second century CE, and traded extensively with the Roman empire, as did kingdoms in South India. Southern Indian ports also maintained a thriving trade with Southeast Asia.

The spread of Buddhist traditions in China inspired some Chinese Buddhists to make pilgrimages to India, where the empire or federation of the Gupta rulers held sway north and east of the Deccan plateau in the fourth and fifth centuries CE. In the sixth century, Gupta power was undermined by yet another invasion spurred by tumult in Central Asia, that of the Hūṇas or Huns. Direct trade between India and Europe decreased with the decline of the Kuṣāṇa and Roman empires, but communication by sea between Southeast Asia and India's east and southwest coasts continued to flourish. In fact, much of Southeast Asia became heavily Indianized, with vigorous Buddhist and Hindu traditions.

After the rise of Islam, southern India's sea trade came to be largely dominated by Muslim Arab traders with commercial ties to West Asia. Arabs also established realms in northern and western India during the Islamic expansion of the early eighth century. At the start of the second millennium, strife in Central Asia once again impelled invaders across the Afghan passes: in this case, Turkic and Persian Muslims who turned from struggles with other Central Asian peoples to raids and conquests in northern India. The resulting Indo-Muslim empires of the mid-second millennium were later supplanted by European colonies, leading to the almost complete political control of the subcontinent by Great Britain in the nineteenth century.

It is plain even from the foregoing brief sketch that India has never been historically isolated from or irrelevant to the rest of Eurasia but rather has constantly exchanged goods and ideas with its neighbors. At the same time, from classical antiquity until the modern period, its multiple strands of influence and innovation were woven into a web of Sanskritized culture and learning that linked the entire subcontinent.

1.3 SANSKRIT LITERATURE AND THE EXACT SCIENCES

Sanskrit texts frequently refer to the "ocean of knowledge," an appropriate metaphor for the vast abundance of subjects covered by the varieties of Sanskrit literature. The sacred Vedas, whose name literally means "knowledge," are often considered the foundation of learning. The genre of "Vedic texts" embraces the four *saṃhitās* or collections of hymns and rituals—namely, the *Ṛg-veda*, *Yajur-veda*, *Sāma-veda*, and *Atharva-veda*—as well as exegetical and philosophical works like the Brāhmaṇas and Upaniṣads. In the first millennium BCE, the divisions of learning included not only the Vedic texts themselves but also the six "limbs," or supporting disciplines, of the Vedas. These were phonetics, grammar, etymology, and poetic metrics, which ensured the proper preservation and comprehension of the archaic verses of the hymns; ritual practice, which specified the details of the various rites; and *jyotiṣa* or astronomy and calendrics, which determined the proper times for performance of the rites. The Vedic texts are generally known as *śruti*, "heard" via divine revelation; the limbs of the Veda, on the other hand, are called *smṛti*, "remembered" from human tradition.

The post-Vedic era of what is known as Classical Sanskrit, beginning in the late first millennium BCE, saw an expansion of the recognized categories within which knowledge was produced and organized. The plethora of Classical literary genres included works treating *dharma*, or religiously mandated law and right conduct; narrative and legend, such as the great epics *Mahābhārata* and *Rāmāyana*, and the Purāṇas; various philosophical, theological, liturgical, and devotional subjects; different types of literary composition, such as stories and poetry, and their aesthetic characteristics; performing arts; building arts; and several sciences, including an enhanced form of *jyotiṣa* that incorporated not only astrology but also computational methods in general, known as *gaṇita*. The exact sciences and most other branches of *smṛti* learning were called *śāstras*, "treatises" or "teachings."

Vernacular languages—Indo-Aryan vernaculars like Pali and Prakrit, as well as classical Dravidian languages such as Tamil—played a large role in the development of Indian literature. Many religious and philosophical works, stories, poems, plays, and grammatical treatises were composed in languages other than Sanskrit. This was especially true among Jains and Buddhists, for whom the ancient Sanskrit Vedas were not as significant as their own sacred canons in Prakrit and Pali, respectively. (A number of Buddhist and Jain scholars in the Classical period, however, wrote in Sanskrit chiefly or exclusively.) The number and variety of surviving texts in vernacular languages increased with the passage of time and included, in the second millennium, many works on astronomy and mathematics. Sanskrit, like Latin in medieval Europe, nevertheless remained central as a widely shared language of scholarship: as the Indologist Sheldon Pollock writes, "There was nothing unusual about finding a Chinese traveler studying Sanskrit grammar in Sumatra in the seventh century, an intellectual from Sri Lanka writing Sanskrit literary theory in the northern Deccan in the tenth, or Khmer princes composing Sanskrit political poetry for the magnificent pillars of Mebon and Pre Rup in Angkor in the twelfth" ([Pol2000], p. 599). But the place of the vernaculars in the culture of learning was never negligible. A view of Indian mathematics drawn almost exclusively from Sanskrit texts, as in the present work, is necessarily partial and incomplete; its only excuse—apology, rather—lies in the limitations on the size of the book and the abilities of the author.

The Vedic veneration of Sanskrit as a sacred speech, whose divinely revealed texts were meant to be recited, heard, and memorized rather than transmitted in writing, helped shape Sanskrit literature in general. The privileged position of orality may have inspired the fascination with, and advanced development of, phonetics and grammar among Indian scholars. Its influence is also visible in the conventional forms of Sanskrit works. Even treatises on secular and technical subjects were ideally considered as knowledge to be learned by heart, not merely kept in a book for reference. (In practice, of course, written manuscripts were crucial to the preservation and transmission of learning, and were produced probably in the hundreds of millions over the last two millennia.) Thus, texts were composed in formats that could be easily memorized: either condensed prose aphorisms (*sūtras*, a word later applied to mean a rule or algorithm in general) or verse, particularly in the Classical period. Naturally, ease of memorization sometimes interfered with ease of comprehension. As a result, most treatises were supplemented by one or more prose commentaries, composed sometimes by the author of the treatise, sometimes by later scholars, either in Sanskrit or in a local vernacular.

In addition to emphasizing the significance of the spoken word, Sanskrit intellectual traditions generally considered knowledge to be founded upon divine teachings. True knowledge of whatever sort was necessarily part of the fundamental truth of the Veda (or, for Buddhists and Jains, of their own sacred principles). Again, it would be misleading to characterize Indian thought simply as "static" or "timeless." It changed over time to accommodate new ideas and new lines of argument, but innovations were generally worked into existing traditions rather than flaunted as revolutionary novelties.

Furthermore, the distinction between *śruti* and *smṛti* did not imply a sharp division of the sacred from the secular; many texts, even on technical subjects like *jyotiṣa*, were ascribed to the revelations of gods or legendary sages. These attributions expunged the historical context of the works to stress the divine importance of their content. Similarly, even historical human authors frequently omitted biographical information and other contextual details as irrelevant or unnecessary to their writings. This sometimes makes it difficult to distinguish reliably between human and allegedly divine authors, a difficulty further compounded by the Indian custom of bestowing on children the Sanskrit names or epithets of gods or sages.

Given this background, we should be prepared to find some substantial differences between mathematics in the Indian tradition and its counterparts elsewhere. To take one example, there are few personal chronicles in Sanskrit literature comparable to the doxographical or biographical accounts of Hellenistic or Islamic scientists. Consequently, several medieval writers whose mathematical works were widely known in India—contemporaries of Theon of Alexandria, Zu Chongzhi, or Thābit ibn Qurra, about whose careers and families at least some evidence survives—are less distinct as historical personages than even the ancient Greek mathematicians Euclid and Antiphon, or Ahmes the scribe of the Rhind Mathematical Papyrus. Educational and professional institutions, libraries, and patrons are also frequently obscure. Consequently, it is hardly surprising that some popular histories filled the resulting void with many pseudobiographical legends about Indian mathematicians.

Another and more fundamental difference is that the Sanskrit tradition does not regard mathematical knowledge as providing a unique standard of epistemic certainty. For many Greek philosophers and their Islamic and European successors, a central concept was the abstraction of universal forms from their sensible manifestations in the same way that numbers and geometrical figures are abstracted from physical quantities and shapes. Hence the validity of mathematical knowledge has had profound implications for the nature of reality in western philosophical thought, from the Pythagoreans on down. It has been suggested that the corresponding role of "paradigmatic science" in Indian thought was filled instead by grammar (*vyākaraṇa*). In Sanskrit philosophy and logic, ideas about reasoning and reality are explicitly linked to the understanding of linguistic statements. What philosophers need to probe in such statements, therefore, is their grammatical interpretation rather than their analogies with mathematical entities.

Mathematics, not being an epistemologically privileged discipline in Sanskrit learning, was generally subject to the same truth criteria as other forms of knowledge. In Sanskrit epistemology, valid ways of knowing include direct perception, inference, analogy, and authoritative testimony. This means that the idea of mathematical proof is somewhat different from the formal chains of explicit deduction mandated in Greek geometry. Mathematical assertions in Sanskrit can be justified in a number of different ways according to philosophical truth criteria, and sometimes they are not explicitly justified at all. This is not to say that rigorous demonstration and formal logic were unknown to Indian mathematicians, nor that Indian mathematicians generally permitted arguments from authority to overrule demonstration. But there was no conventional structure of proof consistently invoked as essential to the validation of mathematical statements. True perception, reasoning, and authority were expected to harmonize with one another, and each had a part in supporting the truth of mathematics.

Chapter Two

Mathematical Thought in Vedic India

2.1 THE VEDAS AND MATHEMATICS

As noted in section 1.3, the earliest extant Sanskrit texts are the ancient religious texts known as the Vedas, which are traditionally grouped into four *saṃhitās* or collections. Probably the oldest elements of these collections, based on comparisons of their vocabulary and grammatical and prosodic forms, are hymns to various deities in some sections of the *Ṛg-veda* or "Praise-Knowledge." The standard model of ancient Indian historiography places their composition sometime in the second millennium BCE. Somewhat later than these Early Vedic hymns are Middle Vedic invocations or mantras used in rituals for performing religious sacrifices, recorded in the *Yajur-veda* ("Sacrifice-Knowledge"). The other two Vedic collections are the *Sāma-veda* ("Chant-Knowledge") and the *Atharva-veda* ("Knowledge of the Atharvan-priest"), containing chants, prayers, hymns, curses, and charms.

This knowledge was shaped into a canonical corpus probably sometime before the middle of the first millennium BCE. The remaining works identified as part of *śruti* or revealed wisdom were composed to interpret and expound the Vedas. Among these, the Brāhmaṇa texts chiefly describe and explain sacrificial ritual. (These texts are not to be confused with the human Brāhmaṇas, or "Brahmins," who were hereditary priests and scholars.) The compositions called Vedānta, or "end of the Vedas," comprising the Āraṇyakas and Upaniṣads, contain teachings on philosophical and spiritual themes.

What do these texts tell us about ancient Indian ideas on mathematical subjects? In the first place, they reveal that by Early Vedic times a regularized decimal system of number words to express quantity was well established. (Most of these number words evidently date back as far as Proto-Indo-European, since they have many cognates in other Indo-European languages.) Some of the most archaic Vedic hymns attest to this system based on decades and powers of ten, including combined numbers involving both decades and units:

> You, radiant [Agni, the fire-god], are the lord of all [offerings]; you are the distributor of thousands, hundreds, tens [of good things]. (*Ṛg-veda* 2.1.8)

> Come, Indra [king of the gods], with twenty, thirty, forty horses; come with fifty horses yoked to your chariot, with sixty, seventy, to drink the [sacred beverage] Soma; come carried by eighty, ninety, a hundred horses. (*Ṛg-veda* 2.18.5–6)

> Three thousand three hundred and thirty-nine [literally "three hundreds, three thousands, thirty and nine"] gods have worshipped Agni ... (*Ṛg-veda* 3.9.9)

Some simple fractional parts such as one-third, using ordinal number forms as in their English equivalents, also occur in Early or Middle Vedic texts.[1] No later than the Middle Vedic period the Indian decimal integers had been expanded to a remarkable extent with the addition of number words for much larger powers of ten, up to at least a trillion (10^{12}). The first record of them occurs among the hymns included in the *Yajur-veda*'s descriptions of sacrificial rites. These hymns invoke not only deities but also aspects of nature and abstract entities, including various sequences of numbers, both round and compound:

> Hail to earth, hail to the atmosphere, hail to the sky, hail to the sun, hail to the moon, hail to the *nakṣatras* [lunar constellations], hail to the eastern direction, hail to the southern direction, hail to the western direction, hail to the northern direction, hail to the upwards direction, hail to the directions, hail to the intermediate directions, hail to the half-years, hail to the autumns, hail to the day-and-nights, hail to the half-months, hail to the months, hail to the seasons, hail to the year, hail to all. (*Yajur-veda* 7.1.15)

> Hail to one, hail to two, hail to three ... hail to eighteen, hail to nineteen [literally "one-less-twenty"], hail to twenty-nine [literally "nine-twenty"], hail to thirty-nine ... hail to ninety-nine, hail to a hundred, hail to two hundred, hail to all. (*Yajur-veda* 7.2.11)

> Hail to a hundred, hail to a thousand, hail to *ayuta* [ten thousand], hail to *niyuta* [hundred thousand], hail to *prayuta* [million], hail to *arbuda* [ten million], hail to *nyarbuda* [hundred million], hail to *samudra* [billion], hail to *madhya* [ten billion], hail to *anta* [hundred billion], hail to *parārdha* [trillion], hail to the dawn, hail to the daybreak ... hail to the world, hail to all. (*Yajur-veda* 7.2.20)

Why did Vedic culture construct such an extensive number system and acclaim it in sacred texts? The computing requirements of everyday life

[1] For example, in *Yajur-veda* 2.4.12.3: "He, Viṣṇu, set himself in three places, a third on the earth, a third in the atmosphere, a third in the sky." (All the *Yajur-veda* cites in this chapter are from the version known as the Taittirīya-saṃhitā recension of the Kṛṣṇa Yajurveda. Note that in these quoted passages and all others throughout the book, text in square brackets represents editorial additions and explanations that are not literally present in the original.)

would not have demanded more than the first few decimal orders of magnitude, as seen among other ancient civilizations, whose known number words reach only into the thousands or tens of thousands. Although infinite speculations are possible about the metaphysical or spiritual implications of these numbers in Vedic thought, there is probably no conclusive solution to the mystery.[2]

The cosmic significance of numbers and arithmetic in ritual reflecting concepts of the universe is brought out clearly in another early first-millennium text, the *Śata-patha-brāhmaṇa* or "Brāhmaṇa of a hundred paths," an exegetical text explaining the symbolism of sacrificial rituals. The following passage refers to sacrificial fire-altars made of baked bricks which symbolize the 720 days and nights of an ideal year. The creator god Prajāpati, representing this year and the concept of time in general, sought to regain power over his creation by arranging these 720 bricks in various ways:

> Prajāpati, the year, has created all existing things.... Having created all existing things, he felt like one emptied out, and was afraid of death. He bethought himself, "How can I get these beings back into my body?"... He divided his body into two; there were three hundred and sixty bricks in the one, and as many in the other; he did not succeed. He made himself three bodies.... He made himself six bodies of a hundred and twenty bricks each; he did not succeed. He did not divide sevenfold. He made himself eight bodies of ninety bricks each.... He did not divide elevenfold.... He did not divide either thirteenfold or fourteenfold.... He did not divide seventeenfold. He made himself eighteen bodies of forty bricks each; he did not succeed. He did not divide nineteenfold. He made himself twenty bodies of thirty-six bricks each; he did not succeed. He did not divide either twenty-onefold, or twenty-twofold, or twenty-threefold. He made himself twenty-four bodies of thirty bricks each. There he stopped, at the fifteenth; and because he stopped at the fifteenth arrangement there are fifteen forms of the waxing, and fifteen of the waning [moon]. And because he made himself twenty-four bodies, therefore the year consists of twenty-four half-months....[3]

(The full sequence of attempted or rejected divisions by all the integers from 2 to 24 is described in the text, although the above excerpt omits some of them for conciseness.)

The final division of 720 into 24×30 is the last possible one that will give an integer quotient. Even more interesting, mathematically speaking, than Prajāpati's ultimate successful division are the divisions that he did

[2]Some inferences about the mystical meaning of numbers are discussed in [BerA1878], vol. 2, ch. 5, in [Mal1996], ch. 14, and in [Mur2005].

[3]*Śata-patha-brāhmaṇa* 10.4.2, [Egg1897], pp. 349–351. I have substituted "divide" as the translation of *vi-bhū* where Eggeling uses "develop." See also the discussion in [Mal1996], ch. 13.

not attempt, which would have produced fractional numbers of bricks. The concept of integer divisibility is thus part of this cosmic narrative. Its sequence of pairs of factors of 720, with the numbers relatively prime to 720 neglected, somewhat resembles Old Babylonian tables of sexagesimal reciprocals or paired factors of the base 60, where 2 is coupled with 30, 3 with 20, and so on, while the relatively prime numbers such as 7 and 11 are omitted.[4] The sexagesimal multiple 720 is also familiar in Old Babylonian texts, being the standard metrological unit called the "brick-sar."[5] Whether these similarities are the result of coincidence or hint at some kind of early transmission remains unclear. Most of the chief characteristic features of Old Babylonian mathematics—sexagesimal place-value numbers, tables for multiplication and division, written numeral forms—have no counterpart in the scanty available evidence for Vedic mathematical ideas.

Late Vedic exegetical texts such as the Upaniṣads, as well as contemporary Buddhist and Jaina philosophy, also offer intriguing possibilities for speculation about the development of some concepts later incorporated in mathematics per se. Examples of these include the synonyms *śūnya* and *kha*, meaning "void," "nullity" (in later mathematical texts "zero") and *pūrṇa* or "fullness."[6] Unfortunately, we have no distinct lines of textual descent from Vedic religious and philosophical compositions on such concepts to their later embodiment in specifically mathematical works. About all we can say is that the Vedic texts clearly indicate a long-standing tradition of decimal numeration and a deep fascination with various concepts of finite and infinite quantities and their significance in the cosmos.[7]

2.2 THE *ŚULBA-SŪTRAS*

Mathematical ideas were explored in more concrete detail in some of the ancillary works classified as Vedāṅgas, "limbs of the Vedas," mentioned in section 1.3—phonetics, grammar, etymology, metrics, astronomy and calendrics, and ritual practice. This section examines mathematics in Vedāṅga

[4] [Hoy2002], pp. 27–30.

[5] [Rob1999], p. 59.

[6] See, for example, [Gup2003] and [Mal1996], ch. 3.

[7] Popular usage of the term "Vedic mathematics" often differs considerably from the mathematical content actually attested in Vedic texts. Some authors use "Vedic mathematics" to mean the entire Sanskrit mathematical tradition in Vedic and post-Vedic times alike, which of course comprises much more than is directly present in these early sources. Most commonly, though, the term signifies the Sanskrit mental-calculation algorithms published in 1965 in a book entitled *Vedic Mathematics*, which the author described ([Tir1992], pp. xxxiv–xxxv) as "reconstructed from" the *Atharva-veda* and which are very popular nowadays in mathematics pedagogy. These algorithms are not attested in any known ancient Sanskrit text and are not mentioned in traditional Vedic exegesis. They constitute an ingenious modern Sanskrit presentation of some mathematical ideas rather than an ancient textual source. The widespread confusion on this topic has been addressed in [DanS1993] and [SarS1989], and a thorough scrutiny of the explicitly mathematical and numerical references that actually appear in the four Vedic collections is presented in [Pandi1993].

texts on ritual practice, which specified the details of performing the various ceremonies and sacrifices to the gods. These texts were classified either as pertaining to *śruti* and describing major ceremonies, or as pertaining to *smṛti* and explaining the routine customs and observances to be maintained in individual households. The former type included the regular fire sacrifices performed at particular times of the year and the month, as well as special rituals sponsored by high-ranking individuals for particular aims, such as wealth, military victory, or heaven in the afterlife.

Some of the ritual practice texts explained how the different types or goals of sacrifices were associated with different sizes and shapes of fire altars, which were to be constructed from baked bricks of prescribed numbers and dimensions. The footprints for the altars were laid out on leveled ground by manipulating cords of various lengths attached to stakes. The manuals described the required manipulations in terse, cryptic phrases—usually prose, although sometimes including verses—called *sūtras* (literally "string" or "rule, instruction"). The measuring-cords, called *śulba* or *śulva*, gave their name to this set of texts, the *Śulba-sūtras*, or "Rules of the cord."

Many of the altar shapes involved simple symmetrical figures such as squares and rectangles, triangles, trapezia, rhomboids, and circles. Frequently, one such shape was required to be transformed into a different one of the same size. Hence, the *Śulba-sūtra* rules often involve what we would call area-preserving transformations of plane figures, and thus include the earliest known Indian versions of certain geometric formulas and constants.

How this ritual geometry became integrated with the process of sacrificial offerings is unknown. Did its mathematical rules emerge through attempts to represent cosmic entities physically and spatially in ritual?[8] Or conversely, was existing geometric knowledge consciously incorporated into ritual practice to symbolize universal truth or to induce a "satori" state of mind in the participants through perception of spatial relationships? No contemporary text can decide these questions for us: the concise *Śulba-sūtras* themselves are mostly limited to essential definitions and instructions, and the earliest surviving commentaries on them are many centuries later than the *sūtras*, which in turn are doubtless later than the mathematical knowledge contained in them.

The rest of the historical context of the *Śulba-sūtras* is also rather vague. The ritual practice text corpora to which they belong are ascribed to various ancient sages about whom no other information survives. The best-known *Śulba-sūtras* are attributed to authors named Baudhāyana, Mānava, Āpastamba, and Kātyāyana, in approximately chronological order. They are assigned this order on the basis of the style and grammar of the language of their texts: those of Baudhāyana and Mānava seem to be roughly contempo-

[8]This is the hypothesis of, for example, [Sei1978], in which a prehistoric ritual origin for Eurasian geometry traditions is reconstructed from ideas of the sky as a circle, the earth as a square, and so on. And [Sta1999] amplifies this thesis for a potential Indo-European ancestor of both Indian and Greek geometry, based on the ritual associations of both *Śulba-sūtra* techniques and the "altar of Delos" legend of the cube duplication problem.

rary with Middle Vedic Brāhmaṇa works composed perhaps in 800–500 BCE, while the *Śulba-sūtra* of Kātyāyana appears to post-date the great grammatical codification of Sanskrit by Pāṇini in probably the mid-fourth century BCE. Nothing else is known, and not much can be guessed, about the lives of these texts' authors or the circumstances of their composition.[9]

The *Śulba-sūtras*, like other manuals on ritual procedure, were intended for the use of the priestly Brāhmaṇa families whose hereditary profession it was to conduct the major sacrificial rituals. But since animal sacrifice and consequently most of the fire altar rituals were eventually abandoned in mainstream Indian religion, and since there are few archaeological traces of ancient fire altars, it is not certain how the prescribed procedures were typically enacted in practice.[10]

The *Śulba-sūtra* texts[11] include basic metrology for specifying the dimensions of bricks and altars. Among the standard units are the *aṅgula* or digit (said to be equal to fourteen millet grains), the elbow-length or cubit (twenty-four digits), and the "man-height" (from feet to upraised hands, defined as five cubits).[12] As early as the *Baudhāyana-śulba-sūtra*, methods are described for creating the right-angled corners of a square or rectangle, constructing a square with area equal to the sum or difference of two given squares, and transforming a square with area preservation into a rectangle (or vice versa), into a trapezium, triangle or rhombus, or into a circle (or vice versa). In the process, it is explicitly recognized that the square on the diagonal of a given square contains twice the original area; and more generally that the squares on the width and the length of any rectangle add up to the square on its diagonal (the so-called Pythagorean theorem).[13] Samples

[9]See [SenBa1983], pp. 2–5. It is suggested in [Pin1981a], pp. 4–5, that the Āpastamba and Kātyāyana *Śulba-sūtras* predate that of Mānava. In [Kak2000a], a much earlier date for *Śulba-sūtra* works is inferred by linking them to astrochronological speculations (see section 2.3).

[10]An archaeological site containing one large brick altar in the traditional shape of a bird with outstretched wings, but differing markedly from the numerical specifications described in the *Śulba-sūtra* texts, has been dated to the second century BCE; [Pin1981a], p. 4, n. 19. And a long-lived South Indian tradition of fire altar construction is attested at the present day in [Sta1983] and in [Nam2002]. But since both of these may have originated in a form of "Vedic revivalism" in some post-Vedic period rather than in a continuous ritual praxis going back to the composition of the *Śulba-sūtras*, we cannot be sure how far either of them represents the original tradition of fire-altar geometry. In [SarE1999], pp. 10–11, such a lapse and revival in the abovementioned South Indian ritual tradition after about the fourth century CE are mentioned.

[11]For an edition and annotated English translation of the four major *Śulba-sūtra* works, see [SenBa1983], on whose edition the following translations are based. Sūtras 1.1–1.2, 1.4–1.13, and 2.1–2.12 of the *Baudhāyana-śulba-sūtra* are quoted and commented on in [Plo2007b], pp. 387–393. An earlier study of *Śulba-sūtra* mathematics is [Dat1993].

[12]See the various metrological *sūtras* in *Baudhāyana-śulba-sūtra* 1.3, [SenBa1983], pp. 17 (text), 77 (translation); *Mānava-śulba-sūtra* 4.4–6, [SenBa1983], pp. 60, 128; *Āpastamba-śulba-sūtra* 15.4, [SenBa1983], pp. 49, 113; *Kātyāyana-śulba-sūtra* 5.8–9, [SenBa1983], pp. 57, 124.

[13]*Baudhāyana-śulba-sūtra* 1–2; [SenBa1983], pp. 17–19 (text), 77–80 (translation). Henceforth the *Śulba-sūtra* citations will be confined to identifying the text and *sūtra* in the edition of [SenBa1983]. The abbreviations used for the text names are listed on

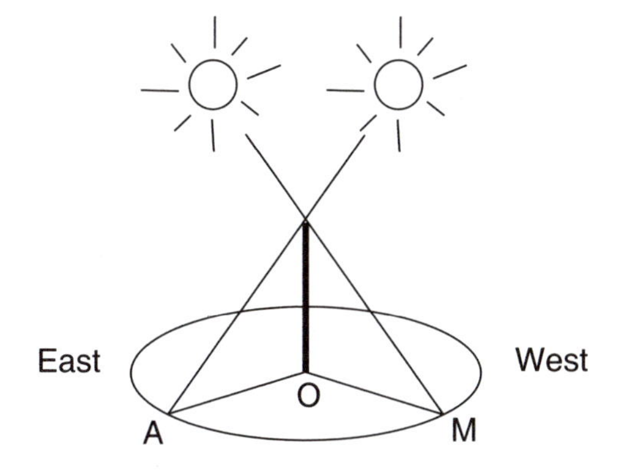

Figure 2.1 Determining the east-west line with shadows cast by a stake.

of such rules from various *Śulba-sūtra* texts are cited in the following part of this section, along with some of their procedures for more elaborate altar constructions.

The preliminary step is the drawing of a baseline running east and west. We do not know for sure how this was accomplished in the time of the early *Śulba-sūtra* authors, but the later *Kātyāyana-śulba-sūtra* prescribes using the shadows of a gnomon or vertical rod set up on a flat surface, as follows:

> Fixing a stake on level [ground and] drawing around [it] a circle with a cord fixed to the stake, one sets two stakes where the [morning and afternoon] shadow of the stake tip falls [on the circle]. That [line between the two] is the east-west line. Making two loops [at the ends] of a doubled cord, fixing the two loops on the [east and west] stakes, [and] stretching [the cord] southward in the middle, [fix another] stake there; likewise [stretching it] northward; that is the north-south line. (*KāSS* 1.2)

The first part of the procedure is illustrated in figure 2.1, where the base of the gnomon is at the point O in the center of a circle drawn on the ground.[14] At some time in the morning the gnomon will cast a shadow OM whose tip falls on the circle at point M, and at some time in the afternoon the gnomon will cast a shadow OA that likewise touches the circle. The line between points A and M will run approximately east-west.

Then a cord is attached to stakes at the east and west points, and its midpoint is pulled southward, creating an isosceles triangle whose base is the east-west line. Another triangle is made in the same way by stretching the cord northward. The line connecting the tips of the two triangles is a perpendicular bisector running north and south. Similar ways of stretching

page xiii.

[14]Note that the text itself is purely verbal and contains no diagrams. This figure and all the remaining figures and tables in this chapter are just modern constructs to help explain the mathematical rules.

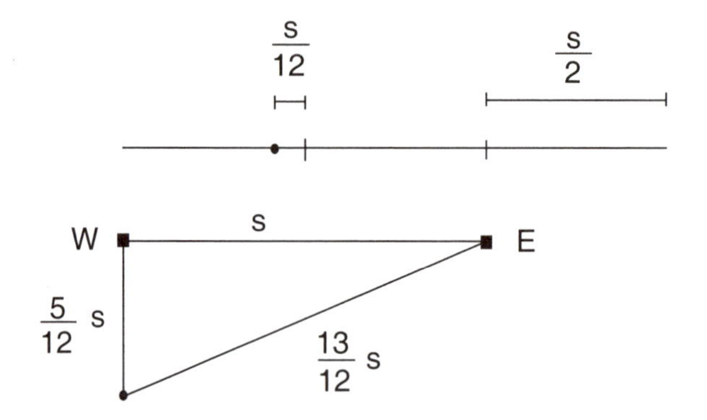

Figure 2.2 Determining the perpendicular sides of a square with a marked cord.

a cord into a triangle are also used for basic determinations of right-angled figures, as in the following construction of a square:

> The length is as much as the [desired] measure; in the western third of [that length] increased by its half, at the [place] less by a sixth part [of the third], one makes a mark. Fastening [the ends of the cord] at the two ends of the east-west line, stretching [the cord] southward by [holding] the mark, one should make a marker [at the point that it reaches]. In the same way [one should stretch the cord] northward; and in the other two directions after reversing [the ends of the cord]. That is the determination. [There is] shortening or lengthening [of the side to produce the desired half-side of the square with respect to] that marker. (*ĀpSS* 1.2)

Here a cord with length equal to the desired side of a square, say s, is increased to a total length of $\frac{3}{2}s$, and a mark is made at a distance of $\frac{5}{12}s$ from one end, as shown in figure 2.2. So when the endpoints are fixed a distance s apart along the east-west line, pulling the mark downwards creates a 5-12-13 right triangle to make the sides perpendicular. The same technique is also used with 3-4-5 right triangles (e.g., in *BauSS* 1.5, *KāSS* 1.4). More general properties of sides and diagonals are stated as well, including versions of what we now call the Pythagorean theorem and a rule for the length of the diagonal of a square with a given "measure" or side:

> The cord [equal to] the diagonal of an oblong makes [the area] that both the length and width separately [make]. By knowing these [things], the stated construction [is made]. (*ĀpSS* 1.4; similarly *BauSS* 1.12)
>
> The cord [equal to] the diagonal of a [square] quadrilateral makes twice the area. It is the doubler (*dvi-karaṇī*, "two-maker") of the

> square. (*ĀpSS* 1.5; similarly *BauSS* 1.9, *KāSS* 2.8)
>
> One should increase the measure by a third [part] and by a fourth [part] decreased by [its] thirty-fourth [part]; [that is its] diagonal [literally "together-with-difference"]. (*ĀpSS* 1.6; similarly *BauSS* 2.12, *KāSS* 2.9)

This rule for the length of the diagonal of a square of side s equates it to $s\left(1+\frac{1}{3}+\frac{1}{3\cdot 4}-\frac{1}{3\cdot 4\cdot 34}\right)$, or about $s\times 1.4142$. Interestingly, the *Kātyāyana-śulba-sūtra* version calls this rule approximate or "having a difference" (from the exact value).

Areas involving multiples of three are also constructed. For example, if a rectangle is made with width equal to the original square side s and length equal to its "doubler" or $\sqrt{2}s$, then the diagonal of the rectangle is declared to be the "tripler," producing a square of three times the original area:

> The measure is the width, the doubler is the length. The cord [equal to] its hypotenuse is the tripler (*tri-karaṇī*). (*ĀpSS* 2.2; similarly *BauSS* 1.10, *KāSS* 2.10)
>
> The one-third-maker (*tṛtīya-karaṇī*) is explained by means of that. [It is] a ninefold division [from the square on the tripler]. (*ĀpSS* 2.3; similarly *BauSS* 1.11, *KāSS* 2.11)

That is, an area one-third of the original area will be one-ninth of the square on the tripler.

Some typical transformations of one figure into another are the following procedures for "combining" or "removing" squares, that is, adding or subtracting square areas:

> The combination of two equal [square] quadrilaterals [was] stated. [Now] the combination of two [square] quadrilaterals with individual [different] measures. Cut off a part of the larger with the side of the smaller. The cord [equal to] the diagonal of the part [makes an area which] combines both. That is stated. (*ĀpSS* 2.4; similarly *BauSS* 2.1, *KāSS* 2.13)
>
> Removing a [square] quadrilateral from a [square] quadrilateral: Cut off a part of the larger, as much as the side of the one to be removed. Bring the [long] side of the larger [part] diagonally against the other [long] side. Cut off that [other side] where it falls. With the cut-off [side is made a square equal to] the difference. (*ĀpSS* 2.5; similarly *BauSS* 2.2, *KāSS* 3.1)

The first of these *sūtras* begins by noting that the previously given definition of the "doubler" or diagonal of a square in essence explained how to make a square equal to the sum of two identical squares. The methods for adding and subtracting two squares of different sizes, again relying on the relations between the sides and hypotenuses of right triangles, are illustrated

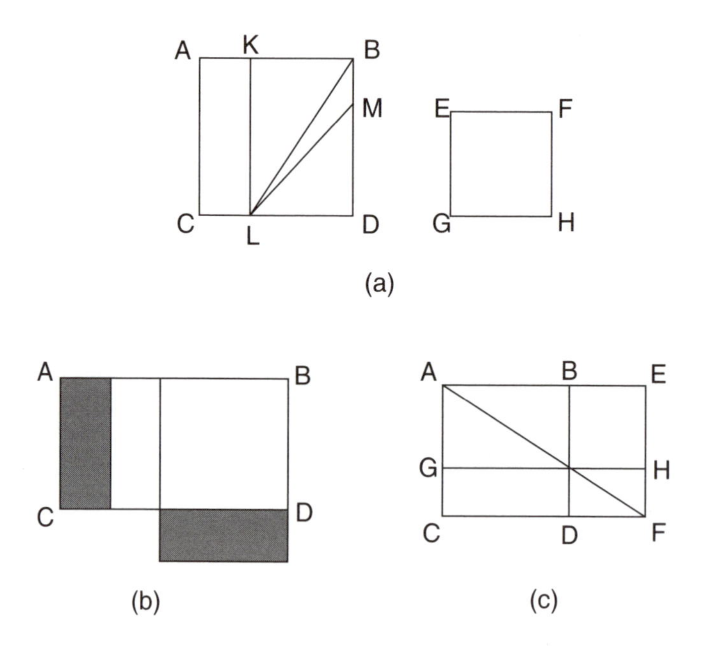

Figure 2.3 Transformations of squares and rectangles.

in figure 2.3a. If $ABCD$ is the larger square and $EFGH$ the smaller, cut off from $ABCD$ a rectangle $KBLD$ with width equal to the shorter side and length equal to the longer. Then its diagonal LB will be the side of a square equal to the sum of the two given squares. But if instead the long side KL is placed diagonally as the segment LM, then the cut-off side MD will be the side of a square equal to their difference. This second technique is employed again in transforming a rectangle into a square:

> Wishing [to make] an oblong quadrilateral an equi-quadrilateral: Cutting off [a square part of the rectangle] with [its] width, [and] halving the remainder, put [the halves] on two [adjacent] sides [of the square part]. Fill in the missing [piece] with an extra [square]. Its removal [has already been] stated. (*ĀpSS* 2.7; similarly *BauSS* 2.5, *KāSS* 3.2)

> Wishing [to make] an equi-quadrilateral an oblong quadrilateral: Making the length as much as desired, put whatever is left over where it fits. (*ĀpSS* 3.1; similarly *BauSS* 2.4)

In the first of these two rules, as shown in figure 2.3b, a square with side BD equal to the width of the given rectangle $ABCD$ is cut off from it, and the remainder of the rectangle is divided into two halves, one of which (shaded in the figure) is placed on the adjacent side of the square. This produces an L shape (also called a gnomon figure—no relation to the vertical stick gnomon for casting shadows) with an empty corner that will have to be filled in with an additional square piece, but the desired square side can then be found by the square-subtraction procedure described above.

It is not quite clear what the *śulba*-priest is supposed to do in the converse case of converting a square into a rectangle. It seems as though a rectangle of the desired width is to be cut off from the square and the remaining bricks of the square's area packed onto the rectangle's end in an ad hoc way. Later commentators have suggested a more rigorous interpretation,[15] illustrated in figure 2.3c, where the given square $ABCD$ is expanded into a rectangle $AECF$ of the desired length AE. Then the intersection of the diagonal AF with the original square side BD defines the side GH of the required rectangle $AEGH$ with area equal to that of the original square. However, this does not seem to be what the *sūtra* actually says, although it is somewhat similar to a simpler transformation rule (*BauSS* 2.3, *KāSS* 3.4) where a square of side s is cut diagonally into three triangles—one half and two quarters—with the quarters then shifted to form a rectangle with dimensions $s\sqrt{2} \times \frac{s\sqrt{2}}{2}$.

Transformations between rectilinear and circular shapes are also tackled:

> Wishing to make a [square] quadrilateral a circle: Bring [a cord] from the center to the corner [of the square]. [Then] stretching [it] toward the side, draw a circle with [radius equal to the half-side] plus a third of the excess [of the half-diagonal over the half-side]. That is definite[ly] the [radius of the] circle. As much as is added [to the edges of the circle] is taken out [of the corners of the square]. (*ĀpSS* 3.2; similarly *BauSS* 2.9, *KāSS* 3.11)
>
> Wishing [to make] a circle a [square] quadrilateral: Making the diameter fifteen parts, remove two. Thirteen [parts] remain. That is indefinite[ly, approximately] the [side of the square] quadrilateral. (*ĀpSS* 3.3; similarly *BauSS* 2.11, *KāSS* 3.12)

In the first of these *sūtras*, the radius of a circle with area equal to a given square is taken to be the half-side of the square, plus one-third of the difference between the half-side and the half-diagonal; that is, the radius is said to equal $\frac{s}{2} + \frac{s\sqrt{2}/2 - s/2}{3}$. To convert instead a given circle into a desired square, one is supposed to use $\frac{13}{15}$ of the diameter of the circle as the square's side; but this is apparently not considered as accurate as the first formula. (See the list in table 2.1 at the end of this section for a comparison of the different values of constants implied by these rules.)

Let us now look at the *Śulba-sūtra* specifications for some actual altar arrangements, starting with the prescribed setup of the traditional three fires used for most sacrificial ceremonies. These are the "householder's fire," which must burn continually under the care of each individual householder,

---

[15]See [SenBa1983], pp. 156–158.

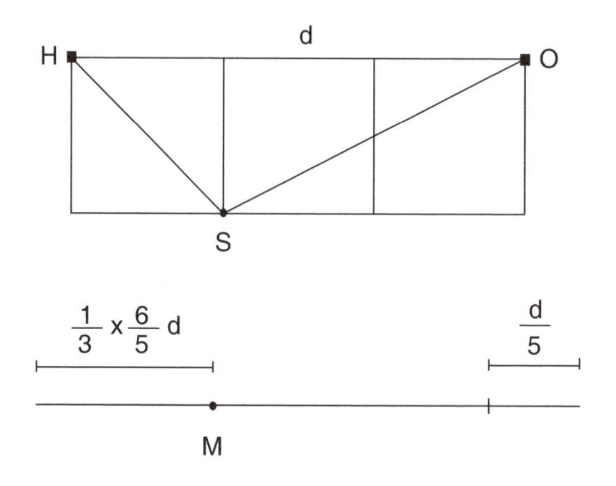

Figure 2.4 Laying out the three sacrificial fires.

the "oblation fire," and the "southern fire." They are to be arranged as follows:

> Now in the construction for setting up the [sacrificial] fires, the distance from the householder's to the oblation [fire]. It is known: The Brāhmaṇa sets [the latter] fire at eight double-paces [where a pace equals 15 *aṅgulas*], the prince eleven, the Vaiśya twelve, [east of the householder's fire]. (*BauSS* 3.1; similarly *ĀpSS* 4.1)
>
> Make three successive [contiguous square] quadrilaterals with [sides equal to] a third of [that] length. In the northwest corner is the householder's [fire]. In the south[east] corner of that same [square] is the [southern] offering fire; in the northeast corner [of the whole] is the oblation [fire]. (*BauSS* 3.2; similarly *ĀpSS* 4.3)

The three fires form a triangle as shown in figure 2.4, with the householder's and oblation fires (H and O respectively) at the western and eastern ends respectively of the east-west line HO; the length of HO depends on the rank of the sacrificer (see section 6.1.2 for a description of the ranks alluded to). The place of the southern fire S (south of the line, as its name suggests) is to be found by laying out the required three squares in a row south of HO. Then S is set in the southeast corner of the western square.

Or else, according to the texts, one can approximate this layout by means of a stretched-cord construction, as follows:

> Dividing the distance [between] the householder's and the oblation [fires] into five or six parts, adding an extra sixth or seventh part, dividing the whole into three, making a mark at the western third, fastening [the ends] at the householder's and oblation [fires and] stretching [the cord] southward by [holding] the mark,

> one should make a marker. That is the place of the southern fire. It agrees with *smṛti*. (*ĀpSS* 4.4; similarly *BauSS* 3.3)

The prescribed cord is also shown in figure 2.4. There it has length $\frac{6}{5}d$, where d is the distance HO between the first two fires (the user may instead choose to make the length equal to $\frac{7}{6}d$). The cord is then divided into three equal parts, and a mark M is made at the eastern end of the western third, that is, at a distance of $\frac{1}{3} \cdot \frac{6}{5}d = \frac{2}{5}d$ (or alternatively $\frac{1}{3} \cdot \frac{7}{6}d = \frac{7}{18}d$) from the western end of the cord.

When the marked cord is attached at the endpoints H and O and stretched toward the south, the mark M is supposed to fall approximately at S, the place of the southern fire. Of course, since the marked length $\frac{2}{5}d$ is somewhat shorter than the actual diagonal of the square $HS = \frac{\sqrt{2}}{3}d$, the triangle produced by the cord will not be exactly congruent to HOS.

An important related construction is that of the Great Altar or "*soma*-sacrifice altar" used in the ceremonies of the sacred ritual beverage *soma* (see section 1.2). The Great Altar is to be set up east of the three fires in the shape of an isosceles trapezium with its base facing west, using prescribed dimensions:

> [The altar] is thirty paces or double-paces on the western side, thirty-six on the east-west line, twenty-four on the eastern side: thus the [dimensions] of the *soma*-altar are known. (*ĀpSS* 5.1; similarly *BauSS* 4.3)
>
> Adding eighteen [units] to a length of thirty-six, [making] a mark at twelve [units] from the western end [and another] mark at fifteen, fastening [the ends of the cord] at the ends of the east-west line, stretching [the cord] south by [holding] the fifteen [mark], one fixes a stake [there]; in the same way northward; those are the two [western] corners. Reversing the two ends, stretching [the cord] by [holding] the same fifteen [mark], one fixes a stake at the twelve [mark]. In the same way northward; those are the two [eastern] corners. That is the construction with one cord. (*ĀpSS* 5.2)

The Great Altar is to be laid out symmetrically about the east-west line as shown in figure 2.5 by means of the now familiar stretched-cord method, utilizing a 15-36-39 right triangle. The height of the trapezium $ABCD$, thirty-six units, is paced off along the east-west line, and its base AB of thirty units is found by stretching the cord twice, to the south and to the north, to form the right triangles WAE and WBE. The same procedure is performed on the eastern side, and the twelve-unit lengths ED and EC are marked off to form the trapezium's top CD.

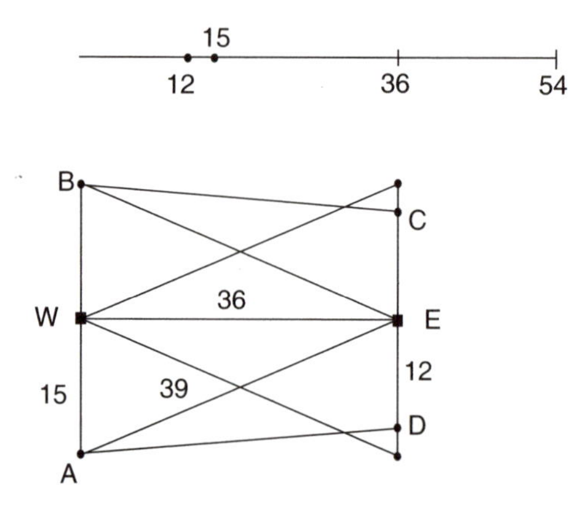

Figure 2.5 Construction of the trapezoidal Great Altar.

The text then describes how to cut and paste this figure into a rectangle—apparently just a mental construction for determining its area of 972 square units:

> The Great Altar is a thousand [square] paces [or double-paces] less twenty-eight. One should bring [a line] from the south[east] corner twelve units toward the south[west] corner. One should place the cut-off [triangle] upside-down on the other [side]. That is an oblong quadrilateral. In that way one should consider it established. (*ĀpSS* 5.7)

For a special sacrifice to the chief of the gods, Indra, the ritual requires a smaller altar with identical proportions to this Great Altar but only one-third of its area. To achieve the desired figure, the linear unit in the Great Altar construction is replaced by its "one-third-maker" (or $\frac{\sqrt{3}}{3}$) described above. Or else the altar dimensions are stated as smaller multiples of the unit's "tripler," $\sqrt{3}$:

> One should sacrifice with one-third [the area] of the *soma*-altar: this is known [for the area] of the Indra-sacrifice altar. The one-third-maker of the double-pace is to be used in place of the double-pace. Or, the widths [eastern and western sides] will be eight [and] ten [times] the tripler, [and] the east-west [length] twelve [times]. The Indra-sacrifice altar is three hundred twenty-four [square] paces [or double-paces]. (*ĀpSS* 5.8; similarly *BauSS* 3.12)

What does this ritual geometry add to our understanding of ancient Indian mathematical thought? For one thing, we see that at least by the time of the *Baudhāyana-śulba-sūtra*, arithmetic (although still not attested

Table 2.1 *Śulba-sūtra* constants

Sūtra	Rule and modern equivalent	Remarks; value
BauSS 2.9, *MāSS* 1.8, *ĀpSS* 3.2, *KāSS* 3.11	Half diagonal of square, minus difference of half diagonal and half side, plus one-third that difference, is radius of circle: $r = \frac{s}{2} + \frac{s\sqrt{2}/2 - s/2}{3}$	$\pi \approx 3.08831$
BauSS 2.10	Seven-eighths diameter of circle, plus one twenty-ninth of remaining eighth, minus one sixth of that twenty-ninth diminished by its eighth, is side of square: $s = \frac{2r}{8}\left(7 + \frac{1}{29} - \left(\frac{1}{29 \cdot 6} - \frac{1}{29 \cdot 6 \cdot 8}\right)\right)$	$\pi \approx 3.08833$
BauSS 2.11, *ĀpSS* 3.3, *KāSS* 3.12	Thirteen-fifteenths of diameter of circle is side of square: $s = 2r \cdot \frac{13}{15}$	Called "approximate" $\pi \approx 3.004$
BauSS 2.12, *ĀpSS* 1.6, *KāSS* 2.9	Side of square plus its third plus a fourth of the third minus one thirty-fourth of the fourth is the diagonal $s\sqrt{2} = s \cdot \left(1 + \frac{1}{3} + \frac{1}{3 \cdot 4} - \frac{1}{3 \cdot 4 \cdot 34}\right)$	*KāSS* says "approximate" $\sqrt{2} \approx 1.4142$
MāSS 11.9–10	$s^2 = \frac{3(2r)^2}{4}$	So interpreted in [Hay1990] $\pi \approx 3$
MāSS 11.15	$r = \frac{4}{5} \cdot \frac{\sqrt{2}}{2}s$	So interpreted in, e.g., [Gup2004b] $\pi \approx 3.125$

in written form) embraced manipulation of arbitrary fractional parts such as one thirty-fourth or two fifteenths, sometimes in quite complicated combinations. Spatial properties of several rectilinear plane figures were well understood, including the relationships among their sides, diagonals, and areas. Properties of the circle were also studied, particularly the challenging task of transforming it into a square of equal area or vice versa. And it was recognized that some of these transformation methods were more accurate than others in terms of preserving area. The transformation rules in fact corresponded to what we would call different values of irrational constants; several of them are summarized in table 2.1.

We have to be cautious about inferring any clear line of chronological development for any of these formulas, since it is perfectly possible that a later text could preserve an archaic rule that was omitted from an earlier text. We are also hampered by the textual isolation of these rules in efforts to understand how they were interpreted, derived or justified by their users. Modern scholars have suggested many ingenious ways to reconstruct their creation and explore their possible implications for other areas of mathematical thought.[16] But none of these is explicitly confirmed by the texts themselves, and there are no known textual traditions directly linking them to extant later works on geometry, which were composed starting around the middle of the first millennium CE. Nonetheless, as we will see in our exploration of those works in chapter 5, we frequently seem to hear in the verses of Classical Sanskrit geometry echoes of the *sūtras* of the ancient *śulba*-priests.

2.3 THE VEDAS AND ASTRONOMY

It has long been debated whether the Vedic corpus, in addition to providing clues about general numeration practices and ritual geometry, also preserves information about an ancient Indian tradition of mathematical astronomy.[17] Since later Sanskrit mathematics is so often closely tied to astronomical texts, it would not be surprising if we found the two subjects linked in Vedic times as well. Certainly there are clear references in Vedic texts to some astronomical and chronometric concepts, as illustrated by one of the Vedic hymns, quoted in section 2.1, which praises not only the sun, moon and constellations but also the directions, seasons, and months.

Vedic texts prescribed periodic sacrifices to be performed at particular times, such as the new and full moon, solstices, and equinoxes. This required keeping track of the passage of seasons and synodic months (synodic

---

[16]For example, see, in addition to the references in table 2.1, [Del2005] and [SenBa1983], pp. 165–168, for intriguing derivations of the $\sqrt{2}$ value in *Baudhāyana-śulba-sūtra* 2.12/*Āpastamba-śulba-sūtra* 1.6, particularly the geometric reconstruction following [Dat1993], pp. 192–194; [Neu1969], p. 34, for speculation on a possible relationship of this value to Old Babylonian mathematics; and [Knu2005] for square root rules and their possible connections to general quadratic problems.

[17]In this section the reader may wish to refer to the glossary in section 4.1 for explanations of unfamiliar astronomical terms.

months are defined as intervals between successive repetitions of a particular lunar phase, such as new moon or full moon). In a late section of the *Ṛg-veda* the year appears to be likened to a twelve-spoked wheel that "revolves round the heavens" with 720 paired children, the days and nights, within it (*Ṛg-veda* 1.164.11). (Presumably this would be an approximate or ideal year, like that represented by Prajāpati's 720 bricks in the *Śata-patha-brāhmaṇa*, rather than a literal estimate of the actual year length as 360 days.) The same section also refers to twelve months and an additional one, evidently an intercalary or "leap" month to keep the year of twelve lunar or synodic months, which is only about 354 days long, aligned with the seasons of the solar year (*Ṛg-veda* 1.25.8). The *Yajur-veda* too mentions the intercalary month (called *saṃsarpa* or "the creeping one," *Yajur-veda* 1.4.14), as well as meteorological names for the twelve regular months and the seasons that they constitute (names like "sweet" and "honey" in the "brilliant" or spring season, "cloud" and "cloudy" in the season called "rains," *Yajur-veda* 4.4.11). The thirteenth month is sometimes said to be inserted twice every five years, although other periods are also attested.[18] Thus, we can tell that by the Middle Vedic period at the latest, attempts were being made to regulate a basic lunisolar calendar arithmetically.

Quantitative astronomy requires some way of identifying positions in the heavens, usually with reference to particular fixed stars or groups of stars. Named constellations are distinctly mentioned in some Vedic texts beginning with the *Yajur-veda*, specifically the band of the twenty-seven or twenty-eight *nakṣatras* or "lunar constellations" that lie in or near the path of the moon's monthly progress through the sky. The identification of a given point in the yearly liturgical cycle depended on what *nakṣatra* the moon or sun occupied at that time, so the sacrificial priests could keep track of where they were in the calendar by observing the positions of the sun and moon with respect to the stars.

At some point, the names of twelve of these *nakṣatras* were applied to the twelve regular months to indicate that the full moon appeared in that *nakṣatra* during that month. (Most of these names are still used for lunar months in the Hindu calendar today.) Table 2.2 shows the sequence of *nakṣatras* along with their associated months and the modern names of stars or parts of constellations that historians have approximately matched to them on the basis of later texts.[19] The order of this cyclic sequence is fixed (except for the occasional omission of the *nakṣatra* called Abhijit), but the way it synchronizes with the liturgical year may vary, as illustrated in the following passage:

> The full moon in [the *nakṣatra*] Phalgunī is the beginning of the year.... That [has] this one defect that the middle falls in

[18] [Pin1973b], p. 6.

[19] See [MacKe1958], vol. 1, pp. 409–420. Note that, for example, the constellation Leo (a group of stars) is *not* identical to the modern zodiacal sign Leo (a 30-degree interval of longitude along the ecliptic measured from the vernal equinox).

> the cloudy season. The full moon in [the *nakṣatra*] Citrā is the beginning of the year. (*Yajur-veda* 7.4.8).

The details of these Vedic calendric concepts are obscured by the brevity of the allusions to them and by the apparent inconsistencies among some of them. Moreover, it is often difficult to tell when a scriptural statement is intended to convey specifically astronomical information and when it should be interpreted otherwise. These characteristics raise a host of perplexing questions. For example, several Vedic texts mention the annual seasons, but were there three of them (as said in, e.g., *Ṛg-veda* 10.90.6) or five (e.g., *Yajur-veda* 7.3.8) or six (e.g., *Yajur-veda* 4.4.11)? We have seen Vedic references to the concept of a 360-day year with twelve months of thirty days, and also to a thirteenth intercalary month. But when the embodied year is described as performing a sacrifice with a ritual of five nights, because "the four-night [ritual] is unsuccessful and the six-night [one] is excessive" (*Yajur-veda* 7.1.10), is that supposed to be interpreted as an astronomical statement about the length of the year? Does it symbolically express that the year is 365 days long as opposed to 364 or 366? Does it signify an intercalation of five leap days annually instead of a leap month less often? And when the *Yajur-veda* states, "Having produced the months with six-day [periods], they cast out a day" (*Yajur-veda* 7.5.6), does that imply an alternative synodic month length of twenty-nine ($6 \times 5 - 1$) days? Why is the beginning of the liturgical year assigned to different points in the astronomical year? Do these seemingly variant versions indicate a change in calendar systems over the course of the development of the Vedic corpus, with references to older features still preserved in later texts?

Similar problems of interpretation arise when we consider other astronomical phenomena besides the basic familiar cycles of the sun and moon. For instance, an early *Ṛg-veda* hymn mentions the sun being "enveloped in darkness" by the demon Svarbhānu and recovered by the sage Atri (*Ṛg-veda* 5.40.5–9). This legend is usually, and plausibly, interpreted to refer to solar eclipses; but does it have a more specific astronomical significance? Also, do the Vedic texts reflect any interest in the five naked-eye-visible star-planets? The uncertainty is compounded by the texts' frequent allusions to deities and other cosmic beings that in later Sanskrit astronomy are explicitly associated with celestial bodies. Should we infer from the later works that this astronomical symbolism was already present in Vedic times?[20]

In short, we can say at least that it is undeniable (and unsurprising) that the Vedic sacrifices depending on periodic lunar and solar phenomena such

[20] For example, it has been suggested that the (male) divine being Vena invoked in *Ṛg-veda* 10.123 represents the planet known to Romans as Venus, indicating by a shared name an ancient astral tradition common to both cultures (see [Kak1996]). However, the Roman identification of the planet Venus with their goddess Venus does not appear to predate the first century BCE: they seem to have adopted it from the Greeks' "star of Aphrodite," which in turn was adopted from the Babylonians' "star of Ištar" (see [PinAVEI]). In this instance, an apparent similarity in later sources is not necessarily a reliable guide for reconstructing earlier history.

Table 2.2 The *nakṣatras* and months

Nakṣatra	Month	Presumed location
Kṛttikās	Kārttika	Pleiades
Rohiṇī		Aldebaran
Mṛgaśiras	Mārgaśīrṣa	Orion
Ārdrā		Orion
Punarvasū		Gemini
Tiṣya/Puṣya	Pauṣa	Cancer
Āśleṣās		Cancer
Maghās	Māgha	Leo
1st Phalgunī	Phālguna	Leo
2nd Phalgunī	Phālguna	Leo
Hasta		Corvus
Citrā	Caitra	Virgo
Svāti		Boötes
Viśākhe	Vaiśākha	Libra
Anurādhā		Scorpio
Jyeṣṭhā	Jyaiṣṭha	Scorpio
Mūla		Ophiuchus
1st Aṣāḍhā	Āṣāḍha	Sagittarius
2nd Aṣāḍhā	Āṣāḍha	Sagittarius
Abhijit		Lyra
Śravaṇa	Śrāvaṇa	Aquila
Śraviṣṭhās/ Dhaniṣṭhā		Delphinus
Śatabhiṣaj		Aquarius
1st Bhadrapadā	Bhādrapada	Pegasus
2nd Bhadrapadā	Bhādrapada	Andromeda
Revatī		Pisces
Aśvinī	Āśvina	Aries
Bharaṇī		Aries

as moon phases, solstices and equinoxes relied on some kind of lunisolar calendar founded on observation of the sky to maintain the pattern of their rites.[21] The fundamental question that has been chewed over by Indologists for the past century and a half is, how quantitatively precise were these observations and the mathematical calendar schemes based upon them?

This question is central not just because a reliable answer to it could furnish valuable details about the practice of ancient Indian exact sciences and their role in the history of ancient science but also because it would determine whether it is possible to use astrochronological methods to date the Vedic texts and consequently the mathematical ideas contained in them. Astrochronology is the application of modern astronomical theories to historical records of celestial phenomena, in order to determine when the events that they describe occurred. Because modern astronomical models are extremely accurate over periods falling in historical time, if a particular celestial event such as an eclipse or nova can be unambiguously identified in a particular historical text as having been accurately observed at a particular celestial location, then astrochronology can often successfully retrodict, or reverse predict, the moment in the past at which the event happened. Consequently, the date of the text recording the observed event, or at least the earliest possible date for it, can be confidently inferred.

The Achilles' heel of astrochronology, as the alert reader will no doubt have guessed, is its requirement of an *unambiguous* identification of an *accurately* observed event. Any astrochronological dating is only as good as the data in its historical source. If the observational record is unclear or imprecise or downright mythical, then any astrochronological inferences drawn from it will be inconclusive or meaningless.

Unfortunately from the astrochronologist's point of view, none of the Vedic texts unambiguously describe known celestial coordinate systems involving, for example, the zodiac or celestial equator, or standardized celestial measuring units such as degrees, or any particular level of observational precision for celestial references in the texts. All of these features do appear in explicitly mathematical-astronomical works from centuries after the Vedic period, and Vedic concepts such as the *nakṣatras* and the half-months are deeply integrated into those later works. But the Vedic corpus itself requires a lot of interpretation, sometimes rather daring interpretation, in order to serve as a source of concrete information about the details of ancient Indian astronomical knowledge or for astrochronological dating of the texts.

A few of the most frequent arguments for astrochronological inferences from the Vedic texts are briefly outlined below. Most of them are based on the "astronomical clock" provided by the precession of the equinoxes caused by the periodic wobbling of the earth's axis. This wobbling makes

[21] Passing references in some Vedic texts to "*nakṣatra*-observers" ([MacKe1958], vol. 1, p. 431) suggest that this was a recognized task or occupation. It was also recognized as part of the broader universe of learning: the Chāndogya-Upaniṣad recounts (7.1.2) that a certain sage, requested to describe the subjects he had already studied, named among them "quantity" or numeration, time, and "*nakṣatra*-knowledge."

the position of the vernal equinox (where the sun rises exactly at the east point of the horizon) shift slowly westward on the ecliptic circle through the constellations of the zodiac. It takes about 26,000 years to complete one cycle of precession, so the equinox moves about one degree of arc every seventy-two years. The vernal equinox point currently lies in the constellation Pisces, while about six thousand years ago it was some 80 degrees further east, in the constellation Taurus.

These facts have been invoked to suggest identifiable dates for occurrences mentioned in Vedic texts. For example, the *Śata-patha-brāhmaṇa* (1.7.4.1–4) recounts a legend that the creator-god Prajāpati in the form of a stag attempted to mate with his daughter Rohiṇī in the form of a doe.[22] Since two adjacent *nakṣatras* in the Vedic constellation lists are named Mṛgaśiras ("deer's head") and Rohiṇī ("reddish"), it is proposed that these names symbolize the shift of the vernal equinox point from the former to the latter *nakṣatra*. In post-Vedic astronomy, these two *nakṣatras* were associated with particular stars, identified by historians as corresponding to a star in the constellation Orion and to Aldebaran (which is reddish) in the constellation Taurus, respectively. Therefore, if the same association existed in Vedic times, and if the *Śata-patha-brāhmaṇa* legend is really intended to symbolize this particular celestial event, we could date its origin astrochronologically on the basis of precession, which would give an approximate time of 4000 BCE.

In another passage (2.1.2.3–4), the *Śata-patha-brāhmaṇa* mentions the *nakṣatra* on the other side of Rohiṇī, the Kṛttikās, identified as a group of six or seven stars that are almost certainly identical to what we call the Pleiades, which also lie in Taurus, to the west of Aldebaran. These stars are said to be married to the stars known as the Seven Sages forming the constellation we call the Big Dipper, but the spouses are permanently separated because the Sages are in the north while the Pleiades "never swerve from the east." If this phrase is interpreted to mean that the Pleiades rise exactly at the east point of the horizon, then it must imply that the vernal equinox was located in the Pleiades at the time when the statement was made, which would put its date around 2950 BCE.[23] If it just means that the Pleiades always rise in the general eastward direction far from the northerly Dipper, however, no particular date can be inferred from it.

Likewise, another Brāhmaṇa text refers to the sun "resting" at the new moon of the month Māgha, "about to turn northward," after which "he goes northward for six months.... Having gone north for six months, he stays, about to turn south."[24] Clearly, this refers to the yearly cycle between the winter and summer solstices, when the sun's rising point on the eastern

---

[22]See, for example, [PinAVEI]. The following astronomical interpretation is described in [Kak2005].

[23]See [Kak2005], pp. 325–326; a nonastrochronological interpretation based on a hypothesized Babylonian influence is suggested in [PinAVEI].

[24]See [Hoc2005], pp. 295–296, on which (and on pp. 297–303 of the same source) much of the discussion in the remainder of this section is based. Many of the astrochronological arguments are presented in [Kak2005], and a summary of earlier attempts at astrochronological dating is given in [MacKe1958], vol. 1, pp. 420–431.

horizon moves from its southernmost position (in the winter of the earth's Northern Hemisphere) to its northernmost one and back again. This means that the sun at the winter solstice was considered to be in conjunction with the moon (that is, at new moon) in the month Māgha that took its name from the *nakṣatra* Maghās, presently identified with stars in the western part of the constellation Leo. Now if the winter solstice month was defined by the full moon lying in Maghās, the sun at the solstice would have been more or less opposite Maghās in the sky, approximately in Aquarius. That would put the vernal equinox in or near the western part of Taurus, pointing to a date around 1400 BCE, though varying interpretations give estimates ranging from at least the mid-third millennium BCE to the late second. Similarly, seasonal interpretations have been suggested for the abovementioned *Yajur-veda* references to the year's commencing with the full moon in Phalgunī or Citrā; if they mean that the winter solstice then fell in the month Phālguna or Caitra, it could imply a date sometime in the third or fourth millennium for the former, and the fourth or fifth millennium for the latter.

These suggested dates are imprecise for a number of reasons. First, we can't infer precise dates from statements that an equinox or solstice is "in" a *nakṣatra*, because a spread of 10–15 degrees of arc along the zodiac for a single *nakṣatra* represents about 700–1100 years of precessional motion. Also, the texts supply no clear definition of *nakṣatra* boundaries or coordinates within them. Nor do we understand precisely how months were demarcated relative to lunar phases at various times in the Vedic period; the synchronization of the calendar year and the astronomical year depends partly on whether months begin at new moon or at full moon. And what about calendar intercalation, that creeping thirteenth month? How far out of sync might the lunar and seasonal periods get before intercalation was applied to re-align them, and did these variations affect the calendar specifications given in the texts?

The hopeful astrochronologist might wonder about the possibility of using Vedic records of more distinctive, noncalendric phenomena to get a better fix on possible dates. This has been attempted for some passages, including the *Ṛg-veda* reference to a solar eclipse mentioned above. Taking the apparently general description in the hymn to imply specific information about the amount of obscuration, location of the observer, and so on, one can hypothetically identify the event with the solar eclipse of 26 July 3298 BCE. Again, though, this depends on assigning a particular technical meaning to the words of the text.

If all of these proposed interpretations are valid, they present us with a rather bewildering and sometimes contradictory array of conclusions to sort through. For example, we have seen that the *Śata-patha-brāhmaṇa* may be astrochronologically interpreted to reveal an internal date around 4000 BCE. But the aforesaid eclipse description in one of the early books of the *Ṛg-veda*, which is unanimously considered on linguistic and textual grounds to be older than the *Śata-patha-brāhmaṇa*, is astrochronologically interpreted to point to a much later date, around 3300 BCE. Furthermore, astrochronology applied to the abovementioned Pleiades reference in the *Śata-patha-brāhmaṇa* gives a different date around 3000 BCE; how do we reconcile the implications of these inferences for the time of its composition of the text? Obviously, astrochronological deductions do not supply a simple or easily understandable structure for Vedic chronology or Vedic astronomical ideas. On the other hand, as discussed in section 1.2, there is no known evidence, textual or otherwise, that indisputably proves any of these dates to be impossible for the composition of Vedic works, or at least for the origin of ideas that were later incorporated into them.

Falling back on a more conservative approach, if we take the explicitly astronomical and calendric references in Vedic texts simply at their face value, they appear to have been intended for the ears of people already familiar with the basic structure of whatever calendar systems were then in use: they were not designed as a systematic exposition of how those systems worked. So it is not surprising that we cannot infer much indisputable data from them, or establish any unique meaning for them. They permit a wide variety of possible interpretations, including speculative interpretations that imply a quite complex or ancient system of observational and computational astronomy.[25]

However, all such interpretations require bringing in substantial assumptions from outside the texts themselves. We may assume, for example, that the celestial references imply a particular minimum level of observational precision, or that some of them are more or less contemporary with the texts in which they appear while others represent remembered traditions from much earlier eras. Or we may read them as corresponding to explicitly astronomical concepts and data attested from texts composed centuries later. But we cannot argue convincingly that any such reconstruction *must* be what the authors of the Vedas intended to convey.

2.4 THE *JYOTIṢA-VEDĀṄGA*

The earliest known explicitly mathematical exposition of astronomy and calendrics, like its counterpart in the case of ritual geometry, is found not in the Vedic corpus itself but in its associated Vedāṅga. In the case of astronomy, the Vedāṅga is considered a single work in various recensions, the *Jyotiṣa-vedāṅga*, also known as the *Vedāṅga-jyotiṣa*. This text is the first available link between the ambiguous celestial and calendric utterances of the Vedas and the full-blown Sanskrit mathematical astronomy of the first millennium CE. The work is known chiefly in two versions, the shorter of

[25]Probably the most comprehensive and ambitious of such interpretations is that outlined in [Kak2000a] and [Kak2005], which combines broad ideas of Vedic epistemology with numerical symbolism involving the prescribed dimensions of sacrificial altars, and the numbers of books and hymns in parts of the Vedic corpus, to argue for a heavily encoded and very detailed esoteric system of astronomical parameters. But no known post-Vedic exegetical texts confirm the existence of such a system.

which is the *Ṛg-Jyotiṣa-vedāṅga* (henceforth *RJV*) consisting of thirty-six verses attributed to one Lagadha or to one Śūci expounding the knowledge of Lagadha, about whom nothing else is known. It is traditionally classified with ancillary works on the corpus of the *Ṛg-veda*. Twenty-nine of its verses are also found, sometimes in slightly different form, in a longer version (*Yajur-Jyotiṣa-vedāṅga*, henceforth *YJV*). It seems likely but is not proven that the *RJV* recension is the oldest, and that the *YJV* expanded and perhaps modified it.[26] We will examine in this section a few of the computational algorithms that the two versions share.

Both recensions, like the *Śulba-sūtra* geometry texts, are somewhat cryptic collections of formulas rather than exhaustive expositions. Both start out by invoking the deity Prajāpati as an incarnation of time cycles: "the one of five years, perceived as [or, the lord of] the *yuga* [literally "yoking," or a periodic cycle], whose limbs are days, seasons, half-years and months" (*RJV* 1, *YJV* 1). The *yuga* here is a five-year intercalation cycle containing two intercalary months. The following rule prescribes how to calculate its progress:

> [The current year] minus one, multiplied by twelve, multiplied by two, added to the elapsed [half-months of the current year], increased by two for every sixty [in the sum], is said [to be] the quantity of half-months [or syzygies]. (*RJV* 4, *YJV* 13)

This formula is based on simple proportion: if two intercalary months are to be inserted every five years or sixty months, then two extra half-months have to be added for every sixty nonintercalated half-months. For example, at the start of the fifth month of the fourth year of a given five-year period, the "current year minus one" is three. Multiplied by twelve and then by two, and added to the eight elapsed half-months of the current year, it gives eighty as the number of regular half-months that have passed since the start of the current *yuga*. By the rule, if two months are intercalated every sixty months, then three half-months must be intercalated in eighty half-months: so the total quantity of elapsed half-months would be eighty-three. This is a somewhat artificial result, because only whole months are actually intercalated in the calendar, not half-months. So what the rule allows us to determine is our approximate current place in the *yuga*.

The starting-point of the *yuga* cycle is defined in the following way:

> When the sun and moon go together in the sky with [the *nakṣatra*] Śraviṣṭhā, then is the commencing *yuga*, the waxing [half-month] of Māgha, [and] the northern half-year. (*RJV* 5, *YJV* 6)

> The sun and moon [going] north enter the beginning of [the *nakṣatra*] Śraviṣṭhā; [going] south, the middle of [the *nakṣatra*]

[26] See [Pin1973b], p. 1, [Ach1997], p. 21 (the latter also discusses a much longer *Atharva-veda jyotiṣa* work), and [SarKu1984]. All the verses cited henceforth are from the edition of the *Ṛg-veda* and *Yajur-veda* recensions in [DviS1908], pp. 1–69. Some of my translations rely on those in [DviS1908], pp. 85–103, and [Pin1973b].

> Āśleṣā; always in [the months] Māgha and Śrāvaṇa [respectively]. (*RJV* 6, *YJV* 7)

As in the Brāhmaṇa text discussed in the previous section, the start of the year at the winter solstice is placed at the new moon of the month Māgha, when the moon and sun are in conjunction at "the beginning of Śraviṣṭhā."

Other verses speak about the changing length of the days in the course of the seasons and the measurement of time during a day. A water clock was evidently used for timekeeping, with the volume of water adjusted by a constant amount every day to account for the changing day length:

> A *prastha* [a particular measure of weight or volume] of water [is] the increase in day [and] the decrease in night in the [sun's] northern motion; vice versa in the southern. [There is] a six-*muhūrta* [difference] in a half-year. (*RJV* 7, *YJV* 8)
>
> Ten *kalās* and a twentieth is [a *ghaṭikā*]; two *ghaṭikās* [are the equivalent] of a *muhūrta*; a day is thirty of those, or six hundred plus three *kalās*. (*RJV* 16, *YJV* 38)
>
> The elapsed [days] of the northern half-year, or the remaining [days] of the southern half-year, times two, divided by sixty-one, plus twelve, is the amount of the daytime. (*RJV* 22, *YJV* 40)

These brief rules tell us that a day including daytime and nighttime comprises thirty of the units called *muhūrtas* (equal to forty-eight of our minutes), or sixty *ghaṭikās*. Each *ghaṭikā* contains $\frac{201}{20}$ of the curious non-round units called *kalās*, of which there are thus 603 in a day.

The total half-yearly change in daytime length is said to be six *muhūrtas* or one-fifth of a day: that is, the summer solstice day will last 18 *muhūrtas* (14 hours and 24 minutes) from sunrise to sunset and the winter solstice day 12. The daylight-length d in *muhūrtas* in, say, the northern half-year is given by a linear proportion depending on the number n of days elapsed since the winter solstice:

$$d = 12 + \frac{2n}{61}.$$

Consequently, the summer solstice day will be reached when d equals eighteen *muhūrtas* or $n = (18-12)\cdot\frac{61}{2} = 183$, meaning that the year is assumed to be $2 \cdot 183 = 366$ days long.

The motions of the sun and moon in shorter time periods are also described:

> The moon is conjoined with a *nakṣatra* [for] one [day] plus seven [*kalās*], the sun [for] thirteen days and five ninths. (*RJV* 18, *YJV* 39)

Here we see that the length of a *nakṣatra* or lunar constellation has evidently become standardized as a constant amount of arc, because it takes the sun or moon a constant amount of time—$1+\frac{7}{603}$ days for the moon, and $13+\frac{5}{9}$ days for the sun—to travel across any one of them in the sky. Each *nakṣatra* must therefore be exactly $\frac{1}{27}$ of the circle of the heavens (although we still don't know exactly where their endpoints fall among the stars).

The above speeds for the sun and moon appear to have been worked out from the year length of 366 days in the following way. Since the sun goes around the complete circle of the *nakṣatras* five times in a five-year *yuga*, being overtaken by the faster moon exactly sixty-two times in that period, the moon must complete sixty-seven laps to the sun's five. Therefore in a *yuga* the moon must traverse $67 \times 27 = 1809$ equal *nakṣatras* while the sun traverses $5 \times 27 = 135$. The speed of each body in days per *nakṣatra*, times the number of *nakṣatras* it traverses in a *yuga*, must be equal to the total number D of days in a *yuga*, taken to be 366×5 or 1830. Dividing D by the number of *nakṣatras* traversed in each case gives the values specified in the verse:

$$D = 366 \times 5 = 1830 = 1809\left(1+\frac{7}{603}\right) = 135\left(13+\frac{5}{9}\right).$$

Another verse explains how to find where in the lunar cycle the current equinox will fall, in terms of the half-month and a smaller unit called the *tithi*. That unit, very important in later Sanskrit astronomy and calendrics, is a sort of artificial "lunar day," equal to exactly one-thirtieth of a synodic month (which is only about $29\frac{1}{2}$ ordinary days long). The verse states,

> The equinox [is] multiplied by two, diminished by one, multiplied by six; what is obtained is the half-months. Half of that is the *tithi*. (*RJV* 31, *YJV* 23)

We can understand the prescribed procedure as follows: If two successive equinoxes are considered to be separated by an interval of six regular, non-intercalary months, the actual (average) number of half-months in that interval, counting intercalation, will be $2 \cdot 6 \cdot \frac{62}{60} = \frac{62}{5}$. And of course, since the *yuga* starts with the winter solstice, the interval from the start of the *yuga* to its first (vernal) equinox will be half of the interval between two equinoxes, or $\frac{31}{5}$ half-months. So the number H of half-months that will have elapsed at the time of the *yuga*'s nth equinox must equal the initial $\frac{31}{5}$ half-months plus $\frac{62}{5} \cdot (n-1)$ half-months. To cast this result in units of half-months and *tithis*, where there are 15 *tithis* in a half-month, we may write

$$\begin{aligned} H &= \frac{62}{5} \cdot (n-1) + \frac{31}{5} \quad = \frac{62}{5}n - \frac{31}{5} \quad = \frac{31}{5}(2n-1) \\ &= \frac{30}{5}(2n-1) + \frac{1}{5}(2n-1) \\ &= \frac{30}{5}(2n-1) \quad \text{half-months} \quad + \frac{15}{5}(2n-1) \quad \textit{tithis} \\ &= 6(2n-1) \quad \text{half-months} \quad + \frac{6(2n-1)}{2} \quad \textit{tithis,} \end{aligned}$$

whence the rule as stated in the verse.

The rest of the *Jyotiṣa-vedāṅga*'s verses describe similar measures and rules for other aspects of this lunisolar calendric astronomy (the five star-planets are not mentioned), in addition to noncomputational details such as the name of the special deity for each *nakṣatra*. Unfortunately, the text is far too brief to tell us everything we would like to know about it. First, when was it composed? The reference to the winter solstice being "at the beginning of Śraviṣṭhā," if this location indicates (as it seems to in later texts) the neighborhood of the star that we call β Delphini, would provide a pretty good astrochronological dating within a century or so of 1200 BCE.[27] Again, though, we cannot be certain how the boundaries of the *nakṣatras* were defined at this period, or exactly how phenomena such as solstices, equinoxes, and new moons were identified. And we still have no unambiguous way to determine which statements and parameters, if any, were established by observation at the time of the composition of the text and which, if any, were reflections of earlier Vedic tradition.[28] So a conclusive date for the text still eludes us. Linguistically it seems to belong to the post-Vedic, pre-Classical Sanskrit corpus, which would probably put it instead somewhere around the fifth or fourth century BCE.[29]

Secondly, where was it composed? The statement that the summer solstice day is eighteen *muhūrtas* from sunrise to sunset while the winter solstice day is only twelve implies a terrestrial latitude of about 33–35° N, well north of most of the Indian subcontinent.[30] It has been suggested that this 18:12 ratio implies, variously, an origin somewhere in the far northwest in the first-millennium urban centers of the Gandhara region; or an origin farther south in the Ganges River basin, where the ratio gives fairly accurate day lengths near the equinoxes, although not near the solstices; or an origin in ancient Mesopotamia, latitude around 33–35°, whence the Indians borrowed

[27] A suggested alternative interpretation would push this date back as far as the early second millennium; see [Kak2005], p. 326.

[28] This is pointed out in, for example, [Ach1997].

[29] See [Pin1973b], p. 1.

[30] The reader is invited to verify this assertion by working through the later Indian trigonometric procedures in section 4.3.4 relating latitude to daylight length.

it via the Persian empire.[31] Finally, do we even understand the most basic parameters of the text's astronomical system? For instance, the *Jyotiṣa-vedāṅga* uses a year length of 366 days, an overly large value which would throw off the determination of the winter solstice by nearly four days in the course of one *yuga*. It has been argued that this parameter should be interpreted instead, at least for the *RJV*, as 366 sidereal days (what we would call diurnal rotations with respect to a particular fixed star) rather than 366 ordinary civil days from, say, sunrise to sunrise. This would imply a somewhat more accurate year length of approximately 365 civil days.[32] The text itself does not definitively resolve any of these uncertainties.

As in the case of the *Śulba-sūtras*, we have to resign ourselves to the fact that ancient mathematical texts in the service of sacred rites leave out a lot of the technical, methodological and historical details we might desire. We can definitely conclude, though, that mathematical practices were considered highly important in this context, as indicated by the following well-known verses of the *Jyotiṣa-vedāṅga*:

> Like the crests of peacocks, like the [head]-jewels of serpents, *jyotiṣa* is fixed at the summit of the Vedāṅga sciences. (*RJV* 35; *YJV* 4 has "*gaṇita*," "calculation," instead of "*jyotiṣa*.")
>
> The Vedas go forth for the sake of the sacrifices; and the sacrifices are prescribed in accordance with time. Therefore, whoever knows *jyotiṣa*, this science of regulating time, knows the sacrifices. (*RJV* 36, *YJV* 3)

2.5 VEDIC INDIA AND ANCIENT MESOPOTAMIA

We have seen in the preceding sections occasional allusions to possible connections between ancient Indian mathematical and astronomical ideas and those of ancient Mesopotamia. As this is one of the crucial (and contentious) issues in the history of Indian science, this chapter concludes with a brief outline of it.[33]

The heart of the problem is that, thanks to the preservation of thousands of ancient cuneiform clay tablets written in Sumerian and Akkadian, we know much more about the development of Mesopotamian mathematical

---

[31] See the explanations of these hypotheses in, respectively, [Fal2000], p. 117; [Oha2000], p. 344; and [Pin1973b], p. 4.

[32] See [Pin1973b], pp. 7–8. [Oha2000], p. 343, disagrees on the grounds that a 366-day year would better preserve the synchronization with the lunar cycle, which was culturally more important than accurate determination of the start of the year with respect to the solstice. This is possible, but if true it undermines the astrochronological assumption of a precisely specified celestial location for the winter solstice.

[33] The standard exposition of the hypothesis of substantial transmission from Mesopotamian to Indian astronomy is [Pin1973b]; a recent re-examination of it appears in [Fal2000]. The history of Mesopotamian astral science is surveyed in, for example, [HuPi1999].

science in the first and second millennia BCE than we do about its Indian counterpart. Consequently, Assyriologists or historians of Mesopotamia have been able to reconstruct quite plausibly the course of the former development. They found that by the Old Babylonian period in the early second millennium, Mesopotamian culture had developed a positive obsession with omens, including celestial omens which were considered particularly fateful for the fortunes of kings and states. The sun, moon, and star-planets were associated with particular gods who sent celestial signs as warnings to humans; lunar and solar eclipses were the most ominous and important of these. In this period scribes already possessed well-developed literate mathematics with a standardized sexagesimal place value number system. But the astral omens were mathematically very simple, merely describing the consequences of a particular event occurring in a particular month (beginning with the first sighting of the lunar crescent in the west) and day (beginning at sunset).

During the second millennium, Mesopotamian omen-observers appear to have figured out a great many things about the periodic cycles of the heavens. Early first-millennium copies of omen catalogues compiled probably in the mid- to late second millennium (although some of their material seems to be older) show that celestial diviners were particularly interested in the so-called "synodic phenomena" of stars and planets, that is, their disappearances and reappearances close to the sun and, in the case of planets, their retrograde motion. All of these had ominous significance, and their periods were very roughly quantified. Days were divided into twelfths and 360ths and the year into twelve months of thirty days each, arranged around the equinoxes and solstices, plus an irregularly intercalated thirteenth month. Circles of constellations were identified for various "paths" in the sky. Time was measured with gnomons and water clocks, with their seasonal cycles tabulated in constant increments and decrements of shadow length or water volume.

In the first half of the first millennium, much of this activity apparently became further standardized. For example, celestial reference circles divided into 360 equal parts and lunar months divided into thirty equal "days" were used. Celestial observers had to make detailed regular records and reports of what they saw. After about 500 BCE the pace of mathematization accelerated: regular, accurate calendar intercalation was established, and the twelve zodiac constellations were assigned equal lengths. Before the turn of the millennium, Babylonian scribes had produced an extremely sophisticated and comprehensive computational system of predictive astronomy, some of which was later adopted by Hellenistic Greek mathematicians.

This grossly oversimplified history nonetheless reveals some of the resemblances that have inspired hypotheses about transmissions from Mesopotamia to India. There are some similarities between cuneiform constellation-lists and Vedic *nakṣatra*-lists; there is the intriguing sexagesimal division of Prajāpati's bricks; there is the schematic 360-day year; and, in the *Jyotiṣa-vedāṅga*, there are also the water clock, the "lunar day" or *tithi*, and periodic daylight-length formulas with a 3:2 maximum-minimum ratio such as cuneiform tables used. (Later Indian texts from the centuries around the turn of the millennium also speak of telling time by gnomon-shadows and of celestial omens, including synodic phenomena of the planets.) Moreover, in the middle of the first millennium BCE the Achaemenid empire, which stretched from West Asia to northwest India, provided a potential conduit for these ideas. The suggestion of a Mesopotamian origin thus furnishes a coherent and plausible explanation for at least some of the features of Indian mathematical astronomy at the close of the Vedic period.

On the other hand, there is nothing in these similarities that necessarily has to be accounted for by transmission, and there are no indisputable traces of transmission such as Akkadian loan-word technical terms in Sanskrit texts. Nor are there any equally clear avenues of transmission for the centuries before the spread of the empire of the Persians. And although formulas for calculating *muhūrtas* and *tithis* are not attested before the *Jyotiṣa-vedāṅga*, the quasi-sexagesimal concept of dividing daytimes, nighttimes and half-months into fifteen parts is at least as old as the *Śata-patha-brāhmaṇa*.[34] Perhaps these similar concepts independently evolved in both cultures from naturally obvious ideas about number and the cosmos. On the third hand, perhaps concepts like sexagesimal time division seem deceptively obvious to us just because they are part of our own Mesopotamian-influenced mathematical heritage. In the absence of conclusive proof either way, and with so many possibilities offered by our many uncertainties concerning the Indian sources, we are ultimately reduced to arguing over probability and plausibility, where no two historians are likely to come up with exactly the same answer.

[34] See, for instance, *Śata-patha-brāhmaṇa* 10.4.2.18–30.

Chapter Three

Mathematical Traces in the Early Classical Period

The middle of the first millennium BCE is generally considered the approximate end of the Vedic period, when the corpus of *śruti* or revealed scriptures was completed and Sanskrit started to shift from a primary language to a learned language.[1] As noted in chapter 1, at about the same time, Buddhism and Jainism emerged as distinct religious movements associated with the teachers Buddha and Mahāvīra, and the Persian Achaemenid empire reached to the northwest of the South Asian subcontinent. Within the next couple of centuries, Sanskrit was linguistically analyzed and standardized by the great grammarian Pāṇini in the form that is now called Classical Sanskrit, the first surviving written documents in an Indo-Aryan language were inscribed, and the incursions of Alexander's armies spawned the first of the Indo-Greek kingdoms.

What was the impact of these various developments on mathematical ideas and methods during the centuries just before and just after the turn of the Common Era? The documents and mathematical texts from this time are exasperatingly scanty, and few definite conclusions can be pieced together from them. However, certain developments can be traced in broad outline to give at least a notion of mathematical thought in this period.

3.1 NUMBERS AND NUMERALS

The widespread use of writing and the resurgence of uban centers fostered the growth of government bureaucracy, which depended heavily on quantification. A text on statecraft called the *Arthaśāstra*, composed in about the third century BCE, lists many kinds of financial transactions that government officials were expected to master, and many detailed regulations about taxation, fines, interest rates, and so on. Such computations of course employed the concepts of gains and losses in revenue; the latter were called *kṣāya*, "decrease," which in later mathematical texts was one of the terms for negative numbers in general.[2] The *Arthaśāstra* also supplied some simple rules for standard timekeeping procedures similar to the arithmetic meth-

[1] New research, however, suggests that parts of the late Vedic corpus may be as much as several centuries younger, coexisting with Classical Sanskrit. See [Bro2007], chapter 3. For an analysis of current theories on the age and origin of Indian writing systems, see [Sal1995].

[2] *Arthaśāstra* 2.7.2, [Kan1960], vol. 1, p. 43. The *Arthaśāstra* is translated in [Sham1960], and the date of the text is discussed at length in [Goy2000].

Figure 3.1 Non-place-value numerals 10 and 7 in a Buddhist cave inscription.

ods found in the *Jyotiṣa-vedāṅga*. These practices doubtless expanded the cultural importance of numeracy and written numbers, but the *Arthaśāstra* preserves no details about how they were used.

The Indian development of the place value decimal system, in which an integer of arbitrary size can be expressed with just ten distinct digits, is such a famous achievement that it would be very gratifying to have a detailed record of it. Unfortunately, the details in the existing evidence are few and far between. The main facts from which the evolution of Indian written numeral forms has been reconstructed are summarized here.

A few non-place-value numerals are found in some of the earliest surviving Indian inscriptions, the monuments of Aśoka around the middle of the third century BCE. These numerals are part of the Brāhmī script, which was in use during Aśoka's time in many parts of India. The origins of Brāhmī are unknown; it is the ancestor of all known subsequent Indian scripts, and of several in Central and Southeast Asia as well. Judging from the evidence of inscriptions and coins, its number system was additive and multiplicative, with separate signs for the values 1 to 9, 10 to 90, 100, and 1000. A multiple of 100 or 1000 was represented by some modification of the sign for that power, for example, by combining it with the sign for the multiplier number.[3] Some typical Brāhmī numerals from a second-century BCE inscription in a

[3][Sal1998], pp. 17–19, 56–57. Various hypotheses have been advanced concerning possible derivations of non-place-value Brāhmī numerals from those of other scripts, including Kharoṣṭhī (see below), hieratic Egyptian, cuneiform, and Chinese, but none of these has been universally accepted; see [Sal1998], pp. 59–60.

Buddhist cave at Nana Ghat in Maharashtra, representing the signs for 10 and 7, are reproduced in figure 3.1.[4] The symbol for 10 is the one somewhat resembling a Greek alpha, while the symbol for 7 may look surprisingly familiar!

Another script represented in Aśokan inscriptions is Kharoṣṭhī, a writing system derived from the Persian Achaemenid bureaucracy's Aramaic script. Kharoṣṭhī was used in Central Asia and northwestern India until about the fourth century CE. It has been suggested as the possible source of the Brāhmī characters, but the argument remains speculative: for one thing, Kharoṣṭhī takes after its West Asian ancestors in being written from right to left, whereas Brāhmī and all its descendants go from left to right.[5]

Kharoṣṭhī has its own additive non-place-value numeral system, presumably also ultimately based on that of Aramaic script. Its most unusual feature is a sort of quasi-quaternary base, where 4 was represented by a character resembling X, to which successive vertical strokes were adjoined to represent 5, 6, and 7. A pair of the X's signifies 8, and is supplemented by a vertical stroke to make 9 (the number 10, however, has a separate sign like a curved stroke). For values between 10 and 100, combinations of the signs for 10 and 20 were used. None of these notational features appear to have been adopted in other numeral systems, and they did not survive the demise of the Kharoṣṭhī script in the early first millennium.[6]

The earliest extant physical examples of decimal place value numerals are found in inscriptions from around the middle of the first millennium CE, written in scripts derived from Brāhmī. (An instance from a late first-millennium inscription in Gwalior, Madhya Pradesh, is shown in figure 3.2.) At present, the first such inscription known in an Indian source may be the one on a certain copper plate from Gujarat (copper plates were a typical medium for records of transactions such as grants and deeds, where the document needed to be portable but permanent). It contains the place value year-number 346 in the so-called Kalacuri era, corresponding to about 595 CE.[7] Decimal place value numbers are also found in some inscriptions from Indianized cultures in Southeast Asia around the same time. For instance, a year-number of 605 in the Indian Śaka era, corresponding probably to 683 CE, is recorded in stone inscriptions in Indonesia and Cambodia.[8] We can certainly infer that if the decimal place value system had been incorporated into epigraphic styles over much of South and Southeast Asia by this time, it must have originated quite a bit earlier.

---

[4]See [Gok1966], pp. 16, 22–24. I am grateful to Professor William Casselman for permission to use the photographs in figures 3.1 and 3.2.

[5][Sal1998], pp. 42–49.

[6][Sal1998], p. 58; [Muk2003].

[7]However, it has been persuasively argued that this particular record is spurious and was actually inscribed at a later date; see [Sal1998], p. 61. Unfortunately, although copper plate property deeds and grants have the advantage of permanence in preserving specimens of medieval writing, their disadvantage is that they are a tempting subject for forgery by later claimants of the property!

[8]See [Sal1998], p. 61.

Figure 3.2 Decimal number 270 (top left) inscribed in a temple in Gwalior.

But we do not need to rely only on such inferences to push back the date of origin of decimal place value beyond the time of its earliest known inscriptional records. The content of some older textual sources includes hints about the writing of numbers that suggest a place value system, although of course the texts themselves are physically recorded only in much later manuscript copies. For example, a commentary from probably the fifth century CE on an ancient philosophical text, the famous *Yoga-sūtra* of Patañjali, employs the following simile about the superficial "changes of inherent characteristics" (*Yoga-sūtra* 3.13):

> Just as a line in the hundreds place [means] a hundred, in the tens place ten, and one in the ones place, so one and the same woman is called mother, daughter, and sister [by different people].[9]

Even earlier, the Buddhist philosopher Vasumitra in perhaps the first century CE used a similar analogy involving merchants' counting pits, where clay markers were used to keep track of quantities in transactions. He says, "When [the same] clay counting-piece is in the place of units, it is denoted as one, when in hundreds, one hundred."[10] Such statements clearly expect the audience to be familiar with the concept of numerical symbols representing different powers of ten according to their relative positions. Due to the brevity of their allusions and the ambiguity of their dates, however, they do not solidly establish the chronology of the development of this concept.

[9] Cited in [SarS1989], p. 63.
[10] Cited in [Plo2007b], p. 395; see also [Fil1993].

A different representation of decimal place value is revealed by a verbal notation called by medieval authors *bhūta-saṅkhyā* or "object-numbers," here designated the "concrete number system." Its function is to provide synonyms for ordinary number words such as "three" or "twelve." Recall that in Sanskrit, at least after the Vedic period, even technical treatises were most often composed in verse. Since Sanskrit (like other languages) generally has only one standard number word for each number, it can be difficult to fit such words into the metrical structure of verses so as to convey a desired mathematical meaning without ruining the scansion of the verse. The concrete number system gets around this problem by allowing the name of any object or being associated with a particular number, in nature or in religious tradition, to stand for that number. For example, the word "hand" in this system means two, "fire" means three (for the three sacrificial fires mentioned in section 2.2), "limb" means six (for the traditional six limbs of the Veda), and so forth.

The concrete number system, to judge from all its extant examples, has apparently always been a place value system, representing large numbers with strings of words that stand for its individual digits or groups of digits, in order from the least significant to the most significant. Thus, if we encounter, say, the verbally expressed concrete number "Veda/tooth/moon," we translate it as "four [for the four Vedic collections]/thirty-two/one," and write it as 1324. These concrete numbers are not combined with number words signifying powers or multiples of ten, so their only unambiguous interpretation is as pure decimal place value. Hence the idea of a positional system for numerals must have been commonplace by the time the concrete number system was invented.

A firm upper bound for the date of this invention is attested by a Sanskrit text of the mid-third century CE, the *Yavana-jātaka* or "Greek horoscopy" of one Sphujidhvaja, which is a versified form of a translated Greek work on astrology. Some numbers in this text appear in concrete number format, as in its final verse:

> There was a wise king named Sphujidhvaja who made this [work]
> with four thousand Indravajra [verses in the "Indravajra" meter],
> appearing in the year Viṣṇu/hook-sign/moon.[11]

The year-number translates as "one [for the deity Viṣṇu]/nine [from the shape of its numeral]/one," and consequently means year 191 of the Śaka era beginning in 78 CE. So it corresponds to 269 CE, or perhaps 270, depending on the (unspecified) month and day. Evidently, then, positional decimal numerals were a familiar concept at least by the middle of the third century, at least to the audience for astronomical and astrological texts.

Exactly how and when the Indian decimal place value system first developed, and how and when a zero symbol was incorporated into it, remain mysterious. One plausible hypothesis about its origin links it to the symbols

[11] *Yavana-jātaka* 79.62; see [Pin1978c] 1, p. 506 and 2, p. 191.

used on Chinese counting boards as early as the mid-first millennium BCE. These counting boards, like the Indian counting pits mentioned above, had a decimal place value structure: they were divided into columns representing successive powers of ten, with units on the right. Small rods were arranged in regular patterns in the columns of the board to designate numbers from 1 to 9, and a column left blank signified a zero. Indians may well have learned of these decimal place value "rod numerals" from Chinese Buddhist pilgrims or other travelers, or they may have developed the same concept independently from their earlier non-place-value system; no documentary evidence survives to confirm either conclusion.[12]

We will see in section 3.3 that there are textual indications of a written symbol for zero in India even before the start of the Common Era, but it is not clear whether the symbol was part of place value notation at that time. The use of zero in decimal numerals and its characteristic round shape may have been reinforced by the round zero markers in sexagesimal place value numerals introduced to India in Greek astronomical and astrological texts.[13] The story behind the transmission of these texts is briefly recapitulated in the following section.

3.2 ASTRONOMY, ASTROLOGY, AND COSMOLOGY

3.2.1 The emergence of Greco-Indian astrology

After the breakup of Alexander's empire at the end of the fourth century BCE, much of the territory he had claimed in western India fractured into small kingdoms under Greek rulers. The control of these "Indo-Greek" states in the centuries around the turn of the millennium largely passed to new invaders, the Scythians or "Śakas." Their name is preserved in the so-called "Śaka era" of the Indian calendar, whose beginning falls in 78 CE and which is still widely used in India. Indian Greeks (called "Yavanas" in Sanskrit, a transliteration of "Ionian," i.e., Greek) remained a powerful group under the Śaka rulers, and their numbers were swelled in the early centuries CE with an influx of Greek traders and settlers who fostered commerce and communication between western India and the Hellenistic culture of the Roman empire. At the same time, many of the Yavanas assimilated into Indian culture, adopting Buddhist or Hindu beliefs, taking Sanskrit names, and intermarrying with Indian dynasties.

This cultural fusion gave birth to Indian horoscopic astrology, largely through the medium of the abovementioned work, *Yavana-jātaka*, of the late third century, whose author describes it as a redaction of a Sanskrit

[12]The Chinese decimal system is described in [Dau2007], pp. 189–199, and arguments for its transmission to India are presented in [LamAn1992], pp. 176–185.

[13][Pin2003c] suggests that these Greco-Babylonian sexagesimal place value numbers may even have originally inspired or helped inspire the Indian development of decimal place value and its zero. For a comparative survey of the role and history of zero in various ancient mathematical traditions, see [Gup1995].

prose translation of a Greek astrological work made in 149 or 150 CE. The *Yavana-jātaka* presents an Indianized version of traditional Greek astrology, including the first known appearance in India of the twelve familiar signs of the Greco-Babylonian zodiac (the Ram, the Bull, the Twins, etc.).[14] It also briefly explains rules for astronomical calculations required in astrology.[15] These combine earlier lunisolar calendric techniques like the ones in the *Jyotiṣa-vedāṅga* with simple arithmetic rules based on astronomical periods or cycles for computing the ascendant or horoscope point and the synodic phenomena of the planets.[16]

The calendric lunisolar astronomy of the *Jyotiṣa-vedāṅga*, with its circle of *nakṣatra* constellations and diverse time divisions, thus coalesced with the Mesopotamian-influenced computational schemes of Greek astrology to form the foundation of all later Indian *jyotiṣa*. To this synthesis Greco-Babylonian sources contributed the twelve signs of the standard zodiac and a wealth of astrological concepts and technical terms. They also apparently transmitted the division of the zodiac into 360 degrees, along with sexagesimal fractions dividing the degree into sixty arcminutes (*liptā*, from Greek *lepton*) and arcminutes into sixty arcseconds.[17] In the *Yavana-jātaka* time units too, namely the *muhūrta* and the *ghaṭikā* or half-*muhūrta* of the *Jyotiṣa-vedāṅga*, were subdivided sexagesimally as far as the fourth sexagesimal place. For instance, the fractional part called a *kalā*, which in the *Jyotiṣa-vedāṅga* meant $\frac{1}{603}$ of a day, now meant a sixtieth of a *muhūrta*.[18] These sexagesimal place value fractions for measurement of arcs and time became standard in Indian astronomy and mathematics, as they had in the Greek tradition.

3.2.2 Early Classical Sanskrit astronomy and trigonometry

A few more of the mathematical details of Indian astronomy in this period can be tentatively filled in from later summaries of some early astronomical treatises. These summaries are preserved in a work called the *Pañca-siddhāntikā* ("Five *siddhāntas* [astronomical treatises]") composed in the sixth century by an author named Varāhamihira.[19] The five treatises are identified individually; some of them are from the early centuries CE, while

[14]See [Pin1978c] 2, pp. 195–198, for a discussion of arguments in favor of an earlier date for Indian use of the zodiac based on iconographic evidence.

[15]The glossary of geocentric-astronomy technical terms in section 4.1 may be useful in reading the following passages.

[16]Such explicit quantitative rules for estimating the positions of the five star-planets first appear in India in a text on divination or interpretation of omens, dating probably from the first century CE; see [Pin1987c]. Apparently the inclusion of the star-planets along with the sun and moon in Indian mathematical astronomy was motivated by their importance in the post-Vedic disciplines of astrology and astral omens, arising from contacts with other cultures.

[17]See, for example, *Yavana-jātaka* 79.23–24: [Pin1978c] 1, p. 408, and 2, p. 188.

[18]For example, in *Yavana-jātaka* 79.11: [Pin1978c] 1, p. 495 and 2, p. 187.

[19]This work is translated in [ThDv1968], in [NeuPi1970] (on whose commentary the following discussion is largely based), and in [SarKu1993].

others are as late as the early sixth century and thus fall into the period of the so-called "standard *siddhānta*" or "medieval *siddhānta*," discussed in the next chapter. Unfortunately, the contents of the works are to some extent mixed together in the *Pañca-siddhāntikā*, so it is not always clear which rule comes from which treatise. This means that the dating of most of its sources and material is uncertain.

The five *siddhāntas*, in the order in which Varāhamihira first names them (not the same as the order in which he cites their contents) are identified as follows: *Pauliśa-siddhānta*, "treatise of Puliśa" (Paulos?); *Romaka-siddhānta*, "treatise of Romans" or westerners; *Vasiṣṭha-siddhānta*, "treatise of [the sage] Vasiṣṭha"; *Sūrya-siddhānta*, "treatise of the Sun"; and *Paitāmaha-siddhānta*, "treatise of Pitāmaha" (the deity Brahman). The summary of the *Paitāmaha-siddhānta* includes an epoch date (or starting point for calendar calculations) that falls in Śaka year 2 or 80 CE, and the *Vasiṣṭha-siddhānta* is mentioned in the third-century *Yavana-jātaka*, so these would seem to be the earliest sources in the collection. Their content, as far as we can tell, concerns arithmetic rules for calendar computations and (in the *Vasiṣṭha-siddhānta*) determining the synodic phenomena of planets, similar to the algorithms in the *Jyotiṣa-vedāṅga* and the *Yavana-jātaka*.

The remaining three texts were composed or revised in the early sixth century, judging from the epoch dates associated with them. They blend numerical algorithms of the sort mentioned above with concepts and methods derived from spherical astronomy, such as terrestrial latitude and longitude on a spherical earth, and celestial latitude and declination for heavenly bodies. Since we cannot tell how much of this mathematics dates from the first few centuries CE and how much of it is contemporary with the fully geometrized mathematical astronomy of the mid-first millennium, which is described in the next chapter, we will not investigate its details here.

One of the geometric features in the *Pañca-siddhāntikā*, however, demands comment. It begins with the following statement:

> The square root from the tenth part of the square of a circumference of three hundred and sixty is the diameter. When one has made four parts [quadrants] in this [circle], the Sine of an eighth part of a zodiacal sign [is to be determined].[20]

This gives the relation between the circumference C of a circle and its diameter D as $D = \sqrt{C^2/10}$, or, as we would say, $\pi \approx \sqrt{10}$. The verse associates Sines[21] with arcs at intervals equal to $30°/8$, or $3°45'$. The subsequent verses give six rules for computing Sines that are equivalent to what we would write as the following, if the radius of the circle is $R = D/2$ and θ is an arbitrary

---

[20] *Pañca-siddhāntikā* 4.1. This rule and the subsequent verses described below appear in [NeuPi1970]. vol. 1, pp. 52–57.

[21] The term "Sine" and the notation $\mathrm{Sin}\, x$ here use a capital letter to signify that they are quantities normalized to a nonunity radius R rather than ratios of quantities normalized to 1, as in modern trigonometry: $\mathrm{Sin}\, x = R \sin x$. Similarly, the capitalized term "Chord" or "Crd" indicates the length of a chord in a circle of radius R.

Table 3.1 Sine values for $R = 120$ in the *Pañca-siddhāntikā*

Arc (°)	Sine	Arc	Sine	Arc	Sine
3;45	7;51	33;45	66;40	63;45	107;37
7;30	15;40	37;30	73;3	67;30	110;52
11;15	23;25	41;15	79;7	71;45	113;37
15;0	31;4	45;0	84;51	75;0	115;55
18;45	38;34	48;45	90;13	78;45	117;42
22;30	45;56	52;30	95;12	82;30	118;59
26;15	53;5	56;15	99;46	86;15	119;44
30;0	60;0	60;0	103;55	90;0	120;0

arc:

$$(1.)\ \mathrm{Sin}\, 30^\circ = \sqrt{R^2/4}, \qquad (2.)\ \mathrm{Sin}\, 60^\circ = \sqrt{R^2 - \mathrm{Sin}^2\, 30^\circ},$$

$$(3.)\ \mathrm{Sin}\, \theta = \sqrt{\left(\frac{R - \mathrm{Sin}(90^\circ - 2\theta)}{2}\right)^2 + \left(\frac{\mathrm{Sin}(2\theta)}{2}\right)^2},$$

$$(4.)\ \mathrm{Sin}(90^\circ - \theta) = \sqrt{R^2 - \mathrm{Sin}^2\, \theta}, \qquad (5.)\ \mathrm{Sin}\, 45^\circ = \sqrt{R^2/2},$$

$$(6.)\ \mathrm{Sin}\, \theta = \sqrt{60(R - \mathrm{Sin}(90^\circ - 2\theta))}.$$

Most of these rules are obvious from modern trigonometric identities; the third can perhaps be more easily understood if expressed as $2\,\mathrm{Sin}\,\theta = \mathrm{Crd}(2\theta)$ $= \sqrt{(R - \mathrm{Cos}(2\theta))^2 + \mathrm{Sin}^2(2\theta)}$, where Crd stands for the Chord of an arc. The last rule is trigonometrically accurate if $R/2 = 60$. This implies that $D = 240$, which is only a rough approximation to $D = \sqrt{(360)^2/10}$ as specified in the quoted verse.

The *Pañca-siddhāntikā* then verbally states values for the Sines in the first quadrant in sexagesimal linear units called by the same names as the Sanskrit terms for arcminutes and arcseconds. These values are listed in table 3.1, which follows the standard convention of using a semicolon as a "sexagesimal point" to separate integer values from sixtieths.

If this material on Sines in the *Pañca-siddhāntikā* is indeed reproduced as it appeared in a text of the early first millennium, then table 3.1 represents the first surviving exemplar of an Indian Sine table. It also bears a strong resemblance to some aspects of Greek trigonometry of Chords; in fact, its Sine values for arcs at intervals of $3^\circ 45'$ could equally well be entries in a Chord table for arcs up to 180° at intervals of $7^\circ 30'$ with $R = 60$. The value $R = 60$ is attested in Hellenistic trigonometry, and so is the division of the quadrant into sixths, that is, arcs of $15^\circ = 2 \cdot 7^\circ 30'$.[22]

[22] [Pin1976], pp. 113–114.

Given these similarities and the other Greek traces in the *Pañca-siddhāntikā*, it seems reasonable to hypothesize (although it remains unproven) that the Indian invention of trigonometry of Sines was partly inspired by Hellenistic Chords. Indian astronomers appear to have been the first to think of replacing the rather clumsy Chord geometry of right triangles inscribed in a semicircle with the simpler Sine geometry of right triangles in a quadrant.[23]

3.2.3 Cosmology and time in the Purāṇas

The early *jyotiṣa* texts that are concerned with calendric computations involving the sun, moon, and *nakṣatras* do not specify a physical configuration for them in the space beyond the earth. The earliest such cosmological system known in Indian texts is a nonspherical model of the universe first fully depicted in the sacred texts called the Purāṇas, part of the scriptural tradition of *smṛti* as opposed to the *śruti* corpus of the Vedas. They recount the exploits of the gods, the creation of the world, and other cosmic events. These texts were shaped over several centuries, but their basic picture of the universe seems to have emerged no later than the early first millennium CE.[24]

In this picture, the world is enclosed in an egg-shaped shell—the so-called "cosmic egg"—as a stack of circular disks. The flat earth is in the center, a disk of immense proportions, corresponding to around half the extent of the solar system in the model of modern astronomy. Its center is occupied by a circular continent, the known world, surrounded by the salt ocean; beyond the ocean is another continent shaped like an annulus or ring, concentric with the central continent and surrounded by another ocean made of sugar-cane juice. These alternating rings of various lands and liquids continue outward, making seven continents and seven oceans in all. In the center (imagined as north of the Himalayas) stands the immense Mount Meru, on whose summit the gods live.

The celestial bodies circle in wheel-like orbits, parallel to the surface of the flat earth, around the "axle" of Mount Meru, which makes them appear to set when they go behind it and rise when they emerge on its other side. The sun's orbit is placed closest to the earth at the bottom of the disk of heaven, the moon above the sun, the *nakṣatras* above the moon, and then the five star-planets in the same order—Mercury, Venus, Mars, Jupiter, Saturn—established in Hellenistic astronomy. Above these, beyond the summit of Meru, are the Seven Sages (our Big Dipper) with the pole star in the middle, which is attached to the orbiting bodies by cords of cosmic wind that keep pulling them in circles around Meru. Above the pole-star are the layers of various higher heavens; below the earth are stacked the various hells. The earth is supported from underneath by, typically, a great serpent or tortoise or other creature.

---

[23]See [Too1994], pp. 7–9, for a detailed discussion of Hellenistic Chord computations.

[24]The details of the following description are discussed in, for example, [Pin1990], [Min2004a], and [Plo2005a]. Sacred texts' allusions to cosmic time cycles are summarized in [Pin2002a] and [Duk2008], and explored in depth in [Gon2002].

This spatial model is accompanied by a temporal one using immense cycles of time. The universe is created and destroyed during one *kalpa* or day-and-night period of Brahman, which lasts for 4,320,000,000 years. There is a shorter period called a *mahāyuga*, or "great age," of 4,320,000 years: it is divided into four smaller intervals in a 4:3:2:1 ratio, during the course of which the world decays from good to bad, as in the Golden, Silver, Bronze and Iron Ages of Greek legend. The last and worst of these sub-periods is the Kaliyuga, which is one-tenth of a *mahāyuga*, or 432,000 years long.[25] There are seventy-one *mahāyugas* in a period called a *manvantara*; fourteen *manvantaras* and some extra years make up a *kalpa*.

This cosmology and its variants became the standard worldview of later Hindu scripture, and profoundly influenced its Buddhist and Jain counterparts as well. As we will see in later chapters, it came into collision sometimes with the quantitative spherical astronomy of the medieval period, which asserted a round instead of a flat earth, a comparatively tiny Mount Meru at the earth's north pole instead of a massive cosmic pillar hiding the stars, and a revised order of distances for the sun, moon, and star-planets. Moreover, the Purāṇas reflect some hostility toward the astrological theories that powerfully spurred the development of mathematical astronomy, perhaps because of their foreign influences from outside the world of *dharma*, perhaps because of their pretensions to knowledge of the future. Already in the mid-first millennium BCE, a sermon of the Buddha had warned the faithful against the sinfulness of divination or foretelling by interpreting omens.[26] Subsequently the Purāṇas showed a similar suspicion toward "calculators" or astrologers, assigning them to particular locations in the subterranean hells along with other professional wrongdoers such as hunters, tanners, and women who sell their hair.[27] Despite these frictions, mathematical astronomy and sacred cosmology retained some concepts in common, particularly the four-billion-year time frame within which all the motions of the heavens were calculated.

3.3 MATHEMATICAL IDEAS IN OTHER DISCIPLINES

Other mathematical developments in these centuries can only be inferred from hints in nonmathematical sources, such as the philosophical texts we have mentioned in section 3.1, and from Sanskrit fields of study tangentially related to mathematics. The two disciplines of this sort that are briefly examined here are the Vedāṅgas of grammar and prosody (poetic metrics).

Grammatical analysis is indisputably the queen and servant of the Sanskrit

---

[25][Pin1990], p. 275, hypothesizes that the length of the Kaliyuga is the fundamental parameter of the system, and is ultimately derived from the sexagesimal $432{,}000 = 2 \cdot 60^3$ years in the antediluvian era of Babylonian cosmology.

[26]See [Pin1992a].

[27][Tie2001], pp. 103–109. My thanks to Johannes Bronkhorst for drawing this point to my attention and supplying the reference.

sciences. As discussed in section 6.4.3, it influenced even the structure of exposition and demonstration in Sanskrit mathematical works. Its canonical text is the *Aṣṭādhyāyī* ("Eight chapters") of Pāṇini, compiled perhaps in the fifth or fourth century BCE. In this work Pāṇini, building on the traditional phonetic structure of Sanskrit (described in section A.1 of appendix A), presented a system for deriving and analyzing phonetically and grammatically valid Sanskrit expressions. Its extremely compressed *sūtra* format relies on notational symbolism and what has been described as an artificial language to explain proper operation on the elements of Sanskrit.[28]

The way Pāṇinian grammatical symbolism works can be (very incompletely) illustrated by examining one of the *Aṣṭādhyāyī*'s rules, which essentially defines what linguists call vowel gradation (i.e., substituting one vowel for another in a word to indicate a change in the word's semantic meaning or grammatical form, as in English "ring, rang, rung"). The *sūtra* states:

> *iko guṇa vṛddhī*
>
> [The replacement] of "ik" [is] the medium (*guṇa*) grade [or] lengthened (*vṛddhi*) grade [of the vowel in question].[29]

But what is the "ik" that is to be replaced here? It doesn't look like a vowel. Indeed, "ik" is not a vowel but a symbol based on Pāṇini's metalinguistic categorization of the sounds of Sanskrit into phonetic classes. He previously defined the first two phonetic classes, for the simple short (or "zero-grade") vowels *a*, *i*, *u* and for the short vocalic forms *ṛ* and *ḷ* of the semivowels, by assigning certain consonants to them as follows:

1. *a i u ṇ*   2. *ṛ ḷ k*

The notation "ik" thus means "all the short vowels from *i* up through the *k*-class, that is, ending with *ḷ*." That is, the quoted *sūtra* means that *i*, *u*, *ṛ* and *ḷ* are the zero-grade vowels that are to be replaced with vowels of higher grades when other grammatical rules decree that replacement is necessary. Although this is not a mathematical formula per se, it is part of a quantitative, symbolic and abstract approach to understanding language that was unique in the ancient world.

It has sometimes been argued that the complex use of abstract symbols and metalinguistic rules found in Pāṇini's grammar stimulated the early development of algebra and of number systems in India.[30] Be that as it may, the *Aṣṭādhyāyī* counts as certainly the most highly developed linguistic analysis of its time, as well as arguably an early instance of formal language, even if its connections to explicitly mathematical subjects are still speculative.

---

[28] [Sta2006], pp. 119–125.

[29] *Aṣṭādhyāyī* 1.1.3, [Vas1962], vol. 1, p. 4. See [Gillo2007] for a concise overview and analysis of Pāṇini's grammar.

[30] See [JosG2000]. More conservatively, it has been suggested that Pāṇini's use of *lopa* or elision constituted a type of "grammatical zero" that directly parallels the concept of mathematical zero: [Pandi2003]. And [Kad2007] draws interesting parallels between Pāṇinian linguistic recursion and place value notation. Finally, the relationship of phonetic science to early Indian music and its quantitative concepts is discussed in [Row1992].

The use of symbolism and mathematical concepts also distinguishes the other technical genre mentioned here, namely, prosody. As explained more fully in section A.2, prosody or metrics consists of determining and identifying the various combinations of Sanskrit syllables in a quarter-verse, where the number n of syllables is given and each syllable may be either heavy or light. Each distinct combination of such syllables is a (potential) poetic meter (although not all theoretically possible meters are actually used in poetry).

In the *Chandaḥ-sūtra* ("Prosody *sūtras*") of Piṅgala, dating to perhaps the third or second century BCE, there are five questions concerning the possible meters for any value of n: (1) What is the arrangement of the "extension" or list of all the possible meters with n syllables in a quarter-verse? (2) What is the serial number m within that list of any given metrical pattern of n syllables? (3) For a given serial number m, what is the corresponding metrical pattern? (4) What is the number of possible metrical patterns with n syllables? (5) What is the number of metrical patterns with n syllables that contain a specified number $1 \leq p \leq n$ of heavy or light syllables? The rules for answering these questions are briefly described below.[31]

(1) The list of metrical patterns containing n syllables is produced by enunciating each possible pattern, starting with the meter having n heavy syllables and ending with the meter of n light ones. The sequence of metrical patterns in between is determined by an algorithm illustrated for $n = 3$ in the following table. Since we don't know what notation was used in Piṅgala's time, we use the symbol | to stand for a heavy syllable and @ for a light one.

1.	|	|	|
2.	@	|	|
3.	|	@	|
4.	@	@	|
5.	|	|	@
6.	@	|	@
7.	|	@	@
8.	@	@	@

By looking at the above list upside down while squinting slightly, the reader will easily see that it corresponds exactly to the numbers 0 to 7 in binary notation (if @'s are read as 0's and |'s as 1's). That is, for a given serial number m in the list, the corresponding metrical pattern read from right to left is equivalent to $n - m$ written in binary. So the algorithm for constructing the list of meters with n syllables is just a trivial variant of binary representation.

(2) and (3) To find out the correct pattern for a given serial number m in the list of patterns, we manipulate m as follows: If m is even, write down @

[31] The following descriptions rely on the more detailed presentations given in [SarS2003] and [Srid2005], both of which draw on [VanN1993].

and divide m by 2. If m is odd, write down | and divide $m+1$ by 2. Apply the same algorithm to the result of the division, and repeat until n syllables have been determined. For example, to find the metrical pattern with serial number $m=2$ in the above list, we perform the following sequence of steps:

2 is even:	$2 \div 2 = 1$,	$(1+1) \div 2 = 1$,
	1 is odd:	1 is odd:
@	|	|

So the second metrical pattern in the list is found to be @ | |, which is correct. By the above analogy with binary representation, it should be clear why this method works. The algorithm for the third problem, finding m from its given pattern, simply reverses the process.

(4) Interestingly, the *sūtra* for determining the total number of possible meters that can be made with n syllables explicitly refers to the word *śūnya*, which, as noted in section 2.1, later became a standard mathematical term for zero:

> When halved, [record] two. When unity [is subtracted, record] *śūnya*. When *śūnya*, [multiply by] two; when halved, [it is] multiplied [by] so much [i.e., squared].

That is, we manipulate the number n of syllables more or less as we did the serial number m in the previous method. But in this case we are instructed to write down a sequence of 0's and 2's, and perform a sequence of doubling and squaring operations to correspond to them.

We will illustrate the rule with the example of $n=7$, as follows: 7 is odd, so subtract unity from it and write down 0. Then $7-1=6$ is even, so halve it and write down 2. Since $6/2=3$ is odd, subtract unity from it and write down 0; since $3-1=2$ is even, halve it and write down 2; since $2/2=1$ is odd, subtract unity and write down 0; since $1-1=0$, stop. Then take 1 as the first operand and retrace the steps from the beginning to the end of the sequence, doubling whenever we encounter a zero and squaring whenever we find a 2. We can imagine a diagram such as the following to represent this process:

$$\begin{array}{ccccccccc} 7 & \rightarrow & 6 & \rightarrow & 3 & \rightarrow & 2 & \rightarrow & 1 \\ 0 & & 2 & & 0 & & 2 & & 0 \\ 64 \cdot 2 = 128 & \leftarrow & 8^2 = 64 & \leftarrow & 4 \cdot 2 = 8 & \leftarrow & 2^2 = 4 & \leftarrow & 1 \cdot 2 \end{array}$$

The answer is $2^7 = 128$, as expected, but instead of seven doublings, the process required only three doublings and two squarings—a handy time-saver in cases where n is large. Piṅgala's use of a zero symbol as a marker seems to be the first known explicit reference to zero (although only as a notational symbol, not as a numerical value) in a Sanskrit text. Of course, it does not conclusively prove that decimal place value notation including zero was established as early as the third century BCE or thereabouts, but it is another piece of evidence pushing back the earliest plausible date for such a system.[32]

(5) Finally, to find out how many of the 2^n metrical patterns contain a specified number p of, say, heavy syllables, Piṅgala prescribes the construction of what later commentators call a "Meru-extension" or mountain-shaped figure, which in fact is just what we know as Pascal's triangle. (The reader may like to try proving that the $(n-p+1)$th entry in the $(n+1)$th row of Pascal's triangle does in fact equal the number of metrical patterns containing p heavy syllables.)

3.4 MATHEMATICS IN JAIN AND BUDDHIST TEXTS

It was not only adherents of the Vedic and Hindu scriptures who left traces of their mathematical ideas in literature of the early Classical Sanskrit period (although of course they had the advantage of numbers), but followers of the Buddha and Mahāvīra as well. The sacred literature of the Buddhists and Jains was mostly composed in the vernaculars of Pali and Prakrit respectively rather than Sanskrit, although they accumulated much sectarian secondary literature in Sanskrit also. The early texts can be very difficult to date: for example, it was not until the fifth century CE that various scriptures and exegeses of Jainism were established in their present canonical form, so some material from various earlier stages is inextricably mingled in them.

Nonetheless, we can extract from such works a partial idea of some aspects of mathematical thought before the middle of the first millennium CE. One of the areas of interest is the intriguing nonbivalent logic used in their philosophy, as in the classic Buddhist "tetralemma" expressed by the philosopher Nāgārjuna (probably in the second or third century CE) as "Anything is either true, or not true, or both true and not true, or neither."[33] Another is the fascination with large numbers similar to that indicated by the list of decimal powers from the *Yajur-veda* quoted in section 2.1. For example, the *Lalita-vistara*, which recounts the life of the Buddha, tells of his prowess in numeracy. As a young prince competing for the hand of the princess Gopā, he outshone all his rivals not only in martial skill, sport, and arts but also in solving astronomical problems and reckoning with large numbers. The number table said to have been recited by him on this occasion includes names for powers of ten going up beyond the fiftieth decimal place.[34]

Cosmology, particularly among the Jains, also provided a fruitful field for the exploration of concepts of the very small and very large. The basic model of the universe in both Buddhism and Jainism was similar to that of the Purāṇas described in section 3.2.3, although with variations: for example, in the Jain universe there are two suns, two moons, and two sets of stars. The description of this universe both qualitatively and quantitatively was one of

---

[32] A case for this hypothesis is persuasively presented in [SarS2003].

[33] See [Lin2005] for a discussion of this doctrine in terms of modern mathematical logic.

[34] *Lalita-vistara*, chapter 12; [Gos2001], pp. 138–143.

the four prescribed Jain *anuyogas* or "subjects of inquiry," namely *gaṇita-anuyoga* or "inquiry into calculation." Calculation, like other practices in Jainism, could be "mundane" (worldly, practical) or "super-mundane" (transcendent). In the branch of calculation called "super-mundane," the problem of defining the ultimate nature of material substance led to the development of sophisticated concepts of the infinite and infinitesimal. Such ideas are illustrated in the following passage from the *Anuyoga-dvāra* ("Doors of inquiry"), explaining the concept of an instant of time:

> [Suppose] there is a particular person, a son of a tailor, who is young, strong, is at a proper period of time, youthful, without disease and with steady forearm; has strong hands and feet ... is capable of physical exercise of scaling, swimming and running quickly; is possessed of internal energy; knows what to do, is clever, learned, proficient, wise, skilful and expert in minute arts. He takes up a big piece of cotton cloth or a silken cloth and quickly tears a cubit of it. Here a questioner asked the teacher thus: Is the time taken by the son of a tailor quickly to tear a cubit ... equal to one [time] instant? [The teacher replied:] Such an assertion is not possible. Why? Because one piece of cotton cloth is produced by the integration of the assemblage of groups of numerable numbers of threads.... The upper thread is cut at a time which is different from the time when the lower thread is cut. Therefore this is not one instant....
>
> Is the time taken by the son of a tailor in cutting the upper thread ... one instant?... No, it is not.... [O]ne thread is produced by the integration of the assemblage of groups of numerable numbers of fibres.... The upper fibre is cut at a time which is different from the time when the lower fibre is cut. Therefore this is not one instant....
>
> Is the time taken by the son of a tailor in cutting the upper fibre ... one instant?... No, it is not.... [O]ne fibre is produced by the integration of the assemblage of groups of infinite numbers of [atomic] conglomerates.... Therefore this is not one instant. Finer still than this, O young ascetic, is the [time]-instant stated to be.[35]

The "numerable" and the "infinite" mentioned in this passage were not the only recognized categories for the size of numbers in Jainism: numbers could be "numerable," "innumerable" (i.e., finite but impossible to denote by a named quantity), or "infinite, endless." There were also different varieties of infinity, described as "infinite in one way," "infinite in two ways," "infinite in partial extent," "infinite in entire extent," and "eternally infinite."[36] Special

[35] *Anuyoga-dvāra* 366, [Han1970], p. 130.

[36] [Kap1937], p. xxxvi; a detailed analysis of a calculation of an unenumerable number in a later Jain text is given in [GupAJM], pp. 185–203.

computational techniques were involved in dealing with these concepts, such as the operation whose name we might translate as "multiple multiplication," or multiplying a number by itself that same number of times—in other words, raising it to the power of itself, as x^x. Successive classes of such numbers denoted repeated applications of multiple multiplication, which produced enormously large numbers very quickly.[37] Historians are still struggling to fully understand and explain the Jain approaches to these mathematical concepts and their possible influence elsewhere.

Returning to the more comprehensible realm of "mundane calculation" in Jain texts, we still find a plethora of mysteries. The first secure attestation of the approximation $\pi \approx \sqrt{10}$ mentioned in section 3.2.2, accompanied by a rule stating that the area of a circle is equal to one-fourth the product of its circumference and its diameter, occurs in a Jain work of about the fourth or fifth century CE.[38] In fact, the $\sqrt{10}$ value of π is so routinely used in Jain texts that it is often known as the "Jain value" for π. This value must have been established somewhere in the gap of several centuries between this attestation and the cruder π values of the *Śulba-sūtras*, but the details of its origin remain obscure.[39]

Even more intriguing is the Jain classification of ten types of computation extending back perhaps to the third century BCE, shown in the following list:[40]

1. "Operation," probably referring to the operations of arithmetic, as the term is interpreted in later texts;
2. "Procedure" or "practice," later used to mean particular subjects of computation, such as series or figures;
3. "Rope," probably meaning cord geometry as in the ancient *Śulba-sūtras*;
4. "Heap," later used to mean either a specific "procedure" involving measurement of heaps of grain or else numerical calculation in general;
5. "Part-classes," later used to denote operations with different types of fractions;
6. "As much as so much" (Sanskrit *yāvattāvat*), a term later designating the unknown quantity in algebra but said by an eleventh-century Jain commentator to mean here multiplication or summation of numbers;
7. "Square" or "product";

[37] Illustrated in, for example, [Gup1992b], pp. xiv–xv.

[38] [Hay1997b], p. 194.

[39] [GupAJM], pp. 113–117, 181–182, discusses some early instances of this value in Jain works and the difficulties in dating them. Various geometric reconstructions of the derivation of the $\pi \approx \sqrt{10}$ approximation are discussed in [Sara1979], p. 65, and [GupAJM], pp. 205–215.

[40] [Kap1937], pp. xi–xii, [Hay2003], p. 120, and [PinGEI].

8. "Cube" or "solid";

9. "Square-square," evidently meaning fourth power;

10. *Kalpa*, the same word used to indicate the lifetime of the universe but not a known technical term in mathematics. Defined by the abovementioned commentator as referring to measurement of sawn lumber, but likely signifying *vikalpa*, "admission of alternatives," that is, permutations and combinations.

As we will see in subsequent chapters, many of the same terms later appeared as standard categories in medieval Sanskrit mathematical texts by Jain and Hindu authors alike. Since there are as yet no known Jain commentaries on this list earlier than the eleventh century and no known early Jain works specifically treating "mundane calculation," we cannot identify with certainty what these terms meant in the early Classical period or what mathematical techniques they involved. Still, these hints are enough to prove that mathematics at this time must have been a rich and broad field of activity, provoking inquiry across disciplinary and sectarian boundaries.[41]

[41]In particular, it is evident that the commonly used synonym "Hindu mathematics" for "Indian mathematics," as in the title of the standard historical survey [DatSi1962], is a misnomer. Vedic-Brāhmaṇic or Hindu mathematical practitioners certainly outnumbered those in other Indian groups, at least from the first millennium BCE onward. But ancient and medieval mathematics in India was equally certainly not only a Hindu development.

Chapter Four

The Mathematical Universe: Astronomy and Computation in the First Millennium

As far as we can tell from the surviving texts, the chief matrix of literate mathematical knowledge in Sanskrit around the middle of the first millennium CE was *jyotiṣa*, a mix of astronomy, calendrics, and astrology. Treatises on *jyotiṣa* bring together in their astronomical computations a wide variety of mathematical knowledge whose complexity and maturity suggest the richness of the mostly unrecorded developments of preceding centuries. Some treatises also include chapters on general mathematical techniques covering a range of subjects that extends far beyond astronomy proper; these are discussed in the following chapter. But the history of Indian mathematics cannot be fully understood without seeing it in the astronomical context where so much of it developed. Hence, this chapter investigates some of the complexities of quantitative modeling and prediction of cosmic phenomena in the Sanskrit tradition.

It is unfortunately impossible to provide a detailed exposition of all—or for that matter of any—of the major medieval *jyotiṣa* works and their commentaries in the space available here. This chapter is therefore limited to describing the general development and content of mathematical astronomy from the middle of the first millennium CE up to the early second, illustrated with excerpts from various texts containing some of its most characteristic and interesting mathematical methods. In the subsequent discussions, the reader may find it helpful to refer to the brief outline of the fundamentals of geocentric astronomy contained in section 4.1.[1]

4.1 AN INTRODUCTION TO GEOCENTRIC ASTRONOMY

The basic configuration of the geocentric universe is shown in figure 4.1. The spherical earth is stationary at the center of the shell of the spherical universe, upon which the fixed stars are placed in an unchanging pattern. The two chief great circles on this celestial sphere are the celestial equator (a projection of the terrestrial equator onto the celestial sphere) and the

[1] A more detailed discussion of geocentric systems can be found in, for example, [Eva1998]. Fuller details of the models and methods described below can be found in [Pin1978a], to which this chapter may also serve as a guide or introduction for those not already familiar with Indian astronomy.

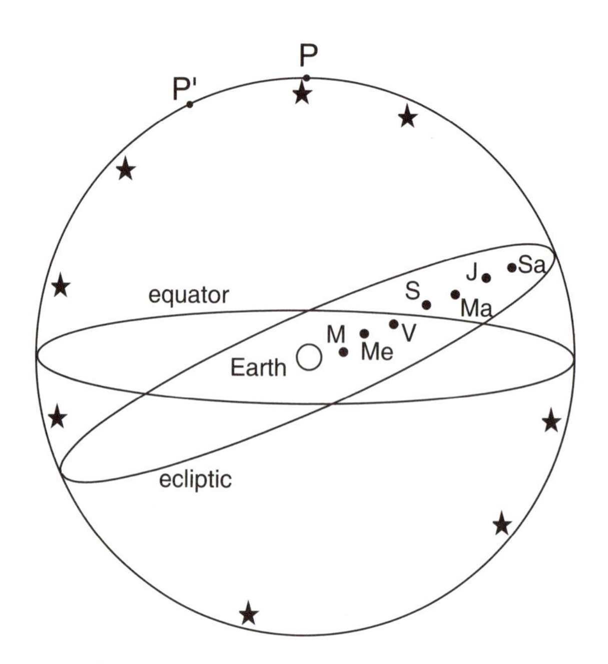

Figure 4.1 The structure of the universe in geocentric astronomy.

ecliptic (a projection of the sun's apparent annual orbit around the earth). The celestial equator and the whole celestial sphere revolve daily from east to west about an axis whose north pole is labeled P in the figure, while P' is the northern pole of the ecliptic. The angle between the planes of the equator and the ecliptic is something under 24 degrees. The daily turning of the celestial sphere represents the passage of time, so time can be expressed either in ordinary time units or as degrees of arc along the celestial equator.

The seven heavenly bodies recognized as "planets"—the moon, Mercury, Venus, the sun, Mars, Jupiter, and Saturn—are laid out in that order between the earth and the fixed stars. All the planets revolve about the earth from west to east in their individual orbits, which are composed of various combinations of circles to account geometrically for phenomena such as their variations in apparent speed and size or their occasional retrogradations. All the planetary orbits lie in or close to the plane of the sun's orbit, the ecliptic. The core concepts of this theory and the symbols used to denote several of its quantities are defined in the following glossary.

Altitude. The angular distance from a point on the **celestial sphere** to the **horizon**, measured along a great circle passing through the **zenith** and its opposite pole; the elevation of a point above the horizon. Forms a coordinate pair with **azimuth**.

Amplitude. Same as **azimuth**, but measured from a different point on the **horizon**, usually the east point.

Anomaly, κ. (More precisely known as "mean anomaly.") The angular distance from an orbit's **apsidal point** traversed by the orbiting body's **mean motion** during a given time.

Apogee (or apsidal point, apsis). The point on a non-**concentric** orbit that is farthest from the earth.

Ascendant. The point of the **ecliptic** appearing on the eastern **horizon** at a given location at a given time. Crucial in the casting and interpretation of horoscopes.

Ascensional difference. The difference between six o'clock (i.e., six hours before or after local noon) and the point of sunrise or sunset, respectively. Expressed either in time units or in degrees of the **equator**.

Azimuth. Angular distance along the **horizon** measured from the north point. Forms a coordinate pair with **altitude**.

Celestial sphere. The notional **concentric** sphere enclosing the universe, on whose surface the reference circles defining celestial coordinates are situated.

Concentric. A sphere or circle whose center coincides with the center of the earth.

Conjunction. The appearance of two celestial bodies at the same or nearly the same point in the sky, or at the same **longitude**.

Day-circle. A small circle on the celestial sphere, parallel to the celestial **equator**, that a celestial body appears to trace out during one daily rotation of the **celestial sphere** about the earth.

Declination, δ. The angular distance from a point on the **celestial sphere** to the **equator**, measured along a great circle passing through the **poles of the equator**. Forms a coordinate pair with **right ascension**. If the given point lies on the **ecliptic**, this angular distance is the **ecliptic declination** of that point.

Deferent. A circle on whose circumference an orbiting body or the center of an **epicycle** is considered to move.

Eccentric. A circle surrounding the earth but whose center does not coincide with the center of the earth. The distance between the two centers is called the **eccentricity** (e).

Ecliptic. The great circle on the **celestial sphere** in the plane of the sun's apparent annual orbit about the earth.

Epicycle. A non**concentric** circle whose center is considered to move along the circumference of another circle, or **deferent**.

Epoch. An arbitrarily chosen zero point of time from which the positions of celestial bodies at subsequent times are calculated.

Equation, μ. The angular distance between an orbiting body's **true longitude** and its **mean longitude** at a given time. (So named because it equates, or equalizes, the mean and true longitudes; also called "equation of center.")

Equator (celestial). The great circle on the **celestial sphere** in the plane of the earth's equator.

Equinox. One of the two points where the celestial **equator** and **ecliptic** intersect. The one through which the sun appears to pass in springtime is called the **vernal equinox**.

Gnomon. A time measurement device consisting of a pointed straight rod positioned so as to cast a shadow, usually in the plane of the observer's **horizon**.

Horizon. The circle where a plane tangent to the sphere of the earth's surface at the observer's location intersects the **celestial sphere**. If the earth's size is considered negligible, this plane of the horizon coincides with a parallel plane passing through the earth's center.

Inequality. A periodic deviation from uniform circular orbital motion, usually interpreted physically by an orbital mechanism such as an **epicycle** or **eccentric**.

Latitude (celestial), β. The angular distance from a point on the **celestial sphere** to the **ecliptic**, measured along a great circle passing through the **poles of the ecliptic**. Forms a coordinate pair with celestial **longitude**.

Longitude (celestial), λ, $\bar{\lambda}$. The angular distance along the **ecliptic** between a given point and the zero point. If that zero point is defined as the **vernal equinox**, this angular distance is **tropical longitude**; if the zero point is defined in relation to some fixed star, it is **sidereal longitude**. The longitude attained at a given time by an orbiting body's **mean motion** is its **mean longitude** ($\bar{\lambda}$); its corrected or true position is its **true longitude** (λ). Forms a coordinate pair with celestial **latitude**.

Mean motion. The average angular speed of an orbiting body in its revolutions about the earth.

Meridian. The great circle on the **celestial sphere** passing through the north and south points of an observer's **horizon** and the **zenith**.

Node. One of the two points where the circle of a **planet**'s orbit intersects the plane of the **ecliptic**. The planet crosses its **ascending node** when moving from south to north of the ecliptic, and crosses the **descending node** going in the opposite direction on the other side.

Oblique ascension. The arc of the celestial **equator**, expressed in units of arc or of time, that rises above an observer's **horizon** simultaneously with a given arc of the **ecliptic**.

Obliquity, ϵ. The angle between the planes of the celestial **equator** and the **ecliptic**.

Opposition. The appearance of two celestial bodies at opposite points on the **ecliptic**, or with **longitudes** separated by 180 degrees.

Parallax. The angular distance between a celestial body's position as seen from a given terrestrial location and its position as seen from the location lying on a straight line between the body and the center of the earth.

Planet. In premodern times, the seven "wandering stars" whose position periodically changes with respect to the fixed stars, that is, the sun and moon as well as the five visible star-planets Mercury, Venus, Mars, Jupiter, and Saturn.

Pole, ecliptic. One of the two points (usually the northern one) where the axis of the **ecliptic** intersects the **celestial sphere**.

Pole, equatorial. One of the two points (usually the northern one) where the axis of the celestial **equator** intersects the **celestial sphere**. The northern pole approximately coincides, at least in historical times, with the fixed star called the pole star (modern Polaris).

Precession. The movement of the **equinoxes** along the **ecliptic** with respect to the fixed stars. (See section 2.3.)

Prime meridian. The **meridian** of an arbitrarily chosen terrestrial location, designated as the zero point of terrestrial longitude.

Prime vertical. The great circle on the **celestial sphere** passing through the east and west points of the observer's **horizon** and the **zenith**.

Retrodiction. The use of quantitative astronomical theories to "predict backward" the positions of celestial bodies at a given time in the past.

Retrograde motion. The occasional apparent backward motion of a planet (other than the sun or moon) from east to west against the background of the fixed stars.

Right ascension. The angular distance along the celestial **equator**, expressed in units of arc or of time, between a given point and the zero point. Forms a coordinate pair with **declination**. **Oblique ascension** for observers on the terrestrial equator reduces to right ascension.

Sidereal. Having its zero point defined with respect to a particular fixed star; see **longitude**.

Solstice. One of the two points on the **ecliptic** 90 degrees distant from the **equinoxes**.

Synodic. Refers to motion or time relative to the position of the sun, as when the moon completes a **synodic month** between two successive conjunctions with the sun. **Synodic phenomena** are special points in a planet's motion relative to the sun, such as **opposition**, starting or stopping **retrograde motion**, or last or first sighting when it is close to the sun shortly before or after **conjunction**.

Tropical. Having its zero point defined as the **vernal equinox**; see **longitude**. (So called from "tropos," "turning," an archaic term for **solstice**, because the sun appears to turn from north to south or vice versa at the solstices; see section 2.3.)

Zenith. The point on the **celestial sphere** directly above a given observer; the upper pole of the observer's **horizon**.

Zodiac. The division of the **ecliptic** into twelve signs—Aries, Taurus, Gemini, Cancer, Leo, Virgo, Libra, Scorpio, Sagittarius, Capricorn, Aquarius, and Pisces—of 30 degrees each. Its zero point may be chosen as **tropical** or **sidereal**.

4.2 EVOLUTION OF THE *SIDDHĀNTA* AND ASTRONOMICAL SCHOOLS

The early first-millennium astronomical texts described in section 3.2.2 were superseded, starting in about the fifth century, by a more comprehensive format called here the "standard *siddhānta* (treatise)" or sometimes "medieval *siddhānta*."[2] The medieval *siddhānta* generally employs the concise verse format typical of Classical Sanskrit didactic genres and thus is not always easy to understand. As in other genres, the verse treatise or "base text" became practically inseparable from the explanatory prose commentaries composed by the treatise's author himself or by later writers. Commentaries generally define or gloss the individual words of the verses, sometimes

[2] This format has been identified elsewhere (e.g., in [Pin1981a], p. 20) as the "classical *siddhānta*," a term that may misleadingly suggest a particular identification with Classical Sanskrit. Medieval *siddhāntas* were indeed composed in Classical Sanskrit, but so were earlier astronomical works.

parsing their grammatical construction, and restate their computational formulas in a more verbose and less ambiguous way. Often the commentary provides one or more worked examples to illustrate the formulas, and sometimes mathematical rationales or demonstrations for them as well. However, the main presentation of both the base text and the commentary is verbal. The detailed geometric diagrams familiar to us in some other premodern mathematical astronomy traditions are rare exceptions, mentioned only in particular astronomical contexts and hardly ever appearing in manuscripts (for an example, see section 4.3.5). The few existing diagrams do not display certain features we now consider standard, such as letter-labeled points. The figures appearing in this chapter, unless otherwise indicated, are merely modern graphical representations of verbal rules. (For an idea of how Indian astronomers themselves used visual aids to understand the rules, see section 4.5.)

The history of the development of the *siddhānta* genre is uncertain at many points, particularly at its beginning. The account presented here attempts to reflect the best estimates constructed from what is known of the texts. The issue of origins is discussed further in section 4.6; in particular, some alternative chronologies are discussed in section 4.6.2.

The first major exemplar of the standard *siddhānta* seems to have been a *Brahma-siddhānta* or *Paitāmaha-siddhānta* (from "Pitāmaha," another name for the deity Brahman) composed probably in the early fifth century, and not to be confused with the first-century astronomical work of the same name that was summarized in the *Pañca-siddhāntikā* (see section 3.2.2). This later *Paitāmaha-siddhānta* now survives only in a fragmentary and corrupted form as part of a Purāṇa text. It is framed as a dialogue in prose between Brahman and the sage Bhṛgu, who seeks instruction in the computation of time. The start of the surviving text echoes the ancient *Jyotiṣa-vedāṅga*'s insistence on the correct ordering of the Vedic rituals:

> Bhṛgu, approaching the Bhagavān [Lord] who causes the creation, continuance, and destruction of the world, the reverend teacher of the moving and the unmoving, said: Oh Bhagavān! The science of the stars is difficult to understand without calculation. Teach me, then, calculation.
>
> The Bhagavān said to him: Hear, my child, the knowledge of calculation. Time, which is Prajāpati and Viṣṇu, is an endless store; the knowledge of this by means of the motion of the planets is calculation....
>
> For the Vedas go forth for the sake of the sacrifices; the sacrifices are established as proceeding regularly in time. Therefore he who knows *jyotiṣa*, this science of time, knows all.[3]

The calculation that Brahman taught, judging from the extant scraps of the text, included the following:

[3][Pin1967], pp. 476–477, 506. See also [Pin1970a], vol. 4, p. 259.

- Use of large integer parameters to calculate
  - mean positions of celestial bodies and their nodes and apsides, and
  - the corresponding measurement of mean time;
- Use of trigonometry to calculate
  - true positions in orbits involving eccentric or epicyclic circles, and
  - determination of terrestrial place and time by shadow-measurement;
- Mathematical prediction of significant planetary positions;
- Computation of eclipses.

Time in the *Paitāmaha-siddhānta* is bounded by the cosmological concepts introduced in section 3.2.3, such as the lifetime of the universe or *kalpa*, equal to 4,320,000,000 years. This duration also equals 1000 *mahāyugas* or 10,000 Kaliyugas, but is arranged as a sequence of fourteen eras called *manvantaras*, which are composed of seventy-one *mahāyugas* apiece and separated by intermediary periods of 1,728,000 years each. All celestial objects—namely, the seven planets including the sun and moon, the apsidal points and ascending nodes of their orbits, and the sphere of fixed stars—are considered to complete integer numbers of revolutions about the earth in one *kalpa*. All of them start together from a great conjunction at the zero point of celestial longitude and latitude, or the beginning of the ecliptic. The comprehensiveness of this system in theory allows an astronomer to predict any astronomical event at any point in the lifetime of the universe.

However, the *Paitāmaha-siddhānta* in its surviving form does not identify any point in this cosmic span with any point in historical time, which makes it useless for computational purposes. Later *siddhāntas* generally assume that the present age is the Kaliyuga of the twenty-eighth *mahāyuga* of the seventh *manvantara* of the *kalpa*, which began with the legendary battle of the rival clans described in the great epic *Mahābhārata*. The traditional date ascribed to it corresponds to Friday, 18 February 3102 BCE.[4] The epoch or zero point of time used for a *siddhānta*'s calculations is sometimes taken to be this initial point of the current Kaliyuga rather than the start of the *kalpa*.

The planets' distances from the earth, and the numbers of their revolutions in a *kalpa*, are thought to be dependent on their observed speeds: faster revolving bodies are closer and slower ones farther away. The sequence of

[4]It is unclear when and how this date became established as the accepted start of the Kaliyuga era; see [Sal1998], pp. 180–181. The standard view (described, for example, in [Pin2002a] and [Duk2008]) holds that astronomers sometime before the middle of the first millennium CE retrodicted an approximate conjunction of the planets around then, which made a handy starting point from which to reckon time. Various attempts to date the *Mahābhārata* battle as a historical event are mentioned in [Kak2005].

planetary orbits, from closest to farthest, is thus Moon–Mercury–Venus–Sun–Mars–Jupiter–Saturn. (Since the mean orbital speed of Mercury and Venus is in fact the same as that of the sun, they are ranked in accordance with the speeds of their synodic motions or *śīghra*-apsides, which are discussed in section 4.3.3.) As the *Paitāmaha-siddhānta* says,

> The orbit of heaven is 18,712,069,200,000,000 [*yojanas*]. For each of the planets which travel [from west] to east, the *yojanas* of the circumference [of its orbit] are obtained by dividing the orbit of heaven by [the number of] its revolutions [in a *kalpa*]; that [planet] which makes many revolutions has an inferior orbit, that which makes few a superior orbit. The diameter of the orbit equals the circumference divided by the square root of ten.[5]

The *yojana* is a standard Indian unit of distance whose exact value in terms of modern units is not known (and may have been somewhat variable), but it is approximately on the order of ten kilometers.[6]

From almost the beginning of its development, the standard *siddhānta* pursued three main mathematical objectives. The most crucial of these was the computation of times, locations, and appearances of future or past celestial phenomena as seen from any given terrestrial location—that is, predictive astronomy. A second purpose was the explanation of the computational astronomy procedures in terms of the geometry of the spherical models underlying them. These explanations are frequently presented in a separate section called *gola*, meaning "sphere." The third goal was instruction in general mathematical knowledge, ranging from basic arithmetic operations to calculation of interest on loans and rules for finding areas, volumes, and sums of series. According to the definition of the tenth-century astronomer Vaṭeśvara,

> [A work] in which are stated all the measures of time, the determination [of the positions] of the planets, [and] calculation in its entirety including the pulverizer [i.e., first-degree indeterminate equations] and so forth, [and] in which the form of the planets, constellations, and earth [is stated] correctly, is indeed called by the most excellent sages a true *siddhānta*.[7]

Not long after its origin around the time of the *Paitāmaha-siddhānta*, the medieval *siddhānta* tradition branched into different schools or *pakṣas*, which

---

[5][Pin1967], p. 479. (Note that this immense number for the circumference of the celestial sphere, like all other numbers quoted from Sanskrit verse treatises, was expressed verbally in the original text, though it is shown here in numerical notation to save space.) The ancient value $\sqrt{10}$ as the ratio of circumference to diameter (see section 3.4) appeared in many medieval *siddhāntas* but was recognized as an approximation, as discussed in section 5.1.2.

[6]The astronomer Āryabhaṭa at the start of the sixth century defined the *yojana* as equal to 8,000 man-heights, and stated that the earth's diameter contains 1,050 of them. *Āryabhaṭīya* 1.7, [ShuSa1976], p. 15.

[7] *Vaṭeśvara-siddhānta* 1.5, [Shu1986], vol. 1, p. 2. See section 5.1 for an explanation of the mathematical technique called the "pulverizer."

are distinguished from one another mostly by the values of the parameters they use for the main divisions of time and the cycles of the heavens. All the *pakṣas*, however, seem to be ultimately based on the system of the *Paitāmaha-siddhānta*, or Brāhma-pakṣa. The sources of competing parameters and authors' reasons for choosing them are not always clear (see also the discussion in section 4.6). A frequently stated motive is the desire to harmonize astronomical calculations as far as possible with *smṛti* traditions about cosmological time, or to bring them into agreement with observed positions.

There seem to have been distinct (although extremely porous) regional boundaries between the various schools, with astronomers in, for example, the south of the subcontinent tending to adhere to a *pakṣa* different from the favorite school in the north. Authors writing in different *pakṣas* often criticized one another's works for being unorthodox (particularly in contradicting *smṛti* texts) or incorrect. But their allegiance to their own particular schools was far from absolute: they sometimes constructed systems of conversion constants (*bījas*, "seeds") for transforming computational results from one *pakṣa* into the corresponding values for a different *pakṣa*.[8] Sometimes an author espousing one *pakṣa* would write a commentary or even a new treatise within a different *pakṣa*, provisionally accepting the parameters that he had elsewhere rejected, and sometimes he would even combine parameters of different *pakṣas* within one work.

The development of the different *pakṣas* by major Indian mathematicians up to about the twelfth century is sketched in the following paragraphs:

The Ārya-pakṣa. This school was initiated around 500 by the *Āryabhaṭīya* of Āryabhaṭa (born 476), which allots 1008 instead of 1000 *mahāyugas* to a *kalpa*. In the *Āryabhaṭīya*, a Kaliyuga is one-fourth of a *mahāyuga* instead of one-tenth, and there is a conjunction of the planets (or at least of their mean positions) at the start of the ecliptic at the beginning and end of each Kaliyuga. Despite introducing these somewhat radical modifications of the *Paitāmaha-siddhānta* system, Āryabhaṭa insists that he is following and restoring the "astronomy of Brahman":

> [The work] called *Āryabhaṭīya* is the ancient *Svāyambhuva* [work of Svayambhu, i.e., Brahman], constant forever.[9]

The structure of the *Āryabhaṭīya* is markedly different from the standard *siddhānta* layout described below, having only four chapters, which are devoted respectively to numerical parameters, computation in general, time-reckoning, and *gola* or the sphere.

The best known author in the Ārya-pakṣa after its founder Āryabhaṭa is his commentator Bhāskara (I), who wrote in the early seventh century two *siddhāntas*, the *Mahā-bhāskarīya* and *Laghu-bhāskarīya* ("Great" and "Lesser [treatise] of Bhāskara," respectively). The *Śiṣya-dhī-vṛddhida-tantra* or "Treatise for increasing the intelligence of students" of Lalla (probably

[8] See [Pin1996a].

[9] *Āryabhaṭīya* 4.50, [ShuSa1976], p. 164.

in the eighth century) also followed the Ārya-pakṣa, although it reverted to the Brāhma-pakṣa's more traditional division of the *mahāyuga*. The *Vaṭeśvara-siddhānta* composed by Vaṭeśvara in 904 combined features of the Ārya-pakṣa and other schools. Ārya-pakṣa texts tended to be most popular in South India (although Āryabhaṭa, Bhāskara, Lalla, and Vaṭeśvara themselves all seem to have been natives of northern or central regions).

The Ardha-rātrika-pakṣa. This school, whose name means "midnight," was also founded by Āryabhaṭa, in a work now lost. It resembles the Ārya-pakṣa except that it chooses its epoch or zero point of time to occur at midnight rather than the traditional sunrise. The Ardha-rātrika-pakṣa partly inspired the Saura-pakṣa (see below), but left no surviving *siddhāntas* of its own.

The Brāhma-pakṣa. The first major work in this school after the *Paitāmaha-siddhānta* is the *Brāhma-sphuṭa-siddhānta* or "Corrected *siddhānta* of Brahman" composed by Bhāskara I's contemporary Brahmagupta in Gujarat. Like Āryabhaṭa, he claims to be restoring the original astronomy of Brahman:

> The calculation of the planets spoken by Brahman, which [has] been weakened by great [lapse of] time, is [here] explained correctly ["*sphuṭa*"] by Brahmagupta, the son of Jiṣṇu.[10]

But he is sharply critical of Āryabhaṭa's deviations from the *smṛti* traditions followed in the original Brāhma-pakṣa:

> Āryabhaṭa defined the Kṛtayuga etc. [i.e., the four ages in a *mahāyuga*] as four equal quarter-*yugas*. None of them is equal to [the ones] stated in *smṛti*.
>
> [The *Āryabhaṭīya*'s] *yuga* and *kalpa* ... are not equal to [those] stated in *smṛti*; hence Āryabhaṭa does not know the mean motions.[11]

Subsequent *siddhāntas* in this school include the *Siddhānta-śekhara* ("Crown of treatises"), composed in the mid-eleventh century by Śrīpati, and the magisterial *Siddhānta-śiromaṇi* ("Crest-jewel of treatises") of Bhāskara II (no relation to his seventh-century namesake) in the mid-twelfth century. The Brāhma-pakṣa's followers were most numerous in the western and north-western parts of the subcontinent.

The Saura-pakṣa. The founding text of this *pakṣa*, from which its name is derived, is the *Sūrya-siddhānta*, or "Sun-treatise," composed or revised in about 800 from an earlier work of the same name[12] and ascribed to direct revelation by the Hindu sun god Sūrya. Its fundamental structure is a compromise among those of earlier schools: it assumes the traditional

---

[10] *Brāhma-sphuṭa-siddhānta* 1.2, [DviS1901], p. 1.

[11] *Brāhma-sphuṭa-siddhānta* 1.9 and 11.10, [DviS1901], pp. 3, 150.

[12] This earlier *Sūrya-siddhānta* was partly preserved in the sixth-century *Pañca-siddhāntikā* of Varāhamihira; see section 3.2.2.

time divisions of the Brāhma-pakṣa along with the midnight epoch of the Ardha-rātrika-pakṣa and the Ārya-pakṣa's mean planetary conjunction at the beginning of the Kaliyuga. These features are reconciled numerically by postulating a long period of immobility at the beginning of the *kalpa* before the planets start to move from their zero point. The concept of a period of immobility, which makes it possible to combine traditional time divisions with planetary period relations that don't conform to them, was adopted by some texts usually assigned to other *pakṣas*, such as the *Mahā-siddhānta* ("Great *siddhānta*") of Āryabhaṭa (II) in the Brāhma-pakṣa in the late tenth or eleventh century.

The Saura-pakṣa itself was most influential in northern and eastern India. Later, it formed the basis of the Gaṇeśa-pakṣa in the sixteenth century, discussed in section 9.3.2.

4.3 ASTRONOMICAL CALCULATIONS IN *SIDDHĀNTAS*

This section describes the primary problems in astronomical computation and the mathematical techniques employed in *siddhāntas* for their solution. The structure of the typical *siddhānta* follows a standard sequence for attacking these problems:

Chapter 1. Astronomical parameters and computation of mean celestial motions and positions.

Chapter 2. Trigonometric methods for finding true celestial positions.

Chapter 3. Computing the apparent direction, place, and time of celestial phenomena as seen from a particular terrestrial location.

Chapter 4. Calculations for lunar eclipses.

Chapter 5. Calculations for solar eclipses.

Subsequent chapters. Various topics of astrological significance, including planetary phases and conjunctions, lunar phases, and the depiction of the moon's crescent.

This sequence is modified in some *siddhāntas*, most radically in the *Āryabhaṭīya*, as noted in the preceding section.

4.3.1 Parameter values and number representation

As in Ptolemaic astronomy, in *siddhānta* astronomy the first task of computing celestial positions is to determine the mean position of a celestial body at a desired time, that is, the position it would occupy at that time if it were revolving about the earth with uniform circular motion on a concentric orbit. These mean motions are presented in the form of simple proportions between the large integer numbers of cycles completed in a *kalpa* or *yuga*.

Consequently, the author of a *siddhānta* needs a way of verbally expressing such large integer numbers—about ten decimal digits long—in verse form. The most usual way to do this is by means of digit sequences expressed in the concrete number system described in section 3.1, where the name of a being or thing can stand for a number conventionally associated with it. Brahmagupta, for example, gives his *kalpa* parameters in this notation, as the following example illustrates:

> In a *kalpa*, the revolutions of the mean sun and Mercury and Venus are seven zeroes, teeth [32], Vedas [4].... [Those] of the moon are equal to five skies [0], qualities [3], qualities, five, sages [7], arrows [5].[13]

Reading these concrete numbers in order from the least to the most significant digit, as usual, they translate into 4,320,000,000 revolutions for the sun and 5,753,300,000 for the moon during the lifetime of the universe.

Other techniques for verbally representing numbers were also devised. For instance, in the first chapter of the *Āryabhaṭīya*, Āryabhaṭa describes an apparently unique alphanumeric encoding system that depends on the phonetic order of the Sanskrit alphabet (explained in section A.1):

> The *varga*-letters [starting] from *k* [are encoded] in the square [places], the non-*varga*-letters, [starting from] *y* which is [equal to] *ṅm*, to the non-square [places]. Nine vowels [are assigned] to the square and non-square [places] in a double nine-tuple of zeros, and [beyond] the square [places] ending with nine.[14]

The basic principle of this cryptic scheme is that consonants are assigned specific integer values and vowels are used to represent decimal powers. Table 4.1 shows the correspondence between consonants and numbers.

The rule's meaning depends on a pun on the word *varga*, signifying either the first five sets of five consonants as grouped in the Sanskrit alphabet or the mathematical concept "square." The values of *varga* consonants, the ones from *k* to *m*, which stand for the numbers 1 to 25, should be understood in the "square" places, that is, in the even powers of ten, which are perfect squares. The succeeding eight non-*varga* consonants from *y* to *h* are to be read in the non-square places, that is, as odd decimal powers, starting with 10^1. We can tell that the numerical values of the non-*varga* consonants must start with 30 because of the statement that *y* is the same as *ṅm*, that is, $5 + 25$.

Each of the nine selected vowel sounds (short vowels and diphthongs) may represent either a square or non-square place, depending on the value of the consonant(s) with which it forms a syllable. The vowels in their sequence indicate successively higher powers of ten, as shown in table 4.2. By "a double nine-tuple of zeros," Āryabhaṭa apparently means that a sequence

[13] *Brāhma-sphuṭa-siddhānta* 1.15–16, [DviS1901], pp. 5–6.

[14] *Āryabhaṭīya* 1.2, [ShuSa1976], p. 3.

Table 4.1 Consonants in Āryabhaṭa's alphanumeric system.

Sanskrit *varga* consonants:							
k	kh	g	gh	ṅ			
1	2	3	4	5			
c	ch	j	jh	ña			
6	7	8	9	10			
ṭ	ṭh	ḍ	ḍh	ṇ			
11	12	13	14	15			
t	th	d	dh	n			
16	17	18	19	20			
p	ph	b	bh	m			
21	22	23	24	25			
Non-*varga* consonants:							
y	r	l	v	ś	ṣ	s	h
30	40	50	60	70	80	90	100

of zeros is to be written down as a template to stand for the consecutive decimal places.

Table 4.2 Vowels in Āryabhaṭa's alphanumeric system.

Vowels following *varga* consonants:								
a	i	u	ṛ	ḷ	e	ai	o	au
10^0	10^2	10^4	10^6	10^8	10^{10}	10^{12}	10^{14}	10^{16}
Vowels following non-*varga* consonants:								
a	i	u	ṛ	ḷ	e	ai	o	au
10^1	10^3	10^5	10^7	10^9	10^{11}	10^{13}	10^{15}	10^{17}

Consonants and vowels are combined into syllables representing numbers. For example, the syllable *gi* would stand for 3 in the second square decimal place, or $3 \times 10^2 = 300$, while *la* would mean $50 \times 10^1 = 500$. Non-*varga* and *varga* consonants can be combined with the same vowel in one syllable, as in *tra* $= 16 \times 10^0 + 40 \times 10^1 = 16 + 400 = 416$.

This ingenious system can represent even very large integers in just a few carefully crafted syllables independently of their order, since their decimal places are fixed by the values of their vowels. On the other hand, the combinations of the consonants are phonetically inflexible, meaning that they (and the words adjacent to them) cannot undergo the euphonic sound changes that are an essential part of correctly formed Sanskrit. Consequently, Āryabhaṭa's astronomical parameters appear as a set of barbarically unpronounceable splutters with no redeeming aesthetic or mnemonic qualities, as seen in the following verse:

> The revolutions of the sun in a [*mahā-*] *yuga* are *khyu-ghṛ*, moon *ca-ya-gi-yi-ṅu-śu-chḷṛ*, earth *ṅi-śi-bu-ṇḷ-ṣkhṛ*, eastward.[15]

We will analyze these sample parameters to illustrate how Āryabhaṭa's alphanumeric encoding scheme is applied. An initial sequence of zeros stands for the necessary decimal places, which are replaced as required by the values derived from the appropriately positioned consonants via table 4.1:

	ḷ	ḷ	ṛ	ṛ	u	u	i	i	a	a
	10^9	10^8	10^7	10^6	10^5	10^4	10^3	10^2	10^1	10^0
Sun:	0	0	0	ghṛ	yu	kh[u]	0	0	0	0
	0	0	0	4	3	2	0	0	0	0
Moon:	0	0	lṛ	ch[ṛ]	śu	ṅu	yi	gi	ya	ca
	0	0	5	7	7	5	3	3	3	6
Earth:	0	ṇḷ	ṣ[ṛ]	khṛ	0	bu	śi	ṅi	0	0
		15	8	2		23	7	5	0	0

So we deduce that the number of the sun's revolutions in a *mahāyuga* is 4,320,000, that of the moon's 57,753,336, and that of the daily rotations of the earth[16] 1,582,237,500. This clever but cumbersome notation may well be Āryabhaṭa's own invention;[17] at least, nobody else seems to have shown any interest in using it, although later commentators on the *Āryabhaṭīya* conscientiously explained how it worked.

A third system for verbal representation of numbers was the so-called *kaṭapayādi* notation, where the thirty-three Sanskrit consonants are mapped not to consecutive numbers but to decimal digits, as shown in table 4.3. Here, the consonants *k*, *ṭ*, *p*, and *y* all stand for the digit 1—hence the name *kaṭapayādi*, "beginning with *k*, *ṭ*, *p*, and *y*." As in the concrete number system, numbers are usually read starting with the least significant digit.

Vowels have no numerical significance in the *kaṭapayādi* convention, and neither do consonants that appear in conjunction with a following consonant or at the end of a word. Only a consonant that is immediately followed by a vowel has a numerical value. The flexibility of the system means that authors can encode digit sequences in actual Sanskrit words. For example, the word *bhavati*, "becomes," is decoded as the sequence 4-4-6, implying the number 644.[18]

---

[15] *Āryabhaṭīya* 1.3ab, [ShuSa1976], p. 6.

[16] Āryabhaṭa's unusual hypothesis about the relative motion of the earth and sky is presented in more detail in section 4.5.

[17] Its basic structure was certainly not without precedent, however. Pāṇini's use of syllables to encode grammatical precepts (see section 3.3) is a related concept, and Āryabhaṭa may also have been acquainted with the somewhat similar *kaṭapayādi* alphanumeric system described immediately below.

[18] This example appears in an inscription cited in [DatSi1962], vol. 1, p. 71.

Table 4.3 The *kaṭapayādi* system of number representation

k	kh	g	gh	ṅ			
1	2	3	4	5			
c	ch	j	jh	ñ			
6	7	8	9	0			
ṭ	ṭh	ḍ	ḍh	ṇ			
1	2	3	4	5			
t	th	d	dh	n			
6	7	8	9	0			
p	ph	b	bh	m			
1	2	3	4	5			
y	r	l	v	ś	ṣ	s	h
1	2	3	4	5	6	7	8

It is unclear when and how the *kaṭapayādi* system originated.[19] We do know that it forms the basis of the so-called *vākya* or "sentence" genre of astronomical texts that were popular in Kerala in southern India (see section 7.4). In these texts, planetary positions at regular intervals were encoded in *kaṭapayādi* sentences. The first such work known is traditionally considered to be the *Candra-vakyāni* or "Moon-sentences" of Vararuci, who is traditionally although somewhat doubtfully assigned to the fourth century CE.[20] Sometime in the early first millennium is probably a reasonable estimate for the date of origin of *kaṭapayādi* notation.

A variant of the *kaṭapayādi* system appears in the *Mahā-siddhānta* of the second Āryabhaṭa around the eleventh century, where it is described as follows:

> The digits starting from unity are the sounds beginning with *k*, *ṭ*, *p*, and *y*, in the order of the sounds. Both *ñ* and *n* are zero.[21]

The following sample of Āryabhaṭa II's *kaṭapayādi* parameters illustrates his approach:

> In a *kalpa*, the revolutions of the sun etc. are *gha-ḍa-phe-na-ne-na-na-nu-nī-nāḥ*, *ma-tha-tha-ma-ga-gla-bha-na-nu-nāḥ*.[22]

[19] A commentator on the *Āryabhaṭīya*, Sūryadeva Yajvan, claimed that Āryabhaṭa's own system was derived from *kaṭapayādi*, but Sūryadeva lived in the thirteenth century and may have been just guessing about the information available to Āryabhaṭa. See [DatSi1962], vol. 1, p. 71, and [SarK1976a], p. 10. A claim that the seventh-century Bhāskara employed *kaṭapayādi* notation ([DatSi1962], vol. 1, p. 71) appears to have been based on an interpolated later verse ([Kup1957], pp. xiv–xv).

[20] See [Pin1970a], vol. 5, p. 558, and [SarK1972], p. 43.

[21] *Mahā-siddhānta* 1.2, [DviS1910], p. 1.

[22] *Mahā-siddhānta* 1.7, [DviS1910], p. 4.

Note that these numbers, which translate to 4,320,000,000 revolutions in the case of the sun and 57,753,334,000 for the moon, are read starting from the most significant rather than the least significant digit, and that they count all the sounds in a conjunct consonant such as *gla*, not just the last one. These innovations never caught on, and neither did Āryabhaṭa II's use of nonsense syllables, as unmemorable (although not quite as unpronounceable) as the syllables of Āryabhaṭa I's alphanumeric system that had failed to catch on centuries earlier.

Even the simpler standard *kaṭapayādi* encoding never attained widespread systematic use, except among the astronomers of south India (see section 7.3) who were familiar with the abovementioned *vākya* genre. Elsewhere, the *kaṭapayādi* system's advantages of conciseness and versatility apparently did not compensate for the effort required to construct meaningful Sanskrit words with the desired digital values. The concrete number system remained by far the most popular technique in *siddhāntas* for expressing numerals in verse.

4.3.2 Mean motions and positions

Once all the basic parameters have been specified in a *siddhānta*'s chosen numerical notation, the astronomer can get on with finding the desired positions of the planets from their ratios. That is, if a body completes R integer revolutions with constant angular velocity in the D days of a *kalpa* or *mahāyuga*, then in the d days between the assumed zero point of time and the desired moment, it must complete r revolutions, where $r = d \cdot R/D$. The fractional part of the current number r of revolutions is the body's current mean celestial longitude, and the value of d is known as the "accumulated days." As Brahmagupta puts it in the first chapter of his *Brāhma-sphuṭa-siddhānta*,

> The result from the accumulated days [since the start of the *kalpa*] times the revolutions of the desired planet, divided by the civil days of the [entire] *kalpa*, is the [accumulated] revolutions etc. The mean [longitude] is [reckoned] from sunrise at Laṅkā.[23]

Finding the value of d requires knowing some accepted value for the time interval between the start of the *kalpa* and some current or historical moment—that is, for the present age of the universe. Brahmagupta follows the convention described in section 4.2 by considering that age at the start of the current Kaliyuga to be 1,972,944,000 years, whereas Āryabhaṭa declares it to be 1,986,120,000 years.[24]

[23] *Brāhma-sphuṭa-siddhānta* 1.31, [DviS1901], p. 9.

[24] *Brāhma-sphuṭa-siddhānta* 1.26–27, [DviS1901], p. 8, and *Āryabhaṭīya* 1.5, [ShuSa1976], p. 9. That is, for Brahmagupta the age of the universe up to the present Kaliyuga is six *manvantaras* plus seven intermediary periods plus twenty-seven *mahāyugas* of the seventh *manvantara* plus the pre-Kaliyuga part of the twenty-eighth *mahāyuga*, or $6 \times 71 \times 4320000 + 7 \times 1728000 + 27 \times 4320000 + \frac{9}{10} \times 4320000$. For Āryabhaṭa, however,

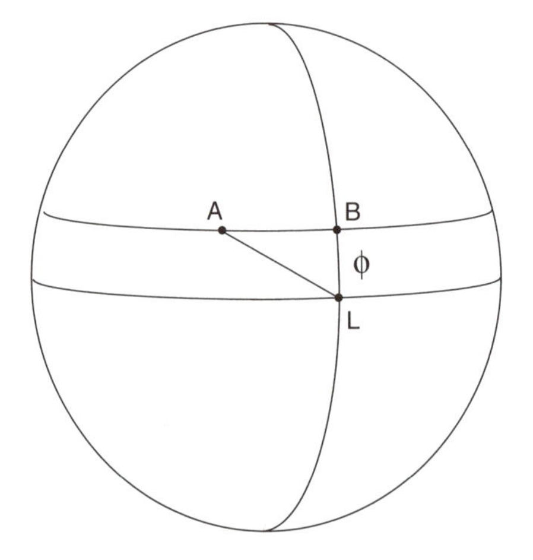

Figure 4.2 The correction for terrestrial longitude.

After the mean celestial longitudes are computed for the given time d by means of these immense integer ratios, they must be adjusted for the observer's own longitude on the earth. The great conjunction of all celestial bodies at the start of the *kalpa* is assumed in the Brāhma-pakṣa to coincide with the vernal equinox at sunrise as viewed from the Indian prime meridian. All *pakṣas* consider this prime meridian to pass through Ujjain (in Madhya Pradesh) and intersect the earth's equator at Laṅkā (i.e., a point in or around the island of Sri Lanka), which is their standard zero point of terrestrial latitude and longitude. Of course, if the observer doesn't happen to be exactly on the Indian prime meridian, there will be a sort of "time zone" discrepancy. That is, the elapsed time d will have to be adjusted by the "location difference," or correction for geographical longitude, which Brahmagupta describes as follows:

> The beginning of a day west or east of the south-north [meridian through] Ujjain is [respectively] after or before sunrise [on that meridian] by the *ghaṭikās* of location-difference.
>
> The circumference of the earth is a line of five thousand *yojanas*. It is multiplied by the degrees of one's own latitude-interval and divided by the degrees of a circle. The square of the distance [of] the location is diminished by the square of the result.... The quotient from the square root [of the difference] divided by the circumference of the earth and multiplied by sixty is in *ghaṭi-*

it is six *manvantaras* plus six intermediary periods of a different size, plus twenty-seven *mahāyugas* of the seventh *manvantara*, plus a differently-sized pre-Kaliyuga portion of the twenty-eighth *mahāyuga*: i.e., $6 \times 71 \times 4320000 + 6 \times 4320000 + 27 \times 4320000 + \frac{3}{4} \times 4320000$.

As we saw in section 4.2, Āryabhaṭa's noncanonical choice of parameters provoked Brahmagupta's criticism.

kās.[25]

As illustrated in figure 4.2, this method uses simple geometry to compute the distance between two points A and B on the earth, which is said to be 5000 *yojanas* in circumference. Point B lies on the prime meridian intersecting the equator at Laṅkā or L, and the observer's location A is west of B on the same latitude circle, ϕ degrees north of the equator. Brahmagupta computes the distance BL along the prime meridian by the ratio $BL/\phi = 5000/360°$. The distance AL in *yojanas* between A and Laṅkā is assumed to be known, and the distance AB is found from $\sqrt{AL^2 - BL^2}$ as though triangle ABL were a plane triangle.

Since the entire 5000 *yojanas* of the equator corresponds to a time difference of one day or sixty *ghaṭikās*, a distance AB on the equator corresponds to a time difference of $60 \times AB/5000$ *ghaṭikās*. (The difference in circumference between the equator and the actual latitude circle of A is apparently considered negligible.) And this time difference is used to correct the time interval d and hence the mean position of the planets as seen from A instead of Laṅkā.

4.3.3 Trigonometry and true motions

The second chapter of a *siddhānta* is usually concerned with the correction of mean planetary positions to true ones. That is, the planet is assumed to be moving with uniform circular motion, but not on a simple concentric orbit. Thus its uniform motion as seen from the earth will appear nonuniform, and the calculated mean positions must be corrected mathematically in order to predict where the planet will actually appear in the sky at a given time. Since these corrections involve combinations of circles, the algorithms for computing them depend largely on Sines[26] relating angular and linear measures. An astronomer's presentation of them generally consists of a versified table of (usually) twenty-four Sines in the first quadrant at intervals of $3\frac{3}{4}^{\circ}$, and sometimes one for the corresponding twenty-four Versines (versed sines), along with a rule for linear interpolation to find Sines and arcs between the tabulated values. There is usually no explanation in the base text of how the Sine table should be derived geometrically. The Cosine or Sine of the complement is also used, but there is no named counterpart to the modern tangent or secant.

The extant *Paitāmaha-siddhānta* describes its Sines and its procedure for interpolating between them as follows:

> The first Sine is a ninety-sixth part of 21,600. If one divides the

---

[25] *Brāhma-sphuṭa-siddhānta* 1.35–38, [DviS1901], p. 10.

[26] As in section 3.2, names of the Indian trigonometric functions are here capitalized to indicate that they are scaled to a nonunity trigonometric radius R (sometimes called the "Radius"), unlike their modern counterparts which are scaled to unity. The value of R is usually an integer on the order of 3000, to provide reasonably accurate integer values of the functions.

> first Sine by the first Sine and subtracts the quotient from the first Sine, one obtains the difference of the second Sine; the sum of the first Sine and the difference of the second Sine is the second Sine. If one divides the second Sine by the first Sine and subtracts the quotient from the difference of the second Sine, one obtains the difference of the third Sine; the sum of this and the second Sine is the third Sine ... [And so on up to the twenty-fourth and final Sine.]
>
> The minutes [in the argument of arc] are to be divided by 225; the Sine corresponding to the resulting serial number is to be put down. One should multiply the remainder by the difference of the next Sine and divide [the product] by 225. The sum of the quotient and the Sine which was put down is the desired Sine.[27]

Here only the first Sine, $\mathrm{Sin}_1 = 225$, is assigned a numerical value. Each subsequent ith tabulated Sine or Sin_i[28] is defined recursively by

$$\mathrm{Sin}_i - \mathrm{Sin}_{(i-1)} = \mathrm{Sin}_{(i-1)} - \mathrm{Sin}_{(i-2)} - \frac{\mathrm{Sin}_{(i-1)}}{\mathrm{Sin}_1}.$$

This Sine-difference formula gets more and more approximate as i increases. It ultimately yields a value for the Sine of 90°—that is, the trigonometric radius R—equal to about 3359, whereas the initial Sine-value $\mathrm{Sin}\left(3\frac{3}{4}^{\circ}\right) = 225$ implies a radius $R \approx 3440$. The *Paitāmaha-siddhānta*'s actual value of R was probably the same as the value first explicitly appearing in the *Āryabhaṭīya* and later accepted as standard in most *siddhāntas*, namely, 3438. This number is the approximate length $C/(2\pi)$ of the radius of a circle with circumference $C = 21{,}600$ arcminutes, if the value of π is taken to be about 3.1416. The advantages of this radius value are similar to those of modern radian measure: it allows the same units to be used for arcs and line segments, and establishes approximate equality between the values of small Sines and their arcs.[29]

The values of the Sine-differences in the *Āryabhaṭīya* are generally correct to the nearest integer, so they cannot have been computed from the above approximate formula (see sections 5.1.1 and 5.1.2 for further discussion of their construction). Āryabhaṭa states their values in his peculiar notation as follows:

---

[27] [Pin1967], pp. 480, 483–484.

[28] The unusual notation Sin_i for the ith Sine value is intended to emphasize the fact that these Sines are regarded as concrete quantities representing line segments in a circle, not ratios of quantities or functions of variables.

[29] It has been debated for several decades whether a trigonometric radius of 3438 may have been used in Hellenistic astronomy and thence transmitted to the Indian tradition. The evidence for this hypothesis is reconstructed from some ratios of astronomical parameters computed by Hipparchus and cited by Ptolemy in the *Almagest*. Arguments for and against the inference that Hipparchus used $R = 3438$ have been recently presented in [Duk2005c] and [Kli2005].

> *makhi-bhakhi-phakhi-dhakhi-ṇakhi-ñakhi-ṅakhi-hasjha-skaki-kiṣga-śghaki-kighva-ghlaki-kigra-hakya-dhāhā-sta-sga-śjha-ṅva-lka-pta-pha-cha* [225, 224, 222, 219, 215, 210, 205, 199, 191, 183, 174, 164, 154, 143, 131, 119, 106, 93, 79, 65, 51, 37, 22, 7] [are the differences of] the Sines in arcminutes.[30]

These differences add up to 3438 for the trigonometric radius. Although the value 3438 was most frequently used, a few other values of R are attested in other *siddhāntas*. For example, the *Brāhma-sphuṭa-siddhānta*, for reasons largely obscure, uses $R = 3270$; the *Siddhānta-śekhara* takes $R = 3415$.[31]

Linear interpolation between the twenty-four entries of a geometrically computed Sine table was the standard method for determining trigonometric quantities in *siddhāntas*, but by no means the only one. Higher order interpolation rules and algebraic approximations also appeared, frequently in later chapters supplying additions and improvements to the texts' basic computational methods. A remarkable example is the following formula first attested in the *Mahā-bhāskarīya* of Bhāskara I and also mentioned in several later works, although its original process of derivation or justification remains mysterious:

> The procedure without [using the tabulated Sine-differences] *makhi* [i.e., 225] and so forth is concisely stated: The degrees of the arc, subtracted from the total degrees of half a circle, multiplied by the remainder from that [subtraction], are put down twice. [In one place] they are subtracted from sky-cloud-arrow-sky-ocean [40500]; [in] the second place, [divided] by one-fourth of [that] remainder [and] multiplied by the final result [i.e., the trigonometric radius].[32]

That is, for a given angle $x°$ whose Sine is sought,

$$\operatorname{Sin} x \approx \frac{R \cdot x(180 - x)}{\dfrac{40500 - x(180 - x)}{4}},$$

[30] *Āryabhaṭīya* 1.12, [ShuSa1976], p. 29. Note that Āryabhaṭa calls the units of Sine length "arcminutes."

[31] *Brāhma-sphuṭa-siddhānta* 2.2–5, [DviS1901], p. 23; *Siddhānta-śekhara* 3.3–6, [Pin1970a], p. 582. Āryabhaṭa's circumference-diameter ratio (see section 5.1.1) is equivalent to a π value of 3.1416, whereas $R = 3415$ implies $\pi \approx \sqrt{10}$ if R is meant to be commensurate with the circumference. It has been speculated ([Ken1983b], p. 19) that Brahmagupta's value 3270 was based on an approximate circumference-diameter ratio of 360/109, but the approximation is so crude that this seems unlikely. See also the discussion of Brahmagupta's remarks on Sine computations at the end of section 5.1.3.

[32] *Mahā-bhāskarīya* 7.17–18, [Kup1957], pp. 377–378. See [Gup1967], [Gup1986], and [Hay1991] for discussions of the various Sanskrit versions of this rule and attempts to reproduce its derivation. (The *Mahā-bhāskarīya*, like the *Āryabhaṭīya*, has an idiosyncratic ordering of subjects; Bhāskara deals with true longitude computations in his fourth chapter and with astronomical parameters, including Sine-differences and this Sine rule, in his seventh.)

which in modern terms, with x in radians, can be rewritten as a very accurate quadratic approximation:

$$\sin x \approx \frac{16x(\pi - x)}{5\pi^2 - 4x(\pi - x)}.$$

Experiments with nonlinear interpolation for trigonometric quantities are first attested in the *Brāhma-sphuṭa-siddhānta* of Brahmagupta, in the context of a simplified Sine table with entries spaced at intervals of 15° or 900′:

> The product of the minutes of half the difference of the previous and current [tabulated] differences, divided by nine hundred, is added to or subtracted from half the sum of those [tabulated differences], [when the half-sum] is [respectively] less or greater than the current [difference]. [The result is the true] current [difference].[33]

This is essentially a means of scaling the current Sine-difference $\mathrm{Sin}_i - \mathrm{Sin}_{i-1}$, or $\Delta\,\mathrm{Sin}_i$, according to where the given arc x measured in arcminutes falls within the ith interval of 900′. Whereas ordinary linear interpolation would prescribe

$$\mathrm{Sin}\,x \approx \mathrm{Sin}_{i-1} + \frac{x - 900(i-1)}{900} \cdot \Delta\,\mathrm{Sin}_i,$$

Brahmagupta's rule states instead that

$$\mathrm{Sin}\,x \approx \mathrm{Sin}_{i-1} + \frac{x - 900(i-1)}{900} \times \left(\frac{\Delta\,\mathrm{Sin}_{i-1} + \Delta\,\mathrm{Sin}_i}{2} + \frac{\Delta\,\mathrm{Sin}_{i-1} - \Delta\,\mathrm{Sin}_i}{2} \times \frac{x - 900(i-1)}{900} \right).$$

(The second term in the scaling factor here is always additive according to Brahmagupta's condition, because successive Sine-differences always decrease, and therefore the average of two successive differences will always be less than the first of them. In computing Versines whose differences successively increase, on the other hand, the second term would be subtractive.) This scaling rule is equivalent to the modern interpolation technique known as Stirling's finite difference formula, up to the second order. Brahmagupta records no information about the mathematical reasoning he used to come up with it, or any other source from which he derived it.

Another approach to increasing the accuracy of Sine values was taken by Govindasvāmin, a commentator on the *Mahā-bhāskarīya* of Bhāskara writing probably in the early ninth century. He recomputed (although he does not explain how) the twenty-four tabulated differences of Āryabhaṭa's table to

[33] *Brāhma-sphuṭa-siddhānta* 25.17, [DviS1901], p. 418. See [Gup1969] for an analysis of this and other second-order interpolation rules.

Table 4.4 Āryabhaṭa's Sine-differences as modified by Govindasvāmin.

1	224;50,23	2	223;52,30	3	221;57,18	4	219;04,57
5	215;16,22	6	210;32,26	7	204;54,26	8	198;23,48
9	191;02,09	10	182;51,27	11	173;52,58	12	164;12,10
13	153;46,49	14	142;42,46	15	131;02,02	16	118;47,38
17	106;02,42	18	92;50,32	19	79;14,31	20	65;18,08
21	51;04,59	22	36;38,41	23	22;03,00	24	7;21,37

the second sexagesimal place, instead of to the nearest integer; his results are shown in table 4.4, which uses semicolons to separate integer from fractional parts and commas to separate the other sexagesimal places. Govindasvāmin also discusses some interesting numerical methods for approximating these tabulated Sine-differences, including the following:

> That final [Sine-difference], successively multiplied by the odd numbers beginning with three, is [each remaining] Sine-difference in reverse order, beginning with the one below that [final difference]. Thus the determination of Sines in the third sign [i.e., in the 30° interval from 60° to 90°].... But the final Sine[-difference] in the last one-eighth of a sign, multiplied by the odd numbers beginning with three, is not equal to [the tabulated differences] *pha* [i.e., 22] and so forth. This is not wrong, because of the incompleteness of the procedure. It is done in *this* way: The final Sine[-difference], decreased by its last [sexagesimal digit] multiplied by the sum of so many [numbers] beginning with 1, multiplied by the odd numbers beginning with three, should be equal to the [Sine-differences] beginning with *pha* in the final sign.[34]

These methods focus on the final 30° of the quadrant, where the size of the Sine-differences changes most rapidly and the accuracy of linear interpolation is poorest. To start with, Govindasvāmin suggests that the seven Sine-differences $\Delta\,\mathrm{Sin}_i$ for $17 \leq i \leq 23$ can be computed from the last one, $\Delta\,\mathrm{Sin}_{24}$, as follows:

$$\Delta\,\mathrm{Sin}_i \approx \Delta\,\mathrm{Sin}_{24} \cdot (2 \cdot (24 - i) + 1).$$

Govindasvāmin's rule would approximate $\Delta\,\mathrm{Sin}_{23}$ by $\Delta\,\mathrm{Sin}_{24} \times 3 = 22;1,37$, then $\Delta\,\mathrm{Sin}_{22}$ by $\Delta\,\mathrm{Sin}_{24} \times 5 = 36;9,37$, and so on, which is no great accomplishment in terms of accuracy. The following intriguing tweak to the procedure, however, notably improves its performance. If the integer in the second sexagesimal place of $\Delta\,\mathrm{Sin}_{24}$ is m, then the new approximation is

[34] These excerpts and Govindasvāmin's corrections to the tabulated Sine-differences are taken from his commentary on *Mahā-bhāskarīya* 4.22, [Kup1957], pp. 199–201. They are discussed in detail in [Gup1971].

defined as

$$\Delta \operatorname{Sin}_i \approx \left(\Delta \operatorname{Sin}_{24} - \frac{m}{60^2} \sum_{i=1}^{24-i} i \right) \cdot (2 \cdot (24 - i) + 1).$$

Applied to Govindasvāmin's value $\Delta \operatorname{Sin}_{24} = 7;21,37$, this rule gives the much more accurate approximations $\Delta \operatorname{Sin}_{23} \approx 22;3,0$ and $\Delta \operatorname{Sin}_{22} \approx 36;38,50$. The approximation gets somewhat worse for larger Sine-differences. How and why Govindasvāmin derived these procedures is unknown, and it is not clear why he chose to compute approximate values for a Sine table where he already possessed more exact values. They may just have been clever puzzle problems solved by numerical experimentation.

Let us now return to the Indian astronomer's immediate motive for studying all these trigonometric methods, namely, the correction of mean planetary positions based on orbital models made up of various arrangements of circles. The displacement of a planet's mean position from its true position depends on its anomaly or angular separation from its apsidal point, or apogee. The sun and moon have only one anomaly, the *manda* or "slow" one, which is derived from what we would call the ellipticity of its orbit. The five star-planets also have a second inequality (*śīghra* or "fast"), which accounts for what we would call the relative orbital motion of the planet and the earth about the sun, causing the planet's apparent retrograde motion.

The physical interpretation in *siddhānta* texts of the relationship between a mean planet and its apsidal points is somewhat ambiguous. The actual path of the planet's orbit with one anomaly could be explained as an eccentric circle, or equivalently as the trace of its motion on an epicycle revolving on a concentric circle, as long as the eccentricity in the first case equals the epicycle radius in the second.[35] The recognition that these two alternatives are equivalent is hinted at by Āryabhaṭa:

> All the planets revolve with their own [mean] motion on the [eccentric] circle [literally *prati-maṇḍala*, "para-circle"] beside the orbit.... [Each] eccentric-circle is equal to its own orbit-circle; the center of the eccentric-circle [lies] beyond the center of the earth's body. The distance [between the centers] of the eccentric-circle and the earth is the radius of its own epicycle [literally *ucca-nīca-vṛtta*, "above-and-below-circle"]. The planets do revolve with mean motion on the epicycle circumference.... The mean planet is at the center of its own epicycle adhering to the orbit-circle.[36]

Another physical interpretation is put forth in the *Sūrya-siddhānta*, in which a planet's displacement from mean longitude is caused by a divine being pulling the planet toward the apsidal point:

> Incarnations of time with invisible forms, located in the ecliptic [and] called the *śīghra*-apsis, *manda*-apsis, and node, are the

[35] A geometric proof of this equivalence is provided in a number of standard references on geocentric astronomy, for example, [Eva1998], pp. 211–213.

[36] *Āryabhaṭīya* 3.17–21, [ShuSa1976], pp. 104–106.

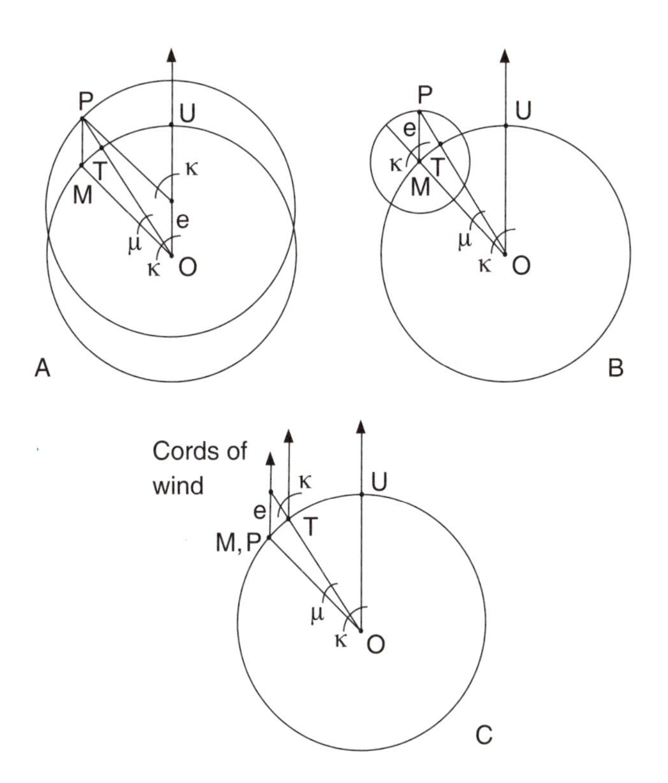

Figure 4.3 Planetary longitude correction models. A. The eccentric orbit. B. The epicyclic orbit. C. The concentric orbit with attraction toward the apsidal deity.

> causes of the motion of the planets. [The planets], attached to their cords of wind, are dragged aside eastward or westward towards them by their left and right hands, in the direction of their own place.…
>
> Because of the greatness of its orbit, the sun is drawn away only a small [amount], [while] the moon is drawn away much more because of the smallness of its orbit. [The star-planets] beginning with Mars, because of their smallness of form, are dragged aside a long way by the deities called the *śīghra*-apsis and *manda*-apsis.[37]

This last model is qualitative rather than mathematically predictive. But since the computational techniques accompanying it are the same as for the eccentric and epicycle models, the results for all are equivalent, as illustrated in figure 4.3. In each case, the earthly observer is at the center O, the planet impelled by its mean motion is at P, and the apsidal point lies in the vertical direction indicated by U. If the planet were moving uniformly on the concentric circle it would be at M, but the orbital setup causes it to appear as though it were at T.

The angular separation between the mean position M and U is the anomaly, κ, and the longitude correction or "equation" μ adjusts the mean position to the true position T.[38] The size of μ depends on κ and on e, the eccentricity or epicycle radius. As the figure indicates, the mechanical differences between these models are mathematically unimportant as long as their parameters are the same. It is not clear how or whether most Indian astronomers made a definite choice in favor of the physical reality of any one of them (this issue is discussed further in section 4.5). The same trigonometric methods applied to all the equivalent models.

A basic trigonometric rule for finding the equation μ is the following one from the *Paitāmaha-siddhānta*:

> The Sine computed from the *manda*-argument is the argument-Sine; that from [the *manda*-argument] subtracted from 90° is the complement-Sine. One should multiply the Cosine and the Sine by the planet's *manda*-epicycle and divide [the product] by 360°; the results are the Sine-result and the Cosine-result [in reverse order]. If the argument is in [the hemisphere] beginning with Capricorn, one should add the Cosine-result to R; if it is in that beginning with Cancer, one should subtract it. [The result] is the

[37] *Sūrya-siddhānta* 2.1–2, 2.9–10, [PandeS1991], pp. 17, 19. The star-planets are typically named in their weekday order rather than in order of geocentric distance; hence they are said to "begin with Mars."

[38] Here we are simplifying the situation by considering only the mean and true "positions" M and T relative to the apsidal line OU, rather than their corresponding mean and true *longitudes* $\bar{\lambda}$ and λ, which are measured from the zero point of the ecliptic. The angle $\angle UOM$ or κ added to the longitude of the apsis would give the mean longitude $\bar{\lambda}$, and $\angle UOT$ plus the longitude of the apsis would give the true longitude λ. Strictly speaking, the planet's longitude is also affected by its latitude or displacement from the ecliptic, but that factor is generally ignored in basic true longitude computations.

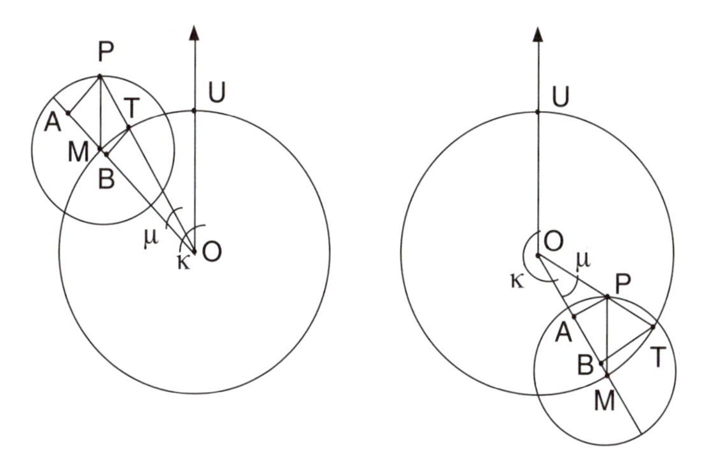

Figure 4.4 Sine-based correction to the mean planetary position.

> Cosine; the Sine is the Sine-result. The hypotenuse is the sum of the squares of the Cosine and the Sine. One should multiply the Sine by R and divide [the product] by the corrected hypotenuse; the arc of the result is the equation. If the argument is in [the hemisphere] beginning with Aries, it is to be subtracted from the [mean] planet; if in that beginning with Libra, it is to be added to it.[39]

Figure 4.4 depicts the quantities used in this rule as they apply to the epicycle model. It shows two different positions of the planet P, which is revolving clockwise on its epicycle while the epicycle center M revolves counterclockwise around the earth. The earth is again at O, the mean position of the planet at M, and the true position at T. The radius of the concentric OM is assumed equal to the trigonometric radius R, and the epicycle radius e is MP. But the epicycle circumference c is assumed in the rule to be smaller than 360°, such that $c/360 = e/R$.

As before, the anomaly is κ and the desired equation to be computed is μ. The "argument-Sine" and "complement-Sine" or Cosine are $\operatorname{Sin}\kappa$ and $\operatorname{Cos}\kappa$ respectively. This Sine and Cosine are then scaled to e instead of R by means of the factor $c/360 = e/R$. The "Sine-result" and "Cosine-result" thus produced are the perpendicular segment AP and the segment AM of the epicycle radius:

$$AP = \operatorname{Sin}\kappa \cdot \frac{e}{R},$$

$$AM = \operatorname{Cos}\kappa \cdot \frac{e}{R}.$$

Then one finds the leg OA and its hypotenuse OP by straightforward geom-

[39][Pin1967], p. 494.

etry, as follows:

$$OA = R \pm AM,$$

$$OP = \sqrt{OA^2 + AP^2}.$$

And from the similar right triangles $\triangle TOB$ and $\triangle POA$,

$$TB = R \cdot \frac{AP}{OP}.$$

But TB is just $\operatorname{Sin}\mu$, since it is the leg opposite μ in a right triangle with hypotenuse R. So the arcSine of TB is the desired correction to the planet's longitude.

Note that in the quoted rule, the references to "Capricorn," "Cancer," and so forth, somewhat confusingly bear no relation to actual zodiacal locations. They are simply a standard way of indicating quadrants of the circle with respect to any given point, which is imagined to be the zero point or beginning of "Aries." In this case, the desired point is U in the direction of the apsis, from which the anomaly κ is measured. So when κ falls in the semicircle beginning with "Capricorn"—that is, within 90° of the apsidal line OU on either side, as shown in the left half of figure 4.4—the distance OA is computed by adding AM to R. If it is in the semicircle beginning with "Cancer," as in the right half of the figure, then $OA = R - AM$.

Similarly, when the anomaly κ is less than 90° or in the semicircle of "Aries," the true position T lags behind the mean longitude at M, so μ is negative. When κ is in the semicircle of "Libra," on the other hand, the equation μ is positive. The left and right sides of the figure respectively show these two cases.

A great deal more is involved in *siddhānta* algorithms for correcting planetary positions than this relatively straightforward trigonometric technique would suggest. Several other astronomical factors are also taken into account. Some are easily explicable, like the solar equation of time, which accounts for the effect of the sun's nonuniform apparent motion on time measurement.[40] Some are more puzzling, such as the numerous rules for applying longitude corrections iteratively, or for varying the size of an orbit or an epicycle depending on the amount of anomaly, the body's altitude above the horizon, or some other quantity. In the treatises these odd procedures are not explained physically or derived geometrically, and many of them have long resisted the best efforts of researchers to explain them historically. However, they sometimes reveal intriguing and creative mathematical approaches, including some of the earliest known uses in Indian texts of iterative approximation techniques.

The following rule from Bhāskara's *Mahā-bhāskarīya* prescribes one such iterative correction. It adjusts the true distance of the sun or moon from the earth (OP in figure 4.4), depending on the size of the anomaly:

[40] These technical issues are beyond the scope of the rudimentary introduction to geocentric astronomy offered in section 4.1. They are dealt with in, for instance, [Eva1998], pp. 235–238.

> When the hypotenuse is multiplied by the Sine-result and Cosine-result and divided by the Radius, [those two quotients] should be the [new] Sine-result and Cosine-result. From these the [new] hypotenuse [is computed] as before. Again, when the [new] hypotenuse is multiplied by the initial [Sine- and Cosine-] results and divided by the Radius, [new Sine- and Cosine-results are produced]. In the same way the hypotenuse is to be derived by this procedure again and again, until the value of the hypotenuse becomes equal to the previously-determined hypotenuse.[41]

That is, once we know from κ and e the "Sine-result" AP and "Cosine-result" AM within the epicycle, and some value for the hypotenuse or planetary distance OP, we recompute new values for them until OP is fixed:

$$AP_1 = AP \cdot \frac{OP}{R}, \qquad AM_1 = AM \cdot \frac{OP}{R},$$

$$OP_1 = \sqrt{(R \pm AM_1)^2 + {AP_1}^2},$$

$$\vdots$$

$$AP_{i+1} = AP \cdot \frac{OP_i}{R}, \qquad AM_{i+1} = AM \cdot \frac{OP_i}{R},$$

$$OP_{i+1} = \sqrt{(R \pm AM_{i+1})^2 + {AP_{i+1}}^2}.$$

If the distance OP in the epicycle model is equal to R (that is, if κ is close to 90°), this procedure will give the same value for it. But if OP is at its maximum (when M is at U so $\kappa = 0$ and $AP = 0$), the iterative procedure will make its value larger, because $OP/R > 1$ and hence $AM_1 > AM$. Similarly, if OP is at its minimum (when $\kappa = 180°$) its value under iteration will also increase, because $AM_1 > AM$ so $(R - AM_1) < (R - AM)$. So this modification will have the effect of elongating the path of the planet along the apsidal line: its orbit will no longer be an eccentric perfect circle but a sort of eccentric oval. It is hard to say for certain what such corrections were intended to accomplish in astronomical theories, but they clearly illustrate the facility with which Indian mathematicians combined numerical and geometric methods.

4.3.4 Direction, place, and time

Once the true positions of the planets have been found relative to the center of the earth, an astronomer must be able to predict their appearance when

[41] *Mahā-bhāskarīya* 4.10cd–12, [Kup1957], p. 189. See [Pin1974], [Yan1997], [Duk2005b], and [DukeIPT] for discussions of this approach and its possible relation to the equant model of Hellenistic astronomy.

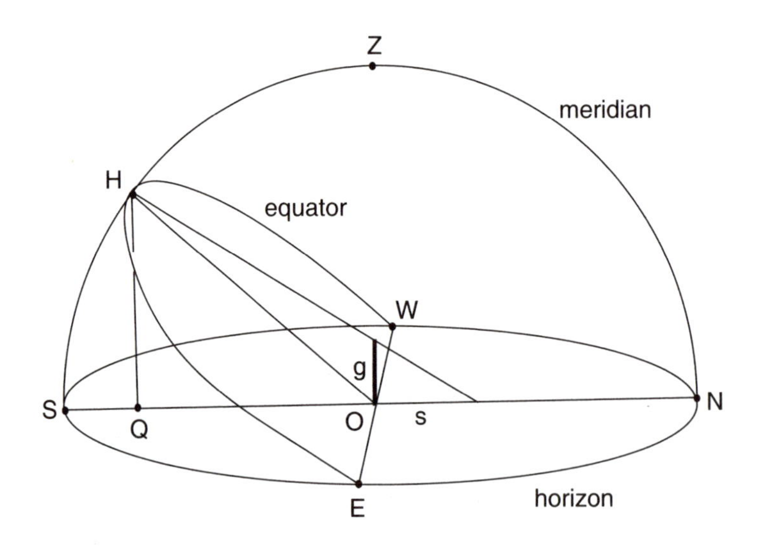

Figure 4.5 Determining local latitude from the noon equinoctial shadow.

observed from a given terrestrial location. These problems are known in *siddhāntas* as the "Three Questions," namely, problems of direction, place, and time. They generally involve Sine-based solutions of triangles determined by gnomon shadows or other plane triangles within the hemisphere bounded by the observer's sky and horizon. Quantities that are typically sought include the time of day, terrestrial latitude, ascensional difference, ecliptic ascendant or horoscope point, and altitude or direction of the sun.

A few examples from the *Brāhma-sphuṭa-siddhānta* will illustrate the chief mathematical techniques employed to answer such questions. First, we consider the problem of finding the local latitude from the shadow of a gnomon when the sun is at true local noon on the day of the equinox, as pictured in figure 4.5. The visible half of the celestial equator extends from the east point E to the west point W, and is bisected by the local meridian NZS at H. We assume that the sun's position is exactly at point H at the time of true local noon—that is, its motion in declination during the equinoctial day is negligible.

A gnomon of height g (whose standard value is twelve digits) is set up in the center of the plane of the observer's horizon, at the intersection O of the north-south and east-west lines, and casts a shadow of length s. The arc ZH between the observer's zenith and the celestial equator is the terrestrial latitude ϕ, and is equal to the angle between the gnomon g and its hypotenuse in the shadow triangle. Consequently,

$$\frac{s}{g} = \frac{\operatorname{Sin}\phi}{\operatorname{Cos}\phi},$$

or as Brahmagupta puts it in the *Brāhma-sphuṭa-siddhānta*,

> At the equinox, the gnomon is [proportional to the Sine of solar] altitude, the shadow [to] the Sine of the latitude, [and] the square root of the sum of the squares of those is the equinoctial hypotenuse.[42]

This assumes that the distance g between point O and the tip of the gnomon is negligible compared to the size of the celestial sphere, so the shadow triangle is similar to the right triangle HOQ containing the Radius OH and the perpendicular HQ dropped on the north-south line.

Now we shall use the latitude ϕ thus derived, plus the current true longitude λ of the sun, to figure out how to tell time. The *siddhānta* procedure is rather long and complicated, involving the computation of many astronomical quantities. But as we will see, it really depends on nothing more than basic trigonometry and the similarity of plane right triangles. All of the formulas involved can be straightforwardly derived from triangles within the celestial sphere, as shown in the explanatory figures below.[43]

Say we want to observe the sun's gnomon shadow at an arbitrary time before noon on an arbitrary day to know the elapsed time t since the start of the day at sunrise. To start with, we have to determine from the sun's current longitude, λ, its corresponding declination δ, namely, its distance north or south of the celestial equator. Then we will compute the so-called "ascensional difference" ω, which measures the time between six o'clock (equinoctial sunrise) and sunrise on the given day. We assume (again neglecting change in declination during the course of the day) that ω also represents the time difference between 6 PM and sunset, so adding 2ω to 12 hours gives us the total time from sunrise to sunset. Knowing this length of the day, we can then use the observed shadow length s_t at the given time to find t.

Brahmagupta's formula for the solar declination δ is as follows:

> The Sine of [the longitude of] the sun is multiplied by the Sine of Jaina-saint [a concrete number meaning 24] degrees [and] divided by the Radius. The quotient is the Sine of the given declination of the sun, north or south of the equinox.[44]

The computation of δ is illustrated in figure 4.6, showing half of the northern

---

[42] *Brāhma-sphuṭa-siddhānta* 3.7, [DviS1901], p. 51. The twelve-digit gnomon length was known as a standard value at least since the time of the *Arthaśāstra* (see section 3.1): *Arthaśāstra* 2.20.10, [Kan1960], vol. 2, p. 138.

[43] The "Three Questions" rules could also be derived from analemmatic plane projection methods, frequently employed by modern researchers to explain their rationales. But since such two-dimensional analemmas are apparently never described or depicted in Sanskrit texts (see [VanB2008], ch. 3), whereas three-dimensional models of the celestial and terrestrial spheres are often mentioned (see section 4.5), this discussion will rely on three-dimensional drawings to illustrate the geometry involved.

[44] *Brāhma-sphuṭa-siddhānta* 2.55, [DviS1901], p. 44. Note that although δ and the ascensional difference ω (see below) are used to solve many of the "Three Questions" problems in the third chapter of Brahmagupta's *siddhānta*, he gives the formulas for them in the second chapter concerning true motions, since they are also needed to determine corrected planetary positions for the times of true sunrise and sunset on a non-equinoctial day.

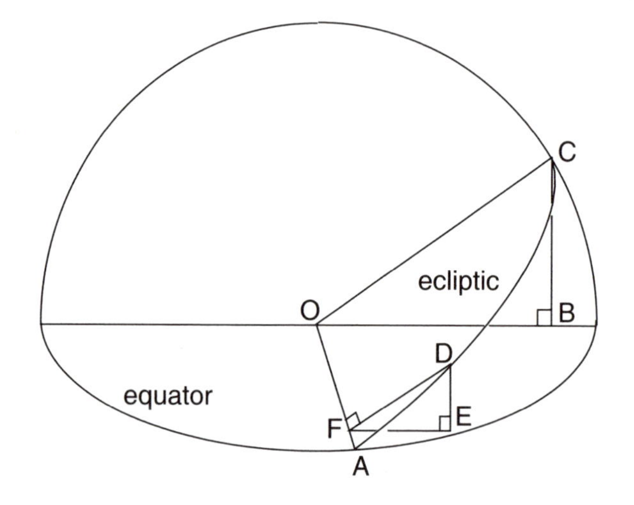

Figure 4.6 Finding the solar declination from similar triangles.

celestial hemisphere, bounded by the plane of the celestial equator and a plane passing through the solstices. The observer on the earth is represented by point O. The equator intersects the ecliptic at the equinoctial point A, and the 90° arc from A to C is the first quadrant of the ecliptic, from the start of the sign Aries to the start of Cancer. The obliquity of the ecliptic with respect to the equator, nowadays called the earth's "axial tilt," is considered in Indian astronomy to be 24°. The radius $OC = OA$ is assumed equal to the trigonometric radius R, so the perpendicular CB dropped from C onto the plane of the equator is Sin 24°.

The point D represents the sun's current position, so the arc AD is the solar longitude λ measured from the beginning of Aries.[45] Its Sine is the segment DF lying in the plane of the ecliptic, and the Sine of its corresponding declination δ is the perpendicular DE. And since the right triangles OCB and FDE are similar,

$$\operatorname{Sin}\delta = DE = DF \cdot \frac{CB}{OC} = \operatorname{Sin}\lambda \cdot \frac{\operatorname{Sin} 24^\circ}{R}.$$

Now we return to the plane of the observer's horizon, half of which is shown within the eastern celestial hemisphere in figure 4.7, to determine the ascensional difference ω. Once again, the observer is at O, with the zenith Z directly above and the north celestial pole P lying on the meridian SZN between the zenith and the north point of the horizon. The arc of the celestial equator HEF intersects the horizon at the east point E and the meridian at H. The arc GCD, parallel to the equator, is part of the so-called "day-circle" representing the sun's apparent motion on the given

[45]We simplify the situation a little by using tropical longitude here. Indian planetary longitudes are generally sidereal, corrected for precession when necessary.

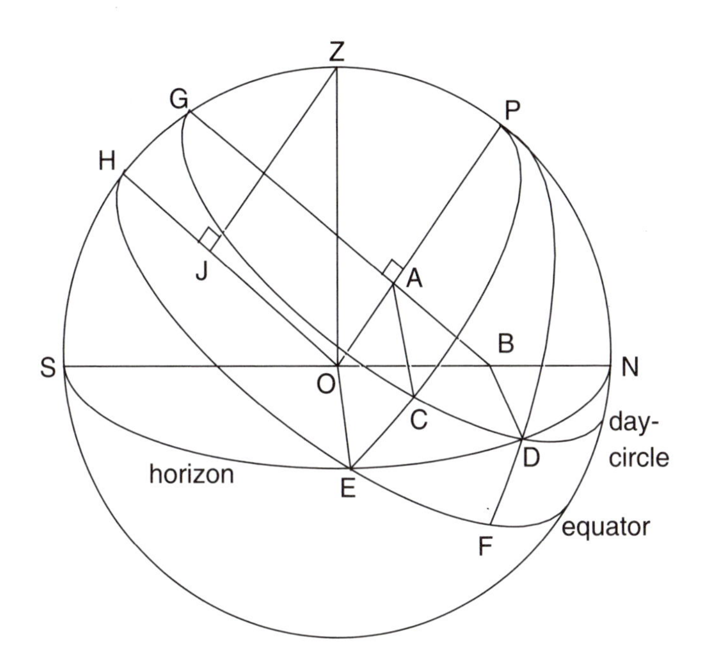

Figure 4.7 The observer's horizon, celestial equator, and day-circle.

day from its rising at point D to its noon culmination at G. (Since this day-circle is north of the celestial equator, the given day evidently falls in the summer half of the year.) Arc EH, a quadrant of the equator, represents one-quarter of a daily solar revolution or six hours of time, during which the corresponding arc CG of the day-circle will rise. So the arc CD from sunrise to local six o'clock is measured by the corresponding time arc EF on the equator, which is the ascensional difference ω.

The arc ZH is again the terrestrial latitude ϕ, and the sun's distance north of the equator, arc GH, is its northern declination δ. The Sine of δ is equal to the segment OA, and its Cosine to the radius of the day-circle, AG. The segment AB, representing the arc CD projected onto the diameter of the day-circle, has a Sanskrit name that translates to "earth-Sine" or "horizon-Sine." Brahmagupta describes how to manipulate all these quantities to compute ω from δ and ϕ, as follows:

> After the square of [the Sine of] the given declination is subtracted from the square of the Radius, the square root of the remainder is the radius of its own day-circle north or south of the equinox. The Sine of the declination, multiplied by the [noon] equinoctial shadow, divided by twelve, is the "earth-Sine," constant on its own day-circle. [The earth-Sine] multiplied by the Radius [and] divided by the half [-diameter] of its own day-circle

> is the Sine of decrease or increase [in day-length].[46]

That is, the radius AG of the day-circle is $\sqrt{R^2 - \mathrm{Sin}^2\,\delta}$, or $\mathrm{Cos}\,\delta$. And since we showed above that the noon equinoctial shadow triangle of a gnomon is similar to the Sine triangle ZOJ of the terrestrial latitude, and since $\triangle ZOJ$ is also similar to $\triangle BOA$,

$$AB = OA \cdot \frac{ZJ}{OJ} = OA \cdot \frac{\mathrm{Sin}\,\phi}{\mathrm{Cos}\,\phi} = \mathrm{Sin}\,\delta \cdot \frac{\mathrm{Sin}\,\phi}{\mathrm{Cos}\,\phi} = \mathrm{Sin}\,\delta \cdot \frac{s}{12},$$

when the gnomon length g equals the standard twelve digits. Then this earth-Sine AB, scaled to the great circle of the equator with radius R instead of the smaller day-circle with radius $\mathrm{Cos}\,\delta$, is just the Sine of the desired ascensional difference arc EF:

$$\mathrm{Sin}\,\omega = \mathrm{Sin}\,EF = AB \cdot \frac{R}{\mathrm{Cos}\,\delta}.$$

The indefatigable astronomer, still wishing to compute the current time t since sunrise for a given moment, now completes the calculation with a procedure for finding the remaining time interval, between local six o'clock and t. We will explain it in terms of the modified hemisphere picture shown in figure 4.8. There, the morning sun is at the point K of its current day-circle, again shown as north of the celestial equator and implying a day in the summer half of the year. The point K is projected onto point L on the day-circle radius AG, so the Sine of the solar altitude above the horizon at that time is LT. We want to find the equatorial time arc EM that corresponds to the sun's progress through arc CK, and the method is as follows:

> The Radius is divided by the shadow hypotenuse [and] multiplied by the [noon] equinoctial hypotenuse. The quotient in the northern or the other [half-year] is decreased or increased [respectively] by the earth-Sine and multiplied by the Radius. The arc of the quotient from division [of that] by the half [-diameter] of its own day-circle is increased in the northern half-year, decreased in the southern, by the arc of ascensional difference.[47]

The first step computes the length of the segment LB from the current shadow hypotenuse. In the case of the noon equinoctial shadow, figure 4.5 showed that the Sine HQ of the sun's altitude SH is the Cosine of local latitude, depending on the shadow length s, the gnomon length g, and their hypotenuse h according to the formula

$$\frac{HQ}{R} = \frac{\mathrm{Cos}\,\phi}{R} = \frac{g}{h}.$$

Consider now the shadow, say s_t, cast by the gnomon at the arbitrary time t when the sun is at K in figure 4.8. The Sine LT of the current solar

---

[46] *Brāhma-sphuṭa-siddhānta* 2.56–58, [DviS1901], p. 44.

[47] *Brāhma-sphuṭa-siddhānta* 3.38–39, [DviS1901], p. 61.

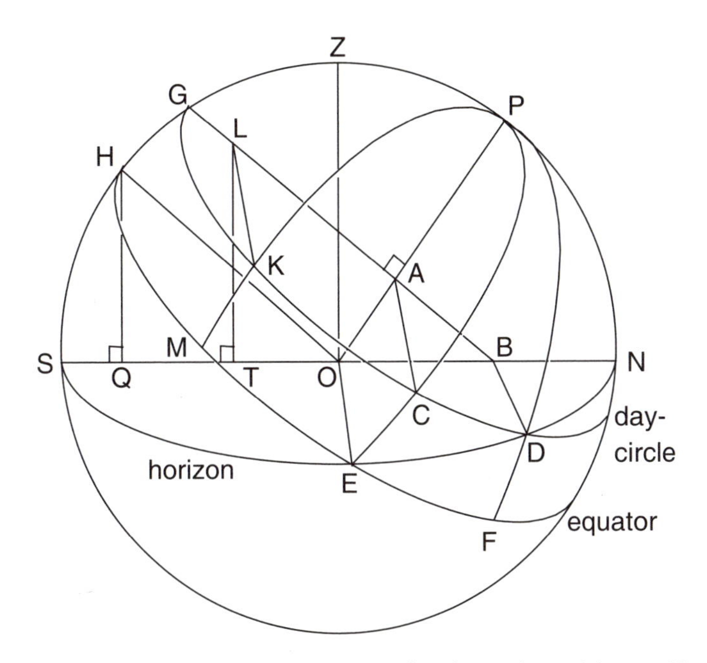

Figure 4.8 Finding the time since sunrise for the sun's position at K.

altitude can be expressed analogously in terms of its shadow hypotenuse $h_t = \sqrt{g^2 + s_t{}^2}$:

$$\frac{LT}{R} = \frac{g}{h_t}.$$

Consequently, owing to the similarity of $\triangle HOQ$ and $\triangle LBT$, "the Radius divided by the shadow-hypotenuse and multiplied by the noon equinoctial hypotenuse" is

$$R \cdot \frac{h}{h_t} = R \cdot \frac{LT}{\operatorname{Cos}\phi} = R \cdot \frac{LT}{HQ} = LB.$$

But LB is composed of just the two segments AB and $AL = LB - AB$, the first of which we already know is proportional to $\operatorname{Sin} EF$ or $\operatorname{Sin}\omega$. Similarly, AL is the projection of the day-circle arc CK onto the day-circle radius AG, and hence it is proportional to $\operatorname{Sin} EM$:

$$\operatorname{Sin} EM = AL \cdot \frac{R}{\operatorname{Cos}\delta}.$$

And EM is the equatorial arc that we were seeking, the interval between six o'clock and the current time t.

Therefore, the sum of EM and ω at last produces our desired equatorial arc FM measuring the time t between sunrise and the sun's arrival at K, a moment probably long past by the time the computation is successfully accomplished! Obviously, this kind of mathematical procedure is not a practical method for routine timekeeping tasks, which were more efficiently (if less accurately) performed by means of water clocks and similar mechanical devices.[48] However, it would permit a single hasty shadow observation at a particularly important moment, such as a birth, to be used at one's leisure for a careful determination of the accurate time in, for example, a nativity horoscope.

The need to find times and celestial positions from scanty data collected at crucial moments may also have inspired the frequent use of iterative techniques seen in several treatments of the Three Questions. The iterative formulas resemble the foregoing closed-form trigonometric rules for quantities such as time, latitude, and ascensional difference, except that they have an extra unknown quantity that must be solved for by successive approximation. Govindasvāmin, the ninth-century commentator on Bhāskara whose computation of Sine values we examined in section 4.3.3, gives several typical examples of such iterative rules. Among them is the following one for finding the sun's rising amplitude, or distance along the horizon between the east point and the place of sunrise (arc DE in figure 4.8). The Sine of arc DE, that is, OB, is initially guessed at and then refined by a computation using the "base distance" OT from the east-west line to the base of the perpendicular dropped from the sun at K on the plane of the horizon. The rule says:

> The Sine of declination increased by some [parts] should be [considered the Sine of] the sun's [rising] amplitude. The [total north-south] distance [adjusted] by that is the base distance.... [Take] the square root of the sum of the squares of that [base distance] and [the Sine of solar] altitude: this should be the line derived from the day-circle, beginning at the horizon and [extending] upward.
>
> After dividing the Sine of declination, multiplied by that desired line, by [the Sine of solar] altitude, [one gets] the Sine of the sun's [rising] amplitude. With that [answer] one should make [the same calculation] again just as before, until [the value of] the Sine of the sun's [rising] amplitude is equal to [the value] produced from it. Subtracting the square of [the Sine of] declination from the square of [the Sine of] the sun's [rising] amplitude, [one gets] a square root [that is] the earth-Sine. He [i.e., Bhāskara] will state the method for [computing] the latitude with [the Sine of] the sun's [rising] amplitude and the earth-Sine.[49]

---

[48] See [SarS2008], pp. 19–34, for descriptions of nonastronomical timekeeping devices used in India.

[49] Govindasvāmin on *Mahā-bhāskarīya* 3.41, verses 4–7ab, [Kup1957], pp. 156–157. The beginning of this passage might also be interpreted as "According to some [people], the increased Sine of declination should be [considered the Sine of] the sun's [rising] amplitude."

In the case depicted in figure 4.8, this rule would require us to know the Sine LT of the sun's altitude a at some point K and its corresponding base distance OT, both of which could be determined from a shadow observation, in addition to the current declination δ or GH. Then we would estimate the Sine of the rising amplitude, OB, to be somewhat greater than $\operatorname{Sin}\delta$ or OA. Adding our estimated OB to OT gives us the base TB of $\triangle LTB$, whose hypotenuse or "line" LB is then found by the Pythagorean theorem. Then a new value of OB can be derived by proportion from the similar right triangles $\triangle LTB$ and $\triangle OAB$:

$$OB = OA \cdot \frac{LB}{LT} = \operatorname{Sin}\delta \cdot \frac{LB}{\operatorname{Sin} a}.$$

Then we repeat the same calculation, recomputing TB from the new OB and LB from the new TB and so on. We can think of this algorithm as iterating the equation

$$OB_{i+1} = \frac{\operatorname{Sin}\delta}{\operatorname{Sin} a}\sqrt{\operatorname{Sin}^2 a + (OB_i + OT)^2}$$

(where we assumed OB_0 equal to $\operatorname{Sin}\delta + \epsilon$ for some small ϵ) until the successive values of OB converge to their fixed point.

Knowing OB and the Sine of declination or OA, we can then compute the earth-Sine $AB = \sqrt{OB^2 - OA^2}$. This is all we need to find the latitude ϕ by the method that Govindasvāmin says Bhāskara "will state,"[50] a simple proportion from the similar right triangles $\triangle OAB$ and $\triangle HQO$:

$$\operatorname{Sin}\phi = OQ = R \cdot \frac{AB}{OB}.$$

In this way, even an astronomer initially unsure of his local latitude could derive all the information he needed to compute the time at the moment of his one shadow observation, as long as he knew the sun's current true longitude and hence its declination. We don't know whether most iterative rules of this sort in Sanskrit texts were developed to satisfy a genuine need of astronomers for algorithms that would work on amounts of data insufficient for closed-form solutions or whether they were sheer effusions of computational creativity. Whatever their purpose, they continued to be developed in *siddhānta* works to a degree apparently unmatched in any other medieval tradition of mathematical astronomy.[51]

4.3.5 Eclipses

Perhaps the most crucial part of an astronomer's knowledge is the computation of eclipses of the sun and moon. As the *Brāhma-sphuṭa-siddhānta* remarks at the beginning of its fourth chapter treating lunar eclipses, "The

[50]And which he indeed does state in *Mahā-bhāskarīya* 3.53cd, [Kup1957], p. 166: "The earth-Sine multiplied by the Radius, divided by [the Sine of] the sun's [rising] amplitude, is the Sine of the latitude."

[51]See section 7.3.3 for a note on some other investigations of iterative techniques.

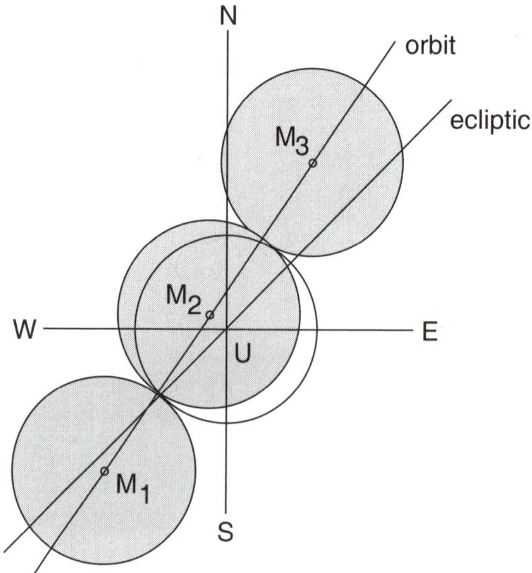

Figure 4.9 First contact, mid-eclipse, and release in a solar eclipse.

knowledge of time is sought by good [astronomers] primarily for the purpose of understanding the syzygies."[52] The ordinary syzygies or occurrences of new moon and full moon are important for the timing of religious rituals and other aspects of the calendar. But when a conjunction or opposition occurs near a node of the moon's orbit so that one of the bodies is eclipsed, it becomes a major astrological event, and its accurate prediction and depiction are matters of great moment.

For a lunar eclipse, it is necessary to know when the true longitudes of the sun and moon will be 180 degrees apart, how close the moon will be at that time to a node of its orbit, and what its latitude will be, in order to predict whether and to what extent it will pass through the conical shadow cast by the earth opposite the sun. The moon's distance from the earth at that time, and thus its apparent diameter, also affect the eclipse computation.

For a solar eclipse, the same basic information is needed about the sun, moon, and node (except that the true solar and lunar longitudes are equal instead of 180 degrees apart). In addition, since the terrestrial locations at which a solar eclipse is visible depend on the moon's parallax or apparent displacement from its true position due to the size of the earth, that parallax must also be computed.

Eclipse computations are one of the very few instances in Sanskrit texts in which a user is instructed to draw a detailed mathematical diagram. This figure is meant to illustrate the progress of the eclipse and the appearance of the obscured body. A modern depiction of an eclipse in a simplified form

---

[52] *Brāhma-sphuṭa-siddhānta* 4.1ab, [DviS1901], p. 72.

might look like the one shown in figure 4.9, for the case of a (partial) solar eclipse. The viewpoint is that of an observer in the Northern Hemisphere, looking down at the diagram drawn on the ground as if it reflected the eclipse in the sky. The disk of the sun is shown centered at point U. The arcs of the ecliptic and the lunar orbit passing through the disk appear as straight lines. As the moon's disk with center M moves along its orbit from west to east, it progresses from the moment when the two disks first touch (first contact at lunar position M_1), to the maximum obscuration (mid-eclipse, or M_2), to release or the end of the eclipse (M_3), when the two bodies separate.

The eclipse diagram described in Sanskrit texts, however, is a somewhat more complicated construction ingeniously devised to apply to both solar and lunar eclipses. The eclipsed body is to be drawn in the center of the figure, and the positions of the eclipsing body moving across it depend on the moon's latitude and the so-called "deflection" of the ecliptic, that is, the angle it makes with the eastward direction at the position of the eclipsed body. These quantities are generally represented in the same linear units that measure the size of the bodies' disks, namely digits. A characteristic description of such a diagram is given in the eighth-century *Śiṣya-dhī-vṛddhida-tantra* of Lalla, as follows:

> For a lunar [eclipse], first contact [occurs] in the eastern part [of the eclipsed body's disk] and release in the western; in a solar [eclipse], they are reversed. For a lunar [eclipse], all the latitudes are [to be drawn] inverted [north to south and vice versa], [but] in a solar [eclipse they are drawn] just as they appear.
>
> One should draw a circle [the size of] the eclipsed body, a circle [the size of] half the sum of both measures [i.e., the disks of the eclipsed and eclipsing bodies] ... with the [cardinal] directions established [in them]. In a lunar [eclipse], place the digits of deflection [at contact] like a Sine in the eastern part of the Radius-circle; in a solar [eclipse], in the western part. In a lunar or solar [eclipse] at release, the last and first are reversed.
>
> After determining the [lunar] latitude, set the mid-eclipse [digits of deflection] east or west of the south or north direction [when the latitude and the deflection have] different or the same direction [respectively], for a solar [eclipse]; for a lunar [eclipse], vice versa. Draw lines [of deflection] from those [three points marking the amount of deflection] to the center.
>
> Extend [each of] the two latitude [values], for contact and release, like a Sine [upon its] line [of deflection] from [its] intersection with the [circle the size of] the sum of the [disks'] measures. [Lay off] the mid-eclipse latitude from the center along the line [of deflection] at mid-eclipse. From those [three points marking the] limits of latitude, [circles with radius equal to] half [the disk of] the eclipsing body are to be produced cutting the eclipsed body;

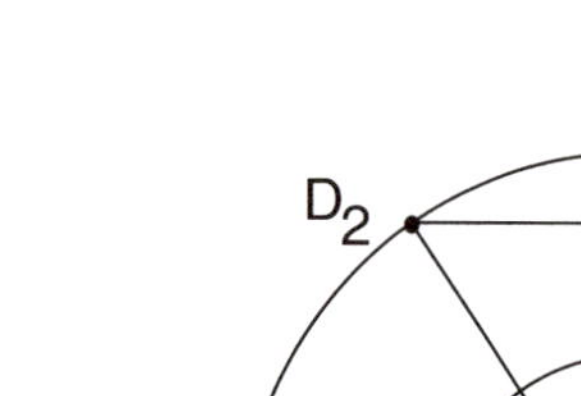
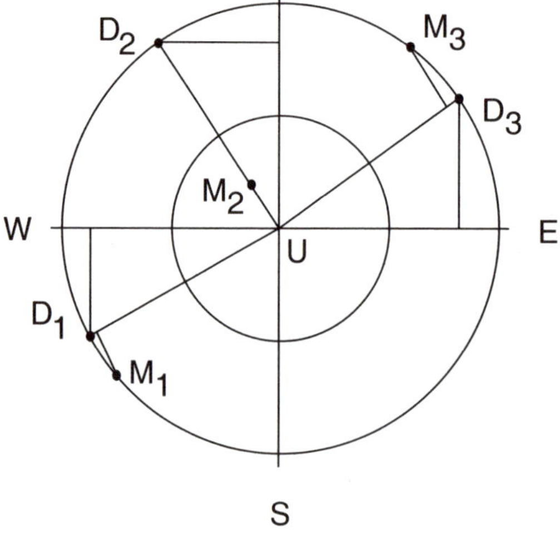

Figure 4.10 Lalla's prescribed diagram for a solar eclipse.

> [they show] respectively [the points of] contact and release [and the amount of] obscuration at mid-eclipse.[53]

What this diagram might look like in the case of the above solar eclipse is shown in figure 4.10. The sun's disk is again centered at U, along with a larger circle with radius equal to the sum of the radii of sun and moon. The positions of the moon's center at contact and release, M_1 and M_3, will fall on the circumference of this outer circle, which we may call the "contact-release circle." The path of the eclipsing moon is diverted from the east-west line by two factors: the inclination or "deflection" of the local arc of the ecliptic with respect to the east-west line, and the offset of the moon itself from the ecliptic when it has a nonzero latitude. These components are represented separately in the diagram, as follows.

The amount of ecliptic deflection in the southern direction at the time of contact is represented by point D_1, showing where the ecliptic intersects the contact-release circle at that time. It is depicted in linear digits "like a Sine." At release, the ecliptic cuts the contact-release circle at point D_3, representing the corresponding amount of deflection toward the north. Thus the line segments UD_1 and UD_3 roughly represent the local arc of the ecliptic during the course of the eclipse. The mid-eclipse deflection, on the other hand, is laid out east-west, 90° from its proper orientation. UD_2 thus represents the perpendicular to the ecliptic at mid-eclipse, deflected westward from the north-south line.

---

[53] *Śiṣya-dhī-vṛddhida-tantra* 5.29–33, [Cha1981], pp. 103–106.

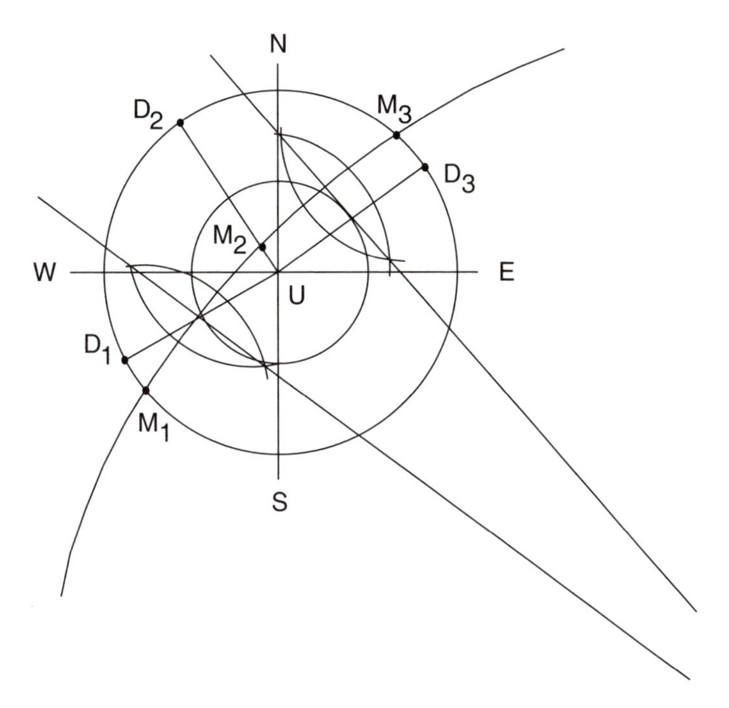

Figure 4.11 Marking the path of the eclipsing body in Lalla's diagram.

Now the moon's latitude, that is, the distance of its center north or south of the ecliptic, can be measured off linearly perpendicular to the ecliptic. The amount of contact latitude is placed perpendicular to UD_1 so that its other endpoint falls on the contact-release circle, at M_1. Similarly, the line segment perpendicular to UD_3, representing the release latitude, touches the contact-release circle at M_3. The mid-eclipse latitude is then laid off along the perpendicular to the ecliptic UD_2. So the three points M_1, M_2, and M_3 represent the positions of the moon's center at contact, mid-eclipse, and release, respectively.

The description then proceeds to the drawing of the apparent path of the center of the moon during the progress of the eclipse, on the assumption that such a path will resemble a circular arc:

> From the intersection of two lines touching the mouths and tails of the two fish [figures] [constructed] with the circles [centered on] the three latitude-points, draw a circle passing through the three latitude [points]. And that is the path of the eclipsing body.. . .[54]

This step is illustrated in figure 4.11. The line segments M_1M_2 and M_2M_3 are to be cut by perpendicular bisectors constructed by means of what Indian

[54] *Śiṣya-dhī-vṛddhida-tantra* 5.34, [Cha1981], pp. 107.

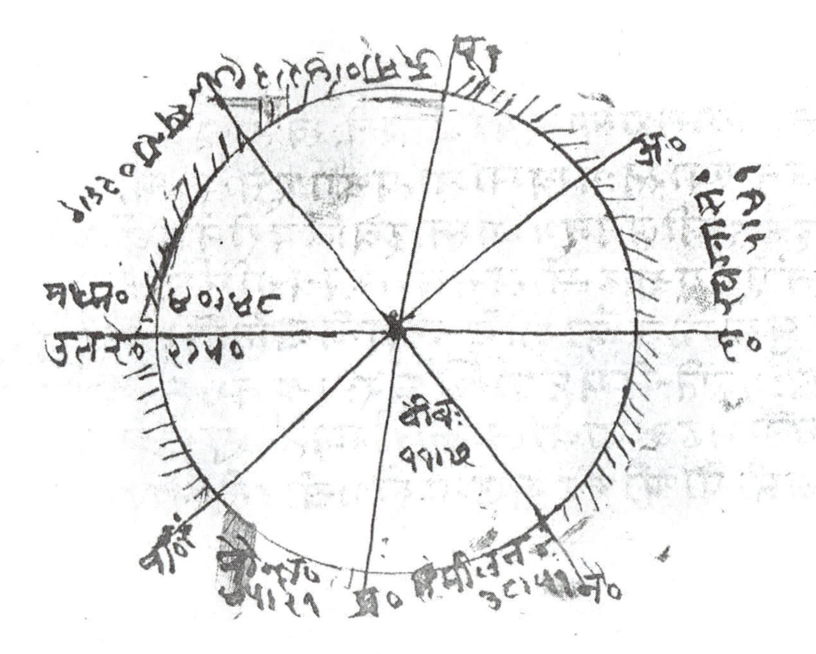

Figure 4.12 A rudimentary eclipse diagram in a Sanskrit manuscript.

mathematicians call "fish figures," or intersecting pairs of circular arcs. The two perpendicular bisectors, running through the "mouths" and "tails" of these fish figures, will of course intersect at the center of the circle that passes through the three points M_1, M_2 and M_3.

An unusual instance of an attempt at such a diagram actually appearing in an astronomical manuscript is reproduced in figure 4.12.[55] The scribe has drawn one circle and the lines of the cardinal directions (with east at the top), along with another pair of perpendicular lines that seems to be intended for marking the deflection. Around the edges he wrote a few numbers including the times of contact and mid-eclipse, but he got no further.

4.3.6 Other typical *siddhānta* topics

There are many other astronomical subjects covered in *siddhāntas*, generally employing similar mathematical techniques to those described in the preceding sections. For the most part, they involve the computation of events and times that have special astrological significance. This section presents a brief overview of the most important of them, as well as some topics concerning

[55]Diagram on f.15v of a manuscript purchased in Jaipur in 2004 and now in the collection of the Jaina Vidya Sansthan, Jaipur.

pedagogy and practice.

Conjunctions and synodic phenomena of the planets. Indian astrologers needed to predict the synodic phenomena or "phases" of not only the moon but also the five star-planets—when they become visible or invisible due to the nearness of the sun, when they start or stop their retrograde motion, and so forth. Conjunctions of planets with one another or with fixed stars were also considered meaningful, especially when they didn't simply coincide in longitude but were actually visible very close to or even occulting another body.

A *siddhānta* therefore gives the coordinates of various named stars in locations on or near the ecliptic where planets can appear, and rules for determining times and distances of conjunctions. There are usually twenty-seven or twenty-eight of these stars, each associated with one of the canonical *nakṣatra* constellations. Their coordinate values are given with varying levels of precision: sometimes to the nearest integer degree, sometimes to the nearest half-degree or arcminute. Some of the values differ from one text to another, although apparently not in a systematic way.[56]

Special features of the phases and positions of the moon. Eclipses and syzygies were not the only lunar matters of interest to astronomers. The size and orientation of the lunar crescent, under a Sanskrit name meaning "elevation of the moon's horns," were computed using a trigonometric model and represented graphically. Another important lunar phenomenon was the astrologically significant appearance of the sun and moon "balanced" about an equinox or solstice point, that is, when they have the same declination and are equidistant from an equinox or solstice on either side of it. These ominous events were predicted using a form of what is now called "Regula Falsi" iteration, or iterated interpolations between two successive time estimates.[57]

Construction and use of mathematical and astronomical instruments. Texts sometimes refer to basic mathematical instruments such as compasses for drawing circles and strings for making straight lines (see section 5.1.2 for an example), and sometimes a *siddhānta* will devote a chapter to describing various kinds of astronomical instruments used for taking measurements. These include, among others, the vertical gnomon and a simple altitude ring dial, along with a sort of armillary sphere (for demonstration rather than observation), which is discussed in section 4.5.[58]

Exercises for students. Some *siddhāntas* include sample problems for students to practice the various computational methods taught in the text; the

---

[56] See the discussion on coordinates used in star lists in [PinMo1989].

[57] The trigonometric computation of the width of the crescent in the *Brāhma-sphuṭa-siddhānta* is discussed in [Plo2007b], pp. 420–421. The method for iteratively computing the moment of symmetrical position about an equinox or solstice is explained in [Plo2002b], pp. 179–182.

[58] See, for example, [Pin1981a], pp. 52–54, [Oha1985], [SarS1992], and [SarS2008], pp. 47–63. Indian texts on instruments in later periods are discussed in [Oha1986].

Mahā-bhāskarīya of Bhāskara, for example, devotes its entire final chapter to this subject. A typical instance reads:

> The terrestrial latitude is one and a half degrees minus eight arcminutes [i.e., 1°22′]. The shadow of a twelve-digit gnomon on a piece of level ground at noon is five digits. Tell [the longitude of] the sun at noon on that day.[59]

Critiques of other treatises. An author may use an occasional verse (or even a whole chapter) to criticize or correct the parameters, models and methods stated in other texts. The boundaries between astronomical and traditional cosmological models are not always distinct in these critiques. A *smṛti* concept such as the flat earth may be disparaged for being irreconcilable with mathematical computation; on the other hand, as we saw in section 4.2, an astronomical parameter may be condemned for being irreconcilable with *smṛti.*

It is not unheard of for the same author to adopt apparently contradictory critical positions at different points in the same text. Brahmagupta, for example, argues variously that traditional cosmology is wrong in placing the moon above the sun but right in denying that the moon eclipses the sun:

> If the moon were above the sun, how would the power of waxing and waning, etc., be produced from calculation of the [longitude of the] moon? The near half [would be] always bright. In the same way that the half seen by the sun of a pot standing in sunlight is bright, and the unseen half dark, so is [the illumination] of the moon [if it is] beneath the sun....[60]

> "The shadow of the earth covers the moon, the moon covers the sun.... Are the sun and [the demon] Rāhu different for each region because of the difference of the magnitude of the solar eclipse? Therefore an eclipse of the sun or moon is not caused by Rāhu": [what is] thus declared by Varāhamihira, Śrīṣeṇa, Āryabhaṭa, and others is opposed to popular [opinion] and is not borne out by the Vedas and *smṛti*.... "[The demon] Svarbhānu or Āsuri has afflicted the sun with darkness": this is the statement in the Veda. So what is said here is in agreement with scripture and *smṛti*.... The earth's shadow does *not* obscure the moon, nor the moon the sun, in an eclipse. Rāhu, standing there equal to them in size, obscures the moon and the sun.[61]

[59] *Mahā-bhāskarīya* 8.7, [Kup1957], p. 388. The student would have to compute the sun's altitude from the shadow, its declination from the altitude and terrestrial latitude, and its longitude from the declination. Notice that Bhāskara does not specify whether the shadow falls to the north or to the south, that is, whether the sun is south or north of the zenith; thus there are two right answers to the problem.

[60] *Brāhma-sphuṭa-siddhānta* 7.1–2, [DviS1901], p. 100. See also [Plo2007b], p. 420.

[61] *Brāhma-sphuṭa-siddhānta* 21.35, 21.38–39, 21.43, 21.48; [DviS1901], pp. 371–374.

4.4 OTHER TEXTS FOR ASTRONOMICAL COMPUTATION

The *siddhānta* was not the only type of text employed by Indian astronomers. Naturally, it would be very cumbersome to use the techniques prescribed by a *siddhānta*, starting from scratch with the immense integers of the *kalpa* parameters, every time one wanted to calculate a planetary position. For greater convenience, astronomers developed the related genres discussed in the following two sections.

4.4.1 The *karaṇa* or handbook

The concise astronomical manuals called *karaṇas* imitate the general structure of the *siddhānta* but simplify its methods. In particular, a *karaṇa* (literally "making") specifies initial planetary positions for a recent epoch, usually some date within the lifetime of its author, rather than using the start of the *kalpa* or Kaliyuga as its zero point. The formulas for calculating current celestial positions are generally simpler approximate versions of the *siddhānta* procedures.

The primary surviving text of the Ardha-rātrika-pakṣa is a *karaṇa* by Brahmagupta with epoch date 665, called the *Khaṇḍa-khādyaka* ("Eating candy," apparently a boast about the ease and pleasantness of using the methods described in the work). Why Brahmagupta should have wanted to write a text employing any of Āryabhaṭa's parameters that he criticized so forcefully in his earlier *Brāhma-sphuṭa-siddhānta* remains unclear. His own rationale is brief and not very informative, but neatly summarizes the purpose of a *karaṇa* work:

> Since in general the procedures [prescribed] by Āryabhaṭa are impracticable in day-to-day [astrological computations for] marriage, birth horoscopes, etc., therefore an easier [system producing] results equal to those is [here] stated.[62]

An example of the "easier" techniques supplied in the *Khaṇḍa-khādyaka* is its procedure for computing the equation of center μ to correct mean longitudes to true ones. Rather than using trigonometry to determine the amount of the equation as described in section 4.3.3, Brahmagupta simply lists in concrete number notation a few computed values of the equation, apparently expecting the user to calculate intermediate values by linear interpolation.

Tabulated Sines do appear in *karaṇas*, but like the computational formulas, they are frequently modified for conciseness and simplicity. The *Khaṇḍa-khādyaka* Sine table has only six entries at 15° intervals, accompanied by the same second-order interpolation method discussed in section 4.3.3. Perhaps

---

See also [Ike2002], pp. 239–248. A fuller discussion of the apparent contradiction in these arguments is given in [Plo2005a]; the attempts of a later Muslim scientist to explain them are described in section 8.2.1.

[62] *Khaṇḍa-khādyaka* 1.2, [Cha1970], vol. 2, p. 1. The equation values and Sine table mentioned below occur in verses 1.16 and 3.6 (vol. 2, p. 10 and p. 94), respectively.

the most extreme example of such simplified Sines occurs in the *Laghu-mānasa* ("Easy thinking") of Muñjāla (or Mañjula), epoch date 932, which has only three tabulated Sine-difference values:

> The sum of [each successive] sign [respectively] multiplied by four, three, and one, [considered as both] degrees and minutes, [determines the Sine] of the arc or its complement.[63]

That is, there are three distinct "difference factors" d_i for the successive 30° intervals of the quadrant: $d_1 = 4$, $d_2 = 3$, $d_3 = 1$. The Sine-difference corresponding to each of those three intervals is the corresponding number of degrees plus the same number of minutes. Thus the Sine of 30° or Sin_1 is $4^\circ 4'$, while Sin_2 is $4^\circ 4' + 3^\circ 3' = 7^\circ 7'$, and the Sine of 90° or Sin_3, the trigonometric radius, equals $7^\circ 7' + 1^\circ 1' = 8^\circ 8' = 488'$.

Of course, such a drastically abbreviated Sine table will not produce accurate results for intermediate values found by linear interpolation. A more complicated interpolation method ascribed to the twelfth-century commentator Mallikārjuna Sūri improves the accuracy:

> Sines of arcs in the [*Laghu-*] *Mānasa* are accurate [when the interpolation factor is] half the sum of the previous and current difference [factors] minus the product [of the arc] with half their difference divided by thirty.[64]

In other words, while ordinary linear interpolation would give the Sine of an arc x° falling in the ith third of the quadrant, between Sin_{i-1} and Sin_i, as

$$\mathrm{Sin}\, x \approx \mathrm{Sin}_{i-1} + \frac{x - 30(i-1)}{30} \cdot d_i{}^\circ d_i{}',$$

Mallikārjuna's rule prescribes instead an interpolation factor based on combining the "difference factors" as follows:

$$\mathrm{Sin}\, x \approx \mathrm{Sin}_{i-1} + \frac{x - 30(i-1)}{30} \cdot \left(\frac{d_{i-1} + d_i}{2} - \frac{d_{i-1} - d_i}{2} \cdot \frac{x - 30(i-1)}{30} \right),$$

with d_i defined as in the previous paragraph. When $i = 1$, d_0 is arbitrarily chosen equal to 4.5. The chief advantage of this more complicated interpolation procedure over simple linear interpolation is manifested in the interval $60^\circ < x \leq 90^\circ$ (see figure 4.13), where the second-order term in x produces a better fit to the sine curve. As usual, no record survives of how Mallikārjuna derived or justified this algorithm, but its structure suggests inspiration from Brahmagupta's second-order interpolation rule.

In general, it appears that the ingenious condensation of astronomical formulas into brief *karaṇa* rules was considered an opportunity to display the author's mathematical expertise and inventiveness. Vaṭeśvara says in the *Vaṭeśvara-siddhānta*:

---

[63] *Laghu-mānasa* 3.2, [Shu1990], p. 65.

[64] Yallaya's commentary on *Laghu-mānasa* 3.2, [Shu1990], p. 65, note. The following interpretation, based on worked examples in one of the manuscripts, is the editor's.

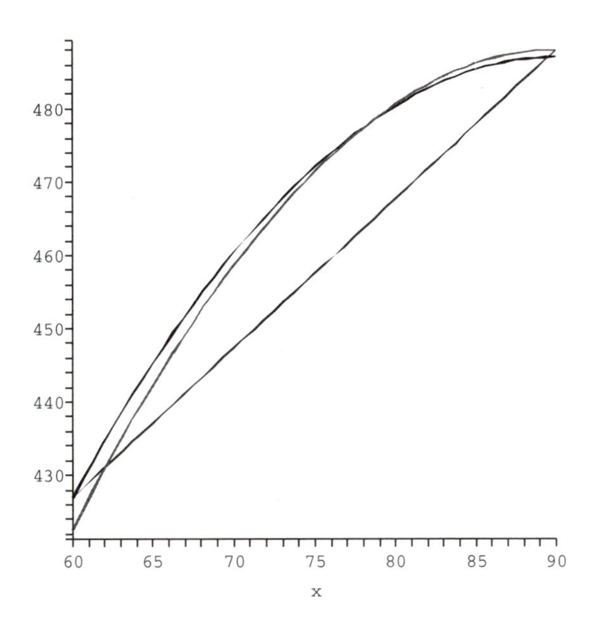

Figure 4.13 Interpolation in the *Laghu-mānasa* Sine table. The modern value of $\sin x$ scaled to $R = 488$, together with its linear approximation and Mallikārjuna's second-order interpolation function, is plotted against x in degrees for the final third of the quadrant. (The true sine curve is the one with the lower y-intercept.)

> A *karaṇa* is to be made quite concise, not apparent to others, easily used by the stupid.[65]

Approximation formulas that are "not apparent to others" attest to the author's originality. A manuscript of a commentary on the *Laghu-mānasa* by Sūryadeva Yajvan recounts an anecdote illustrating the use of original *karaṇa* techniques as a sort of trademark of mastery:

> Muñjāla, having studied ... many other works on astronomy and having summarized in sixty verses in *anuṣṭubh* meter the astronomy stated there, wrote the *karaṇa* entitled *Laghu-mānasa* and got it copied neatly by one of his pupils to show it to the king of the region. That wicked pupil, taking those verses to the king, told him that he himself wrote the *karaṇa* entitled *Laghu-mānasa*. The wise king advised him to show it to his teacher. Thereafter, when Muñjāla happened to visit the king, the king asked him whether his pupil had written the *karaṇa* entitled *Laghu-mānasa*. At this Muñjāla laughed. The king asked him the reason of his laughter. Muñjāla replied: "Your majesty being the king, all people are competent to do all things: this is why I laughed." The king then said: "Tell me how to know the fact." "Let this *karaṇa* remain in your custody and let both of us under your protection be asked to write another *karaṇa* this very day," said Muñjāla. Both of them were then put under guard by the king.... Muñjāla then in a short time, remembering the earlier multipliers and divisors etc. and summarizing them in an unusual way, composed another *Laghu-mānasa* and showed it to the king. His pupil, on the other hand, could not do anything. Thereupon the king, getting angry with that pupil for the wrong done by him to his teacher, banished him from the kingdom. Muñjāla he gratified with presents and honours.[66]

Sūryadeva hammers home the point in his commentary on the final verse of the *Laghu-mānasa*, where Muñjāla proclaims that "those who denigrate (or copy) this work shall lose [their] reputations":

> Those who will produce counterfeit works in imitation of this work shall earn infamy. For, nobody can know the rationale etc. of the rules given in this *karaṇa* work written by me, and therefore the learned people will easily know that such-and-such a person has forged another work on the same subject by stealing the contents of this work. Thus such authors shall certainly earn a bad reputation.[67]

---

[65] *Vaṭeśvara-siddhānta* 1.6.10, [Shu1986], vol. 1, p. 56.

[66] Translated by K. S. Shukla, [Shu1990], pp. 98–99. I have ventured to substitute "*karaṇa*" where the translator says "calendrical work", since the quoted excerpts from this commentary on p. 25, and in the translation below, make it clear that Sūryadeva calls the *Laghu-mānasa* a *karaṇa*.

[67] Translated by Shukla, [Shu1990], p. 192. Original base text verse on p. 77.

4.4.2 Tables and the calendar

Even simpler to use than the abbreviated formulas of a *karaṇa* are the entries in a table of precomputed values of a desired function. Detailed tables of this sort, along with instructions for how to use or compute them, were known as *koṣṭhakas* (literally "storehouse"). They seem to have come into widespread use in the second millennium, probably originally in imitation of Islamic astronomical tables, and were frequently geared to the needs of calendar makers. The content and construction of *koṣṭhakas* are discussed more fully in sections 8.3.2 and 9.3.2.

The ultimate goal of Sanskrit calendrics in general is the production of the annual calendar, the *pañcāṅga* or "five elements." This ephemeral text identifies the moments that demarcate years, months, days, and their various subdivisions, as well as the stated times for performance of ritual observances, and those that are auspicious and inauspicious. The "five elements" of its name are the five basic time units:

- The weekday starting at sunrise;

- The *tithi* or one-thirtieth of the synodic month, during which the difference in the eastward motions of the sun and moon amounts to 12°;

- The *karaṇa* or half-*tithi* (not to be confused with the identically named genre of astronomical handbooks);

- The *nakṣatra* or "constellation," a twenty-seventh part of a sidereal month, or the time required for the moon to traverse an interval equal to 13°20′, or one of the twenty-seven *nakṣatra* constellations comprising its orbit;

- The *yoga*, a sort of "anti-*tithi*," a time interval in which the sum of the eastward motions of the sun and moon amounts to an increment of one *nakṣatra* length or 13°20′.

The weekday and *tithi* are useful for ordinary timekeeping and the scheduling of rituals that depend on the lunar synodic phases, particularly on the times of new and full moon. The remaining three time units have mostly auspicious and astrological significance. In addition to keeping track of these five quantities, a *pañcāṅga* also records the beginnings of solar months (when the sun enters a zodiacal sign), and the occurrence of intercalary months when required to keep the solar year more or less synchronized with the seasons.

4.4.3 Texts for astrological computation

The various astrological applications of Indian mathematical astronomy are beyond the scope of this book. However, the importance of astrology in the context of Sanskrit science justifies a brief overview of its literature. Texts on astral sciences are traditionally divided into omens, horoscopic prediction, and *gaṇita* or astronomical calculation.[68] So, according to this classification, all astronomical and mathematical texts might be seen as essentially a subset of astrology.

The primary astrological topic in the medieval period is *jātaka* or nativity horoscopes. Their computation requires the use of Three Questions-type formulas to convert observed appearances of celestial bodies at the moment and location of birth into celestial longitudes and latitudes that can be evaluated by criteria of auspiciousness. In particular, the horoscope point or rising point of the ecliptic circle on the eastern horizon must be identified, for the astrological "houses" are reckoned from it.

The concept of the horoscope, and the calculations required for it, also motivate the other major types of astrological practice. A horoscope may be cast not just for a birth but for any desired moment at which information about a particular subject is sought; the implications of the horoscope at that moment are interpreted to provide the desired information. Conversely, a propitious time for some special activity may be selected based on the qualities of the horoscope in effect at that time. Subspecialties of this latter technique were developed for certain especially momentous events, namely military campaigns and marriage ceremonies.

4.5 GEOMETRIC MODELS IN ASTRONOMY

The subject of *gola* or the sphere in a *siddhānta* is frequently treated apart from *gaṇita* or astronomical computations. The *gola* part may consist of an individual chapter or even a separate section composed of multiple chapters, explaining the geometric models of the celestial and terrestrial spheres on which the computations are based. Generally it describes the construction of a sort of armillary sphere to illustrate the models.

Presumably, this separation of *gola* and *gaṇita* was motivated by the practical aim of making the computational formulas as compact as possible, unencumbered by physical rationales. In addition, the spherical cosmology underlying *siddhānta* algorithms may have seemed unfamiliar or controversial (given that Indian popular and sacred ideas of the universe involved a flat earth), thus needing its own part of the treatise for its full explication. Authors such as Lalla who devoted sections to *gola* stressed its importance (perhaps concerned that pragmatic-minded students would be tempted to skip it?):

> A year-computer [i.e., astrologer or calendar maker] ignorant of *gola*, like a debater without [knowledge of] grammar, a [ritual] sacrificer deprived of scriptures, or a physician entirely without experience, will not shine. Those who really know [this] science call *gola* [necessary] for calculation. The ignorant who do not

---

[68] [Pin1981a], p. 1.

> know *gaṇita* and the essence of *gola* [can] in no way know the motion of the planets.[69]

The first known *siddhānta* author to allot a chapter specifically to *gola* is Āryabhaṭa, although he does not clearly separate spherical models from computational algorithms. The *gola* chapter of the *Āryabhaṭīya* treats basic spherical cosmology, the spherical astronomy and geography usually subsumed in later works under "Three Questions," and the configuration and prediction of eclipses. Its most unusual feature is doubtless its insistence on the daily rotation of the earth rather than the sky. This was implied earlier by Āryabhaṭa's reference in his opening chapter to the earth's rotations in a *yuga*,[70] but the implication was skillfully dodged by his commentator Bhāskara:

> The stars bound to the wheel of the constellations move to the western direction because of the [cosmic] wind drawing that constellation wheel. The stars see [the earth] as if [it were] turning with its own motion towards the east, like a planet. Thus [the rotation of the sky is equivalently] named the turning of the earth, for this reason.[71]

But in his *gola* chapter Āryabhaṭa makes the rotating-earth hypothesis much more explicit, though he also balances it with a statement that could be interpreted otherwise:

> In the same way that someone in a boat going forward sees an unmoving [object] going backward, so [someone] on the equator sees the unmoving stars going uniformly westward. The cause of rising and setting [is that] the sphere of the stars together with the planets [apparently?] turns due west at the equator, constantly pushed by the cosmic wind.[72]

At the very least, these verses are arguing that the models of a rotating earth and a stationary earth are equivalent from the point of view of the

---

[69] *Śiṣya-dhī-vṛddhida-tantra* 14.3–4, [Cha1981], vol. 1, pp. 197–198.

[70] *Āryabhaṭīya* 1.3ab; see section 4.3.1. Of course, the hypothesis of the earth's rotation is not to be confused with the heliocentric hypothesis, the idea that the earth revolves once annually about the sun instead of vice versa. As the verse in question illustrates, the *Āryabhaṭīya* assigned daily rotations to the earth but annual revolutions to the sun. Popular claims that Āryabhaṭa propounded heliocentrism seem to be based on an overinterpretation of the concept of the *śīghra* anomaly, which for the superior planets is computed with respect to the mean sun. But a synodic anomaly (i.e., one that depends on the position of the sun) does not have to imply a physically heliocentric orbit, and neither Āryabhaṭa nor any other medieval Indian astronomer is known ever to have interpreted it as such. (See section 7.4, however, for allegations concerning a modified, potentially heliocentric system in a later Indian work.) Note that the concept of a synodic anomaly is not unique to Indian astronomy; it was present in the numerical algorithms of Late Babylonian astronomical texts, and modeled by an epicycle in, for example, the *Almagest* of Ptolemy, as discussed in [Eva1998], pp. 337–346.

[71] Bhāskara on *Āryabhaṭīya* 1.2, [ShuSa1976], p. 20. See also [Plo2007b], p. 417.

[72] *Āryabhaṭīya* 4.9–10, [ShuSa1976], p. 119.

observer. However, the idea of the earth's rotation was universally rejected by later astronomers on physical grounds, as in Lalla's rebuttal:

> And if the earth turns, then how do birds reach their nests [again after leaving them]? Arrows released up [toward] the sky should fall down in the west if there is an eastward rotation of the earth, [and] a cloud should move toward the west. Now [if it is argued that these effects do not happen] because the motion [is] slow, then how [can] there be a full rotation in one day?[73]

Although Āryabhaṭa's notion of the motion of the earth remained unpopular, his claims for its sphericity were universally accepted and stoutly defended in *siddhānta* texts against the objections of traditional flat earth cosmology:

> The sphere of the earth [made of] earth, water, fire, and air, in the middle of the cage of the constellations [formed of] circles, surrounded by the [planetary] orbits, in the center of the heavens, is everywhere circular.
>
> [Bhāskara comments:] "In the center of the heavens": The earth is not at all above [the center], and not below, hence it is not falling.... Now others think [that] the earth is supported by [the cosmic serpent] Śeṣa or [something] else: that is not rational.... Now if they [i.e., Śeṣa etc., can] stay fixed by their own power, why cannot this power be assumed for the earth?[74]

This spherical configuration of the cosmos was supposed to be demonstrated by a physical model that could be manipulated. Bhāskara's commentary on Āryabhaṭa's *gola* chapter lays out the procedure for its construction:

> So with a pair of well-joined half-circles or with three well-joined [piecewise] circular strips, one should form a single circle.... Then join the circular strips to one another with copper pegs. Then, setting down the one circle thus constructed [as] the east-west [circle], [make] a second, north-south vertical [one].[75]

These are the great circles passing through the four cardinal points on the horizon at zero latitude; the "east-west circle" is the celestial equator, with a circular scale marked on it in *ghaṭikās* to show the passage of time in the revolving of the sky, and the "north-south circle" is the meridian perpendicular to it. Bhāskara goes on to describe the addition of a third circle

---

[73] *Śiṣya-dhī-vṛddhida-tantra* 20.42–43, [Cha1981], vol. 1, p. 238.

[74] *Āryabhaṭīya* 4.6, [Shu1976], pp. 258–259.

[75] Bhāskara on *Āryabhaṭīya* 4, [Shu1976], pp. 240–243. See also the similar explanation by a commentator on *Brāhma-sphuṭa-siddhānta* chapter 21 in [Ike2002], pp. 162–166. One wonders if astronomers or their students ever really constructed the theoretically complete spherical model with dozens of circles to represent all the reference circles and all the planetary orbits, or if it was more of a thought experiment. See [SarS2008], p. 61, for a reference to powering the automatic rotation of such an armillary sphere with a mercury-driven perpetual motion machine—evidently a thought experiment!

perpendicular to the first two to represent the horizon, and a fourth inclined to the equator and marked with 360 degrees, to represent the ecliptic. An iron rod is to be inserted through the north and south intersections of the circles, perpendicular to the equator, to represent the polar axis, and the sphere of the earth is to be represented by a ball of mud or clay clumped onto the center of that rod. Other circles can be added to represent the orbits of the planets, the local horizon, day-circles, and so forth.

4.6 THE PROBLEM OF ORIGINS

As in the case of the pre-Classical quantitative astronomy discussed in sections 2.3–2.5, a clear and detailed narrative of the formation of mathematical astronomy in medieval *siddhāntas* would shed light on the often perplexing chronology of Indian mathematics as a whole. The question of how its parameters and models arose—that is, the empirical underpinnings of *siddhānta* astronomy—has long been debated. Unfortunately, we cannot settle the issue by direct appeal to the sources, because the overwhelming majority of medieval Sanskrit astronomical sources say nothing about how astronomical systems were constructed. The conventions of the genre maintain that an astronomical system is a revelation from a deity or part of a truthful authoritative tradition, but also that it produces predictions in accordance with observed positions of celestial bodies in the sky. Innovation by an individual author, although a great deal of it may have gone into the composing of a text, is not explicitly boasted of, at least not as it affects the fundamental parameters of the system.[76]

The historiographic debate is inevitably somewhat skewed by our modern preconceptions about the nature of science, especially by our core narrative of a heliocentric revolution based on the work of Copernicus, Galileo, Kepler, and Newton. We now take it for granted that mathematical astronomy involves a planned program of controlled observational experiments whose goal is to modify a single, self-consistent geometric model of celestial motions in order to maximize the model's predictive accuracy. As a consequence, we naturally think it important to find out whether Indian mathematical astronomers conceived and carried out such an observational program well or badly. But first we must inquire whether Indian astronomers even considered such a program necessary to their science, or whether they used different methods for reconciling calculation and observation.

4.6.1 The mystery of Indian observational astronomy

The major stumbling block to investigating these questions is that we have no detailed record of the observational practices of Indian astronomers from this period. Observational practices in other major traditions in ancient

[76] See, however, sections 7.4 and 9.3 for a discussion of some different perspectives in second-millennium texts.

astronomy are at least somewhat easier to reconstruct from contemporary texts. For example, we have the surviving letters sent to kings by neo-Assyrian star observers reporting on astral omens and the cuneiform tablets preserving observational data in the Babylonian astronomical diaries; there are also Ptolemy's descriptions in the *Almagest* of observations made by himself in the early second century CE to test particular features of his models.

Such specific records of data gathering or references to them do not appear in Indian astronomical texts of the first millennium CE or thereabouts, and are found only rarely in later texts. For the most part, all we have in the way of data in Indian texts are the orbital parameters themselves, celestial coordinates for some twenty or thirty fixed stars, mean longitude positions for epoch dates in *karaṇas*, and occasionally some *bīja*-corrections (see section 4.2) to modify the given parameters.

Of course, from the many Sanskrit descriptions of observing instruments and the passing references to their use, as well as the plethora of predictive techniques that require an observed value of some astronomical quantity, we know that Indian astronomers must have frequently looked at the sky. And authors often boasted that the calculations produced by their systems agreed with observation. But how did these concepts interact with astronomical theory? What, if anything, did Indian astronomers do with the measurements they made, beyond feeding them into existing algorithms for predicting particular events? And how and why did they derive different sets of parameters for those algorithms?

Were observational data never systematically collected by early medieval Indian astronomers as part of a conscious project for empirically constructing and refining quantitative geometric models? Or were they collected and used in this way but never published in the form of a disseminated text? If the former hypothesis is true, that would imply that *siddhānta* parameters were not independently derived from an organized observational program but instead were originally borrowed from other sources. They might then have been modified in an ad hoc way to conform to isolated observations (much as some early Islamic astronomers confirmed or adjusted a constant here and there in Ptolemaic astronomy without comprehensively testing its whole structure) or to various other criteria. On the other hand, if the latter hypothesis is true, it implies that Indian parameters were routinely derived from Indian observations. There is no existing evidence on which to base a conclusive historical proof of either position.

The historical arguments in favor of the first hypothesis are based on the similarities between early *siddhānta* astronomy and Hellenistic Greek theories; the arguments in favor of the second are based on their differences. We have seen in section 3.2 that Sanskrit astronomical and astrological texts from around the turn of the first millennium used units and concepts, such as the zodiacal signs and the horoscope, evidently derived from Greek sources. The Sine tables present in those early texts also have apparent connections with Hellenistic trigonometry of chords. Indian geometric cosmologies in medieval *siddhāntas* involving epicycles and eccentrics likewise suggest some familiarity with Greek models.

However, the parameters and orbital models that we see in standard *siddhānta* texts beginning with Āryabhaṭa's are not identical to the ones in Ptolemy's *Almagest*, the earliest surviving systematization of Greek mathematical astronomy. So we know that *siddhānta* astronomy could not have been simply borrowed from Ptolemy's. But there were different pre-Ptolemaic and quasi-Ptolemaic models in Hellenistic astronomy employing many of the same concepts and methods that Ptolemy used, which either preceded the *Almagest* or coexisted with it for a while. In the long run these variant systems went extinct and left only scanty traces in the Greek record as currently known. It has been suggested that these variants were the chief source of many of the characteristic parameters or "elements" of *siddhānta* astronomy.[77] The remainder of this chapter summarizes the basic arguments both for and against an independent origin for mathematical models in Indian *siddhāntas*.

4.6.2 Inferences from statistical analyses of parameters

The most detailed and best known case for the independent Indian development of *siddhānta* mathematics is based on a type of astrochronological argument comparing *siddhānta* parameters with modern estimates of what their values should have been. According to current astronomical theory, the motions of celestial bodies change in many ways over time scales of centuries and millennia, owing to various gravitational effects. Consequently, the values of parameters such as mean motions or epicycle sizes that give reasonably accurate results at a particular time will produce less accurate predictions for times much earlier or later. And it seems reasonable to suggest that the time for which the parameter is most accurate was approximately the time at which the observations were made from which the parameter was derived. On this assumption, parameters of ancient texts can be roughly dated by investigating their deviation from modern retrodicted values.

This error analysis approach was first applied to various *siddhānta* elements by the British Indologist John Bentley near the turn of the nineteenth century. Many of Bentley's statistical deductions clashed with conclusions drawn by other historians trying to establish the chronology of Indian astronomy based on textual records; for example, he assigned the astronomer Varāhamihira (see section 3.2.2), who is now universally agreed to have worked in the sixth century, to a time around 1528.[78] As a result, Bent-

[77]See [JonA2004] and [JonAMA] for discussions of the extinct Hellenistic astronomical systems, and [Pin1976], [Pin1974], [Duk2005a], [Duk2005b], and [DukeIPT] for arguments relating them to later Indian material.

[78]Although Bentley's approach involved some plausible ideas, his motivations now appear absurd if not somewhat deranged. He was persuaded that Indian astronomical eras stretching back thousands or billions of years constituted a deliberate anti-Christian hoax perpetrated by medieval and modern Brāhmaṇas, who wished to undermine the "Mosaic account" of the age of the universe (i.e., Biblical chronology) by laying claim to astronom-

ley's theories were received with skepticism and ultimately discarded.[79]

More recently, Bentley's method was revived in a more mathematically sophisticated form by the late French Indologist Roger Billard.[80] Billard's analysis involves evaluating the differences over time between planetary mean longitudes as retrodicted by a set of *siddhānta* or *karaṇa* elements and the same mean longitudes retrodicted by modern astronomy.[81] The differences thus evaluated usually yield a fairly short time interval during which the total variance of all the errors is minimized. This can be interpreted to mean that the elements were derived from observations made within that time interval, which is presumably the approximate date of the text in which they are found (or at least the lower bound of the set of its possible dates, since a text might be based on observations older than itself). In fact, almost all the sets of elements tested were interpreted by Billard as based on single sets of observations clustered fairly closely around particular epochs, "none of which is ever mentioned in any text."[82]

The reason why such dates can be reconstructed in this way, according to Billard, is that owing to Indian astronomers' unrealistic assumption of an accurate conjunction of all planetary positions at the start of one or another *smṛti* era, initial planetary longitudes in their systems were somewhat displaced from the real ones. So, mean motion values based on that assumption, which were intended to produce accurate mean longitudes for some chosen epoch, will generally be too fast or too slow compared to the real mean motions. Consequently, the Indian parameters generally are "extraordinarily far from reality," and the chronological window within which they can be used to produce accurate results is narrow. The inference is that the short interval of accuracy is when the parameters were actually derived from contemporary observations.[83]

Using this reasoning, Billard reconstructed a chronology of Indian *siddhānta* astronomy containing many conclusions that conflict with the mainstream historical narrative outlined in section 4.2. Some of these conclusions are plausible (if sometimes rather boldly revisionist), while others contradict inferences from textual evidence. A few of the points of conflict are listed below.

---

ical evidence predating Judeo-Christian estimates of creation. Bentley's assertions about the alleged hoax of the astronomical era and about the date of Varāhamihira are found in [Ben1970] on, for example, p. 79 and pp. 158–163, respectively.

[79] See, for example, [BurE1971], pp. 23–24.

[80] Billard of course did not thereby endorse or excuse Bentley's paranoid and chauvinistic motives in employing this method; see his reference to Bentley in [Bil1971], p. 2.

[81] [Bil1971], pp. 47–50, and the brief summary of Billard's approach in [Wae1980]. Note that the longitudes computed from the Indian systems are sidereal and those computed from the modern system are tropical.

[82] [Bil1971], pp. 31, 51.

[83] [Bil1971], pp. 60–61. Of course, it is usually not possible to directly observe a mean longitude, unless it happens to coincide with the true longitude at some time when the body is at its apsidal point and its anomaly is zero. And it is also not possible to directly observe an apsidal point. Mean positions and motions, and the parameters for correcting them to true ones, have to be worked out together from observations of true positions.

- Billard concludes that the errors of the Ārya-pakṣa parameters are minimized shortly after 500, while those of the Brāhma-pakṣa parameters are minimized in the mid- to late sixth century and those of the *Sūrya-siddhānta* around 1100. He infers from this that the Ārya-pakṣa parameters are based on very accurate observations made in the early fifth century, presumably by Āryabhaṭa himself, and those of the Brāhma-pakṣa are based on observations shortly prior to Brahmagupta, from the middle and end of the sixth century. The Ārya-pakṣa thus predates the Brāhma-pakṣa in the development of *siddhānta* astronomy.

- It follows that the *Paitāmaha-siddhānta* containing the Brāhma-pakṣa parameters cannot be as old as the early fifth century, but must date to the end of the sixth century or later. Therefore, that *Paitāmaha-siddhānta* cannot be the ancient "astronomy of Brahman" mentioned by both Āryabhaṭa and Brahmagupta, who must be referring instead to some other, unknown treatise.[84]

- This chronology would imply that the Ārya-pakṣa and Ardha-rātrika-pakṣa of Āryabhaṭa were not variant forms of a recently established *siddhānta* tradition incorporating *smṛti* concepts such as the *kalpa*. Rather, Billard argues that Āryabhaṭa himself probably originated the "speculative" notion of blending a hypothesized early Classical tradition of "scientific" astronomy with the structure of the *yugas* and the notion of the Kaliyuga conjunction.[85] He suggests that most of Āryabhaṭa's successors in Indian astronomy took his synthesis somewhat too seriously, "sacralizing" the requirement that *siddhānta* parameters should conform to the temporal structure of *smṛti* cosmology. Thus they had to keep revising the parameters to obtain a transient accuracy during their own lifetimes.[86]

- Although the astronomer Lalla mentions *bīja*-corrections referring to the year 748, Billard finds that the parameters of Lalla's *Śiṣya-dhī-vṛddhida-tantra* are optimized for about 898. The discrepancy is explained by hypothesizing an unknown "Proto-Lalla" of the eighth century whose data were partly incorporated into the *Śiṣya-dhī-vṛddhida-tantra* a century and a half later.[87]

- Since the *Vaṭeśvara-siddhānta* is explicitly dated by its author to 904 but Billard finds its parameters optimal for a date around 1100, he infers that the work must have been corrupted by a later scribe or

---

[84] [Wae1980], pp. 57–58. In [Pin1980], p. 61, it is argued in response that only the abovementioned *Paitāmaha-siddhānta* is a plausible candidate for this role.

[85] [Bil1971], pp. 18–19, 122.

[86] [Bil1971], pp. 19, 26–28.

[87] [Bil1971], pp. 139–146.

redactor, "Pseudo-Vaṭeśvara," who substituted for its original parameters corresponding ones from the *Sūrya-siddhānta*.[88]

- The *Mahā-siddhānta* of Āryabhaṭa II, which is mentioned in the twelfth century by Bhāskara II, contains parameters which Billard found to be optimal for the early sixteenth century.[89]

On the whole, it appears that the Bentley-Billard error analysis approach provides some intriguing information and perspectives, but not consistently reliable historical conclusions. A narrowly focused statistical test of quantitative accuracy, although it may agree with our modern ideas of how astronomical texts ought to be evaluated, is not adequate to answer the complex questions about the interaction of computation with observation in the work of medieval Indian astronomers.

4.6.3 The transmission hypothesis and construction of parameters

If we are not convinced that Indian astronomical parameters can be satisfactorily explained as the products of observational programs for testing and refining specific geometric models, then how do we explain where they originally came from and why they were modified by later authors? If we postulate foreign sources, we are once again forced to rely largely on conjecture. There are no explicit references in medieval *siddhānta* or *karaṇa* texts to non-Indian astronomy, other than occasional mentions of the few "Yavana"-derived treatises that we surveyed in section 3.2. We cannot determine with certainty when and how the spherical astronomy of the medieval *siddhāntas* was formulated, but by the time of its first securely datable texts, it was already fully integrated into the Indian *jyotiṣa* tradition.

Bearing these caveats in mind, we will briefly state the case for the "transmission hypothesis" that argues that the fundamentals of the *siddhānta* system were derived from Hellenistic sources around the fourth century CE, and that although they were modified many times thereafter, they were not systematically tested and revised in accordance with any comprehensive observational program. The basic arguments for this view are as follows.[90]

There are explicit records (in the *Yavana-jātaka* and other astrological works) of a transmission, via Indo-Greek intermediaries, of Greek astrology

[88] [Bil1971], pp. 147–151. Billard notes in support of this hypothesis that the *Vaṭeśvara-siddhānta* was at that time known only from a single manuscript, concluding that the reliability of the manuscript "doubtless leaves much to be desired" (p. 147). Since Billard's publication, however, the *Vaṭeśvara-siddhānta* has been re-edited using an additional manuscript, which agrees with the first in its parameter values; [Shu1986] 1, pp. xix–xxiv and 3–5.

[89] [Bil1971], pp. 161–162; for the mention by Bhāskara, see [Pin1980], p. 61. We cannot account for this discrepancy by hypothesizing that Bhāskara must have been referring instead to Āryabhaṭa I, since he mentions a feature peculiar to the *Mahā-siddhānta* of Āryabhaṭa II; [DviS1910], p. 23, [Pin1992b].

[90] Fuller explanations can be found in, for example, [Pin1976], [Pin1993], and [Pin1980]. A detailed reconstruction of how planetary parameters in different *pakṣas* might have been derived according to this hypothesis is given in [Duk2008].

and astronomical computations applied to astrology into the Sanskrit tradition during the first few centuries of this era. Shortly afterward, there appeared Indian astronomy texts containing features previously not attested in the *jyotiṣa* tradition but familiar in Hellenistic astronomy, such as trigonometry, spherical models of the earth and heavens and the application of trigonometry to them to solve astronomical problems, and geometric devices such as eccentrics and epicycles to explain mathematically the unevenness of circular motions exhibited by the celestial bodies. There are no detailed records of Indian observations in medieval texts, and the catalogues of star coordinates that they contain are too small and inconsistent to serve as a basis for detailed planetary observations.

Therefore, we hypothesize that the basic models and parameters of the earliest astronomical *pakṣas* were taken over from some unidentified Greek sources, including values of planetary mean motions and contemporary mean longitudes. These features were adapted and developed with original approaches in the Indian tradition: for example, trigonometry of chords was changed to the more efficient trigonometry of Sines, and mean planetary motions were expressed in terms of very long periods conforming to Indian concepts of cyclic time. The planetary periods were derived from Greek mean motions by use of the so-called "pulverizer," a mathematical technique for solving linear indeterminate equations (discussed in sections 5.1.1 and 5.1.3), more or less in the following way.

The mean motion for any planet can be expressed in terms of some integer number A of revolutions in some very large integer number Y of years; the ratio A/Y is the planet's yearly mean motion in longitude. Over some given smaller time interval $y < Y$ years, the mean motion will produce some integer number x of complete revolutions plus a fractional revolution c equal to the mean longitude for the planet at the end of y years; in other words, $\frac{A}{Y} \cdot y = x + c$. Suppose, however, we want the planet's mean longitude at that time to be a different value, say, b, but we don't want to change the value for the mean motion significantly. That is, we need to find a new large integer number of revolutions R such that $R/Y \approx A/Y$, but $\frac{R}{Y} \cdot y = w + b$ (where w is some integer). Then we just need to solve this latter (indeterminate) equation for R and w, choosing an R that is not too different from our original A.

This method, it is argued, allowed Indian astronomers to tweak the borrowed planetary parameters in order to fit them into the chosen Indian astronomical eras without impairing their accuracy too much. Following the original adaptation and adjustment of the parameters, they were subsequently modified in various ways by *bījas* or conversion constants, for reasons not well understood. The role played by observations in this process is unclear, but does not seem to have been fundamental in determining parameter changes and revisions in the medieval period.

This hypothesis, like the previous one, is unsupported by any direct confirmation from the authors of Indian texts. And it too fails to explain completely some puzzling aspects of the Sanskrit mathematical astronomy tradition. In particular, since there is no textual evidence for any significant transmissions of Western astronomy into India between the early first millennium and the early second, what was inspiring medieval astronomers in the interval to make changes to their parameters, and how did they compute what those changes ought to be?

The divergence of the above two hypotheses about the foundations of *siddhānta* spherical astronomy, together with the vast number of details left unexplained by both of them, indicates how much work still needs to be done before the issue can be confidently resolved, if it ever can. In the meantime, the most plausible assessment seems to be something like this: medieval Indian mathematical astronomy was originally largely inspired by the models and parameters of its Hellenistic counterpart, but developed many original methods as well. In particular, the relations between numerical parameters, observations, and geometric models were far from simple, and cannot be interpreted naively in terms of modern ideals of "scientific method." As one historian of Indian astronomy recently expressed it, "The Indian astronomers, unlike Greeks, were not locked into any particular geometrical model.... They had the freedom to reconcile their mathematical predictions with observations in different ways as necessary. Theirs was a case of what can be called 'computational positivism'—like the Babylonian rather than the Greek practice."[91]

[91]Dr. S. Balachandra Rao, lecture, "A comparative study of the procedures of three astronomers of the early sixteenth century," World Sanskrit Conference 13, Edinburgh, 14 July 2006. Another perspective on Indian "computational positivism" is presented in [NarR2007], p. 532: "It is not that models are banished but they are not primary; models should not be 'preconceptions' (∼axioms?), but should be *inferred*—from observation and computation."

Chapter Five

The Genre of Medieval Mathematics

In the previous chapter, we saw that the term "*gaṇita*" was used in Indian texts in a variety of ways, to refer to the computational algorithms of astronomy as opposed to its geometric models, or more broadly to mean quantitative astronomy as opposed to the more descriptive disciplines of astrology. However, *gaṇita* is also frequently used in an even wider sense, to mean any type of computational or quantitative practices: what we might call "mathematics" in general. It is this sense of *gaṇita* that now comes to the fore as we explore some chapters in medieval *siddhānta* texts, and some independent works, that are explicitly devoted to calculation beyond the context of astronomy.

Instead of following a topical structure highlighted by mathematical material plucked from various *siddhāntas*, as happened in the previous chapter for reasons of economy, in this chapter I examine some major exemplars of more general mathematical works one by one, in more or less chronological order. This presentation is thus a compromise between a broad historical survey emphasizing highlights and the complete translations and analyses that would be necessary to give a thorough understanding of the individual works. Although not every verse of every text can be presented in detail, the goal is to convey at least an idea of the range of topics and methods covered in each of the texts, in the order that the author (presumably, unless he was traduced by later scribes) chose for them. Perhaps somewhat unfairly, the first known appearances of these topics and methods in earlier texts are treated more fully, while variations and repetitions of them in later works are skimmed over.[1]

The earliest known *siddhānta* chapters on *gaṇita* appear to predate by a few centuries the earliest known independent treatises on *gaṇita*. The fact that such chapters cover topics such as cube root extraction, rules for interest and barter, volume formulas for stacks of grain and lumber, and so forth, seems to imply that astronomy was the chief educational context for mathematics as a whole until the abovementioned mathematical treatises established it as a separate subject.[2]

---

[1]Furthermore, several other *siddhānta* chapters and independent treatises on *gaṇita* as a general topic in this period have had to be omitted from this discussion. See the entries in appendix B for information on the medieval mathematicians Āryabhaṭa (II), Śrīdhara, and Sripati.

[2]A strong form of this view is adduced in [Rao2004], p. 11: "[M]athematics mostly served as the hand-maid of astronomy.... The credit of happily divorcing mathematics from astronomy for good and giving it an independent status and treatment goes par-

But some indications suggest that mathematical instruction also took place outside the context of astronomy, even before the middle of the first millennium. For example, chapters on mathematics in general are never placed at the beginning of *siddhāntas*.[3] We can infer that teaching general computation was not considered the first duty of a textbook on mathematical astronomy. So perhaps the first independent texts on *gaṇita* in its broader sense developed in parallel with early works on mathematical astronomy, or even before them, but simply failed to survive. In any case, even if mathematics as an intellectual or pedagogical category started out strictly as a subfield of mathematical astronomy, it soon gained a separate existence. (See chapter 6 for further discussion of the place of mathematics within the medieval Sanskrit intellectual tradition.)

It is clear from the sheer volume and variety of mathematical material discussed, and the sophistication of its presentation, that even the earliest surviving *siddhāntas* were building on a highly developed mathematical tradition, albeit one that continued to grow and mature in succeeding centuries. The scanty and fragmentary sources that we examined in chapter 3 only hint at how that tradition must have flourished for at least several centuries before the middle of the first millennium. Consequently, it must be stressed that little, if any, of this material can be confidently described as an original discovery made by the author of the earliest known text in which it appears. Medieval *siddhānta* authors hardly ever discuss the provenance of a particular algorithm or explicitly claim one as their own invention, making it generally impossible to tell the difference between the rules they discovered themselves and the ones they compiled from other sources.[4] But at least it can be positively asserted that none of them were inventing their subject from scratch. Despite the extravagant claims of this sort in some popular histories—such as that Āryabhaṭa himself invented decimal place value notation or that Brahmagupta personally developed the arithmetic of negative numbers and zero—there is no historically plausible reason to believe that Indian mathematicians were first formulating such fundamental concepts as late as the middle of the first millennium.

5.1 CHAPTERS ON MATHEMATICS IN *SIDDHĀNTAS*

5.1.1 Mathematics in chapter 2 of the *Āryabhaṭīya*

We begin with the earliest surviving *siddhānta* chapter on *gaṇita*, the second chapter of the *Āryabhaṭīya*, composed around 500. This chapter consists of

---

ticularly to the Bakshāli manuscript as also to Śrīdhara (8th Cent.) and Mahāvīra (9th Cent.)."

[3]We might except the ever idiosyncratic *Āryabhaṭīya*, whose chapter on *gaṇita* is preceded only by one short introductory chapter stating parameter values for its astronomical models.

[4]See, however, the preceding discussion in section 4.4 of original rules as a sort of authorial trademark in *karaṇas* or astronomical handbooks.

thirty-three verses, which contain brief rules beginning with basic arithmetic and geometry and ending with first-degree indeterminate equations.[5] After an invocatory verse praising Brahman, the earth, and the celestial bodies, the remainder of the chapter consists of the statements described below. For convenience in explaining them we have assigned them topical subdivisions and drawn various inferences about their context and purpose, but in Āryabhaṭa's actual text they are just a stream of very elliptical versified algorithms.

Verse 2. Decimal notation. First of all Āryabhaṭa sets out the names of the places or powers for decimal place value notation, as follows:

> One and ten and a hundred and one thousand, now *ayuta* [ten thousand] and *niyuta* [hundred thousand], in the same way *prayuta* [million], *koṭi* [ten million], *arbuda* [hundred million], and *vṛnda* [thousand million]. A place should be ten times the previous place.

Many of these decimal place names have evidently come down unchanged from the forms in much older texts (see section 2.1), although some of the larger ones have been transposed or changed.

Verse 3. Arithmetic and geometric meanings of square and cube. Without any explanation of basic operations on place value numbers, which the student was evidently expected to understand already, Āryabhaṭa proceeds to squares and cubes in both their numerical and geometric senses. He defines a square (*varga*, see section 4.3.1) as an "equal-four-sided" figure, or equi-quadrilateral, and its area (or the result of squaring) as the product of two equal (sides or quantities in general). Similarly, a cube is called the product of three equals, as well as a "twelve-edged" solid.

Verses 4–5. Algorithms for square roots and cube roots. The procedures for extracting square and cube roots of numbers in decimal place value notation are concisely explained as follows:

> One should divide, constantly, the non-square [place] by twice the square root. When the square has been subtracted from the square [place], the quotient is the root in a different place.
>
> One should divide the second non-cube [place] by three times the square of the root of the cube. The square [of the quotient] multiplied by three and the former [quantity] should be subtracted from the first [non-cube place] and the cube from the cube [place].

[5] Verses 1–11, 16–17, 19, 25–28, and 30–33, along with parts of the commentary on them by Bhāskara I, are translated in [Plo2007b], pp. 401–417, where they are reproduced from a recently published annotated English translation of the whole of the *Āryabhaṭīya*'s second chapter and Bhāskara's commentary on it, [Kel2006]. The translations here, which do not differ substantially from the earlier ones, are based on the text in [ShuSa1976], pp. 33–84.

Table 5.1 Finding the square root of 341,056 by Āryabhaṭa's method.

		□		□		□
Set out the number:	3	4	1	0	5	6
Subtract [largest] square from square place:	−2	5				
		9	1			
Divide non-square place by 2× root: $91 \div (2 \cdot 5) = 9$						
Subtract product $9 \cdot (2 \cdot 5)$ *from non-sq. place*:		−9	0			
			1	0		
Subtract square [of 9] from square place:			−8	1		
Oops. Start over.						
		□		□		□
Set out the number:	3	4	1	0	5	6
Subtract [largest] square from square place:	−2	5				
		9	1			
Divide non-square place by 2× root: $91 \div (2 \cdot 5) = 8$						
Subtract product $8 \cdot (2 \cdot 5)$ *from non-sq. place*:		−8	0			
		1	1	0		
Subtract square [of 8] from square place:			−6	4		
			4	6	5	
Divide non-square place by 2× root: $465 \div (2 \cdot 58) = 4$						
Subtract product $4 \cdot (2 \cdot 58)$ *from non-sq. place*:			−4	6	4	
					1	6
Subtract square [of 4] from square place:					−1	6
Final square root: 584.						

The algorithms outlined by these rules are very similar to the modern long-division-style root extraction rules that were ubiquitous in the days before electronic calculators, but a brief explanation of them may be helpful. The first procedure assumes that the digits of the given number have been alternately designated "square" and "non-square," that is, even and odd powers of ten, respectively, as described in section 4.3.1. The rest of the graphical manipulations to be used in the course of applying the algorithm are not specified; only the essential steps are sketched. One way the procedure may have been carried out is illustrated by the example for the square root of 341,056 in table 5.1, where the square places are marked with □.[6] Steps not explicitly stated in the rule are shown in italics.

As the example shows, some of the steps may require a little trial and error to ensure that the remainder from a subtraction will be large enough to accommodate the next step. But the procedure is essentially straightforward, as will be seen if we consider our sample square root 584 as the sum $500 + 80 + 4$, or more abstractly $(a + b + c)$, whose square is

$$(a + b + c)^2 = a^2 + 2ab + b^2 + 2ac + 2bc + c^2.$$

The repeated subtractions required by Āryabhaṭa's algorithm in our example are just successively removing from the given square number its components $a^2 = 250000$, $2ab = 80000$, $b^2 = 6400$, $2(a + b) \cdot c = 4640$, and $c^2 = 16$.

Similarly, the cube root algorithm apportions the digits of a given number—let us think of it as $(a + b)^3$—into "cube," "first non-cube," and "second non-cube" places. Then it subtracts out the successive components a^3, $3a^2b$, $3b^2a$, and b^2. This is illustrated by our example for the cube root of 103,823 in table 5.2, where the cube places are marked with •.

Verse 6. Area of a triangle and volume of a pyramid. Āryabhaṭa's formula for the area of a triangle is the familiar "half the base times the height" rule, but his corresponding rule for the volume of a related solid is puzzling:

> The bulk of the area of a trilateral is the product of half the base and the perpendicular. Half the product of that and the upward side, that is [the volume of] a solid called "six-edged."

This seems to imply (and is interpreted by Āryabhaṭa's commentators to mean) that the volume V of a tetrahedron or pyramid with triangular base of area A and height or "upward side" h is given by $V = \frac{Ah}{2}$, instead of $\frac{Ah}{3}$. The source and rationale of this approximate rule are not known; perhaps it was considered appropriately analogous to the product of one-half the base and height for the area of a triangle.[7]

Verse 7. Area of a circle and volume of a sphere. Āryabhaṭa says:

[6] Note that Āryabhaṭa himself has not specified any particular marking for the square versus non-square places; this is just an editorial addition to help explain the procedure. For more details on the layout of Indian square root and cube root operations, see [DatSi1962], vol. 1, pp. 169–180.

[7] It has been suggested by a modern scholar, as noted in [Sara1979], p. 199, that this volume formula should instead be interpreted as a correct rule for a different quantity,

Table 5.2 Finding the cube root of 103,823 by Āryabhaṭa's method.

			•			•
Set out the number:	1	0	3	8	2	3
Subtract [largest] cube from cube place:		−6	4			
		3	9	8		
Divide 2nd non-cube place by $3 \cdot (\text{root})^2$: $398 \div (3 \cdot 4^2) = \not{8}\ 7$						
Subtract product $7 \cdot (3 \cdot 4^2)$ *from 2nd non-cube:*		−3	3	6		
			6	2	2	
Subtract product $7^2 \cdot (3 \cdot 4)$ from 1st non-cube:			−5	8	8	
				3	4	3
Subtract cube [of 7] from cube place:				−3	4	3
Final cube root: 47.						

> Half of the even circumference multiplied by the semidiameter, only, is the area of a circle. That multiplied by its own root is the volume of the circular solid without remainder.

That is, if the radius of a circle is r and its circumference is C, its area A is $\frac{C \cdot r}{2}$. Then $A\sqrt{A}$ is said to be the volume of the corresponding sphere, which again is only an approximate rule—it is equivalent to $\pi r^3 \sqrt{\pi}$ instead of $\frac{4}{3}\pi r^3$, if $C = 2\pi r$. Again, it may be that the incorrect volume rule was considered loosely analogous to the correct area one.

Verse 8. Segments of the altitude of a trapezium and its area. Here Āryabhaṭa says that the two sides of an unidentified figure times its height, divided by the sum of the sides, equal what we may translate as "two lines at its own intersection." The area is given as the height times half the sum of the two lengths. From considering the relations between the line segments shown in figure 5.1, we can infer that the subject of this rule is an isosceles trapezium with altitude h. If its two "lengths" or "sides" are the horizontal top and bottom t and b, respectively, then indeed its area A is equal to $h \cdot \frac{t+b}{2}$. If its altitude h is divided into the segments h_t and h_b at the intersection of the diagonals, the values of h_t and h_b can be found from similar

namely six times the volume of a pyramid whose base area is one-fourth that of the given triangle. Although this interpretation would indeed make the rule exact, there is no known basis for it in the Sanskrit mathematical texts. The corresponding argument that the incorrect rule for the volume of a sphere in Āryabhaṭa's next verse ought to be interpreted instead as a correct rule for its surface area (see [Sara1979], p. 209) is also unsupported.

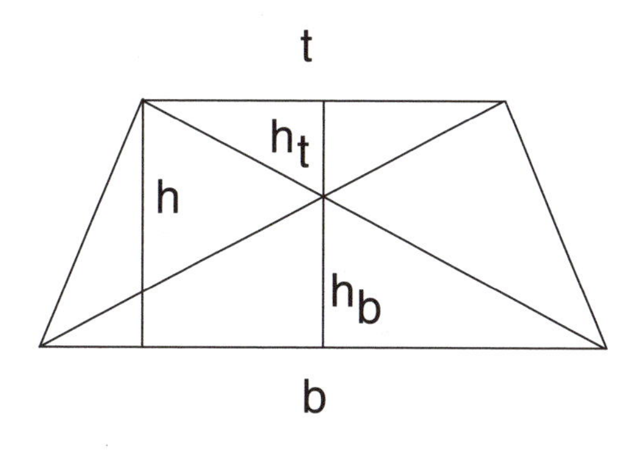

Figure 5.1 The line segments of the trapezium.

right triangles as follows:

$$\frac{h_t}{t/2} = \frac{h_b}{b/2} = \frac{h}{t + (b/2 - t/2)} = \frac{h}{(b+t)/2}, \quad \text{so}$$

$$h_t = \frac{th}{b+t} \quad \text{and} \quad h_b = \frac{bh}{b+t}.$$

Verse 9ab. Area of an arbitrary plane figure. A general rule for the area of unspecified (presumably rectilinear) figures is the following:

> For all figures, when one has acquired the two sides, the area is their product.

This could presumably be an approximation to be used for roughly rectangular quadrilaterals whose area is more or less the product of two adjacent sides.

Verses 9cd–12. Chords and circumference of a circle and construction of a Sine table. Āryabhaṭa now goes on to explain a little more about the Sines whose values he stated without discussion in his initial chapter (see section 4.3.3).

> The Chord of a sixth part of the circumference is equal to the Radius. A hundred increased by four, multiplied by eight, [added to] sixty-two thousands [i.e., 62,832], [is] the approximate circumference of a circle with diameter two *ayuta* [i.e., 20,000].
>
> One should divide up a quarter of the circumference of a circle. And from triangles and quadrilaterals, as many Sines of equal arcs as desired [can be found for a given] Radius.
>
> The Sine of the first arc, divided and [then] diminished [by itself] is the second difference. The remaining [differences] are [succes-

sively] diminished by those [cumulative differences] divided by that first Sine.

This value $\frac{62832}{20000} = 3.1416$ for the ratio of the circumference of a circle to its diameter is the first such value to appear in Indian mathematics after the standard approximation $\pi \approx \sqrt{10}$ from the early first millennium CE or earlier.[8] However, the standard trigonometric radius value $R = 3438$ for a circle with circumference 21,600 arcminutes must have been derived from $\pi \approx 3.1416$ or a similarly accurate value. So if the standard radius $R = 3438$ is indeed as old as the early fifth century, it implies Indian knowledge of such an accurate π value prior to Āryabhaṭa's work. How the value was derived is not explained by Āryabhaṭa; modern reconstructions base it on computing the sides of inscribed regular polygons, up to a 384-sided figure.[9] This seems likely, and would explain why Āryabhaṭa introduces the subject with the comment that the Chord of a 60-degree arc is equal to the radius, since that provides a starting point for determining the successive polygon sides.

The process of geometrically constructing Sines is only sketchily described: after dividing a quarter of the circumference into the desired number of equal arcs, one derives the Sines in an unspecified way "from triangles and quadrilaterals" (see section 5.1.2 for an elaboration by the commentator Bhāskara). The remaining rule looks like a vague restatement of the *Paitāmaha-siddhānta*'s approximate formula for the ith Sine-difference discussed in section 4.3.3, and that is the way Bhāskara interprets it. But it may instead have been intended to convey a different, more complicated meaning corresponding to a more accurate formula for Sine-differences, as follows:

$$\Delta \operatorname{Sin}_{(i-1)} - \Delta \operatorname{Sin}_i = (\Delta \operatorname{Sin}_1 - \Delta \operatorname{Sin}_2) \frac{\operatorname{Sin}_{(i-1)}}{\operatorname{Sin}_1},$$

where Sin_i as before is the ith Sine in the Sine table, $\Delta \operatorname{Sin}_i = \operatorname{Sin}_i - \operatorname{Sin}_{(i-1)}$, and $\Delta \operatorname{Sin}_1 = \operatorname{Sin}_1$. This relationship between the successive differences of Sine-differences and the corresponding Sines is geometrically exact, up to the equivalence of the first Sine and its Chord.[10]

Verse 13. Physical construction of figures. The manual work of mathematics is touched on here with the direction that circles are to be drawn by turning a compass, and triangles and quadrilaterals with pairs of diagonal lines (the commentator Bhāskara's explanation of this is discussed in section 5.1.2.) Horizontal and vertical surfaces respectively are to be made with water—that is, by water leveling, or pouring water on the surface to detect bumps

[8] [HayKY1989], p. 11.

[9] [Sara1979], p. 156.

[10] However, this interpretation appeared in the commentarial tradition only in the fifteenth century, so it is not certain whether it reflects what Āryabhaṭa actually meant. For a full explanation of this and other interpretations of Āryabhaṭa's Sine-difference rule, and its possible relationship to the corresponding rule in the *Paitāmaha-siddhānta*, see [Hay1997a].

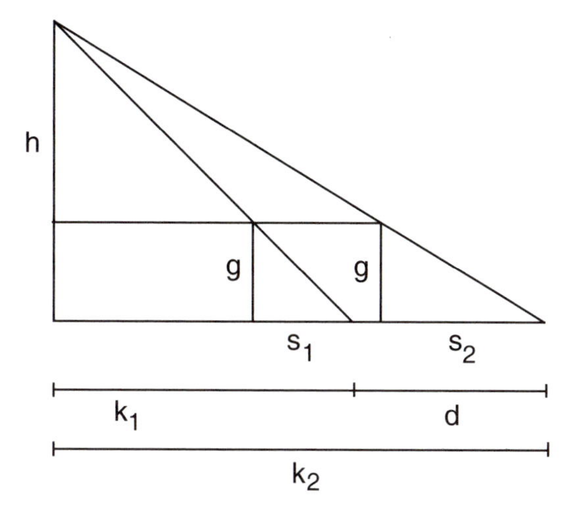

Figure 5.2 Computing with gnomon shadows.

and hollows—and with a perpendicular or plumb line. It may be that these procedures are described here because true horizontals and verticals are necessary for working with gnomons, the topic of the next verses.

Verses 14–16. Shadows of gnomons. The shadow of a gnomon is considered to determine a "shadow circle" whose radius is defined as the square root of the sum of the squares of the gnomon length and the shadow—that is, the hypotenuse of the shadow triangle. The next two verses describe the relationships between shadows of gnomons and the heights and distances of light sources, which are depicted in figure 5.2. The first deals with a single gnomon of height g illuminated by a light source of given height h and casting a shadow of length s_1; the distance between the base of the light source and the base of the gnomon is $(k_1 - s_1)$. The unknown shadow length s_1 is determined from h, g, and $(k_1 - s_1)$ by a simple proportional rule which we may express as $s_1 = g \cdot \dfrac{k_1 - s_1}{h - g}$.

Āryabhaṭa's last shadow rule involves using two equal gnomons to find the height and distance of a light source when they are not directly measurable. The gnomons are set up in line with the light source whose height h is now unknown, and cast shadows of known length s_1 and s_2, respectively. Then the distance k_1 or k_2 from the tip of either shadow to the base of the light, as well as the height h of the light, are to be found as follows:

> The upright is the distance between the tips of the [two] shadows multiplied by a shadow divided by the decrease. That upright multiplied by the gnomon, divided by [its] shadow, becomes the arm.

The terms "arm" and "upright," which are standard in Sanskrit mathematics to refer to the horizontal and vertical legs of a right triangle, are a little confusing in this case because in the figure, the "upright" k_1 or k_2 is actually horizontal. From the two pairs of similar right triangles in the figure, the ratios of the "uprights" k_1 and k_2 to their corresponding shadows s_1 and s_2 must be equal:

$$\frac{s_1}{k_1} = \frac{g}{h} = \frac{s_2}{k_2}.$$

Consequently $s_1 = k_1 \cdot \frac{g}{h}$ and $s_2 = k_2 \cdot \frac{g}{h}$. So, in terms of the distance $d = (k_2 - k_1)$ between the shadow tips and the "decrease" or difference $(s_2 - s_1)$ of their shadows,

$$s_2 - s_1 = (k_2 - k_1) \cdot \frac{g}{h} = d \cdot \frac{g}{h}, \qquad \text{and}$$

$$k_1 = s_1 \cdot \frac{h}{g} = d \cdot \frac{s_1}{s_2 - s_1}, \qquad k_2 = s_2 \cdot \frac{h}{g} = d \cdot \frac{s_2}{s_2 - s_1}.$$

Then the height h, which Āryabhaṭa calls the "arm," is derived from either of the above ratios:

$$h = \frac{g \cdot k_1}{s_1} = \frac{g \cdot k_2}{s_2}.$$

Verses 17–18. The right triangle and segments of chords. The sum of the squares of the arm and the upright—evidently referring to an arbitrary right triangle—is stated to be equal to the square of the hypotenuse. The remainder of these verses deals with the relationships of chord segments in circles. First, the product of the "arrows [or Versines] of two arcs" is said to be equal to the square of the half-chord or Sine. We can understand this as applying to a circle cut by a diameter d and by a chord $2s$ perpendicular to that diameter, as in the left side of figure 5.3, where a and $(d - a)$ are the Versines or "arrows" of the smaller and larger bow arcs, respectively. Their product will indeed equal s^2 because, from the similar right triangles in the upper semicircle, $\frac{d-a}{s} = \frac{s}{a}$.

Then Āryabhaṭa gives a rule involving a similar configuration for two intersecting circles:

> [The diameters of] two circles decreased by the obscuration are multiplied by the obscuration. Divide [them] separately [by] the sum of [the diameters] decreased by the obscuration; the two quotients are the arrows of the [arcs] intersecting each other.

This situation (pictured on the right sight of figure 5.3) is reminiscent of the geometry of an eclipse, and the word "obscuration" or "eating" is used in astronomy for the obscured part of an eclipsed body; but this particular rule has no known astronomical purpose. The algorithm may be expressed

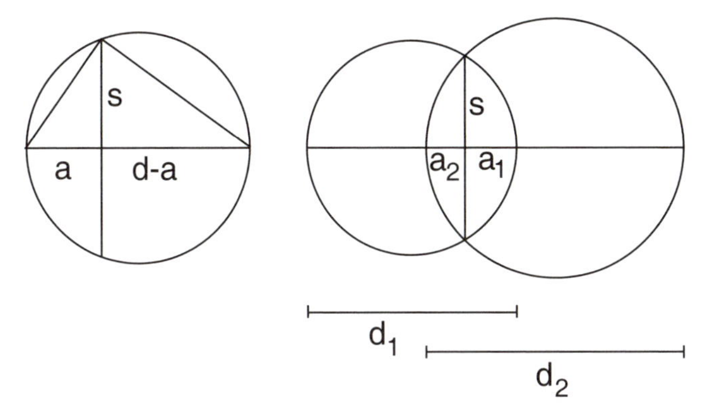

Figure 5.3 "Arrows" in circular segments.

in terms of our figure, where the two circles have diameters d_1 and d_2 and the obscuration a is the sum of the corresponding "arrows" a_1 and a_2, as follows:

$$a_1 = \frac{(d_2 - a) \cdot a}{(d_1 - a) + (d_2 - a)}, \quad a_2 = \frac{(d_1 - a) \cdot a}{(d_1 - a) + (d_2 - a)}.$$

Although Āryabhaṭa himself gives no explanation of this rule, we can justify it by means of modern algebraic manipulations, as follows. From the previous rule we know that $a_1(d_1 - a_1) = s^2 = a_2(d_2 - a_2)$, so replacing a_2 by $(a - a_1)$, we may write

$$\begin{aligned} a_1(d_1 - a_1) &= (a - a_1)(d_2 - a + a_1), \\ a_1 d_1 - {a_1}^2 &= ad_2 - a^2 + aa_1 - a_1 d_2 + aa_1 - {a_1}^2, \\ a_1(d_1 - a + d_2 - a) &= ad_2 - a^2, \quad \text{so} \\ a_1 &= \frac{(d_2 - a) \cdot a}{(d_1 - a) + (d_2 - a)}, \end{aligned}$$

and we can similarly derive the corresponding expression for a_2.

Verses 19–22. Series. Āryabhaṭa now turns to the subject of "progression" or "successive sequence," usually translated as "series." His very compressed and multivalent rules concerning them invoke the concepts of the first term and last term of a sequence or subsequence, its middle term or mean, the "increase" or constant difference between successive terms (in an arithmetic progression), the given number of terms, and the sum of the terms. We will briefly restate his formulas in modern notation for the terms of a sequence $(a_1, a_2, \ldots, a_{n-1}, a_n)$, where the constant difference $(a_i - a_{i-1})$ for $1 < i \leq n$ is called d, the mean of the last k terms $(a_{(n-k+1)}, \ldots, a_n)$ is m_k, and the sum of the last k terms is s_k. It will be left to the reader to verify, if desired, that Āryabhaṭa's rules are equivalent to modern formulas for these quantities.

Mean of the last k terms:

$$m_k = a_1 + \left(\frac{k-1}{2} + (n-k)\right) \cdot d.$$

Sum of the last k terms:

$$s_k = m_k \cdot k, \quad \text{or} \quad s_k = (a_{(n-k+1)} + a_n) \cdot \frac{k}{2}.$$

Number of terms:

$$k = \frac{1}{2} \cdot \left(\frac{\sqrt{8ds_k + (2a_{(n-k+1)} - d)^2} - 2a_{(n-k+1)}}{d} + 1\right).$$

All these rules can also be interpreted to apply to the specific case where $n = k$, thanks to the flexibility of the meaning of Āryabhaṭa's expression "first term" for what we have designated $a_{(n-k+1)}$.

Rules for finding the sums of the following series are also given:

Sum of the "pile solid" $\sum_{j=1}^{n} \sum_{i=1}^{j} i$, or $1 + (1+2) + (1+2+3) + \ldots + (1 + 2 + \ldots + n)$:

$$\sum_{j=1}^{n} \sum_{i=1}^{j} i = \frac{(n)(n+1)(n+2)}{6}, \quad \text{or} \quad \sum_{j=1}^{n} \sum_{i=1}^{j} i = \frac{(n+1)^3 - (n+1)}{6}.$$

Sum of the "pile solid of squares" $\sum_{i=1}^{n} i^2$ and of the "pile solid of cubes" $\sum_{i=1}^{n} i^3$:

$$\sum_{i=1}^{n} i^2 = \frac{(n+1)(2n+1)(n)}{6}, \qquad \sum_{i=1}^{n} i^3 = \left(\frac{n(n+1)}{2}\right)^2.$$

Verses 23–24. Factors and products. These are two rules for manipulating factors of a product, say $a \cdot b$. If the factors a and b are known, then their product is said to be half the difference of the square of their sum and the sum of their squares, or as we would write it:

$$ab = \frac{(a+b)^2 - (a^2 + b^2)}{2}.$$

On the other hand, if the product ab of two unknown factors is known, and also their difference $a - b$, then the factors themselves are to be found by a procedure that we can express in modern notation as follows:

$$a = \frac{\sqrt{4ab + (a-b)^2} + (a-b)}{2}, \quad b = \frac{\sqrt{4ab + (a-b)^2} - (a-b)}{2}.$$

If we think of the known product and difference as $p = ab$ and $d = a - b$, respectively, then $a = \frac{p}{b} = b + d$ while $b = \frac{p}{a} = a - d$. So we can view the above procedure as just a version of the quadratic formula solving the equations $p = a^2 - ad$ and $p = b^2 + bd$ for a and b, respectively.

Verse 25. Interest. This rule computes the monthly amount of interest i earned by a given principal p. Also given is the sum $q = i + e_i$, where e_i is the earnings accrued by i invested at the unknown rate $\frac{i}{p}$ for a given number t of months: thus $e_i = i \cdot \frac{i}{p} \cdot t$. Āryabhaṭa's rule for finding the monthly interest i from q, p and t can be expressed as follows:

$$i = \frac{\sqrt{ptq + \left(\frac{p}{2}\right)^2} - \frac{p}{2}}{t}.$$

As in the previous rule, we can understand this as a form of the quadratic formula solving for i in the equation $q = i + e_i = i + \frac{i^2 t}{p}$. These rules indicate that Āryabhaṭa was quite conversant with what we call quadratic equations in general and methods for solving them, although he nowhere states a general formula for quadratic solutions.

Verse 26. Rule of Three Quantities. As we will see in subsequent sections, the core concept called the *trai-rāśika*, or "having three quantities," is considered the very foundation of mathematics in Indian texts. It deals with what we would call simple proportion: given some ratio $a : b = x : d$, if the three quantities a, b, d are known, find x. The three given quantities a, b, d are known by Sanskrit technical terms generally translated as the "fruit" (or "result"), the "measure" (or "argument"), and the "desire" (or "requisition"), respectively; x is the "desired result" or "fruit of the requisition." Āryabhaṭa presents the first known statement of the *trai-rāśika* in which all the above names appear, so this terminology either was already standard in Āryabhaṭa's time or soon afterward became so. Āryabhaṭa's solution procedure is also standard, corresponding in our notation to $x = ad/b$.[11]

Verse 27. Manipulating fractions. Āryabhaṭa speaks of fractional quantities as having "multipliers" and "divisors," or what we would call numerators and denominators. His rules for dividing fractions and reducing two fractions to a common denominator are familiar enough to require no explanation beyond his own words:

> The divisors of [fractions] in division are multiplied by each other's multipliers. [Each] multiplier and divisor is multiplied by the other's divisor: that is [reduction] to the same kind [literally "same-coloring"].

Verse 28. Inverse operations. This verse describes the method of "inversion" or "[working] backward" in cases where the result of certain arithmetic operations on some number is known but the number itself is not.

[11] See [SarS2002] for a detailed discussion of the development of the Rule of Three Quantities in Indian mathematics and its use elsewhere.

To borrow a typical example from Bhāskara's commentary, "What [number when] multiplied by three, decreased by one, halved, increased by two, then divided by three, then decreased by two, becomes one?" Āryabhaṭa's rule prescribes changing every operation to its inverse to get the unknown from the result: "multipliers become divisors, divisors multipliers; what is added is subtracted, and what is subtracted is added in inversion." Working backwards in this way from the given result 1 in our example, we get 1 plus 2, times 3, minus 2, doubled, plus 1, divided by 3: namely, 5.

Verses 29–30. Finding unknown quantities from their sums. Two rules deal with basic problems of separating out unknown quantities from given combinations of them. We can express the first case as involving n unknowns $a_1, a_2, \ldots, a_n$ and their unknown sum $S = \sum_{i=1}^{n} a_i$. What is known are the so-called "less-[one]-quantity" sums $S_i = S - a_i$ for $1 \leq i \leq n$. The prescribed rule to find the desired total sum S adds up all the given "less-[one]-quantity" sums and divides the resulting sum (in which each unknown a_i appears $n-1$ times) by $n-1$:

$$S = \frac{\sum_{i=1}^{n} S_i}{n-1}.$$

Presumably, although Āryabhaṭa does not explicitly say so, each unknown a_i is then easily found from $S - S_i$.

The second rule determines the value of a "bead" or unknown quantity from two equal combinations. Namely, if the unknown value or "bead" is x, and one property consisting of b_1 beads plus c_1 units equals another property of b_2 beads plus c_2 units, then the rule prescribes the equivalent of $x = \dfrac{c_2 - c_1}{b_1 - b_2}$.

Verse 31. Distance, rate and time. Like the problem on intersecting circles, this formula for determining the time of conjunction of two objects moving toward or away from each other suggests a possible astronomical context involving celestial conjunctions. If the objects are moving in opposite directions, then the time interval remaining before their meeting (or after it, if they have already passed each other) is the distance between them divided by the sum of their speeds. If they are moving in the same direction, that time interval is the distance divided by the difference of their speeds.

Verses 32–33. The "pulverizer" or linear indeterminate equations. Āryabhaṭa's mathematical chapter culminates in the earliest surviving description of the *kuṭṭaka* ("pulverizer" or "grinder"), a technique for finding quantities that satisfy certain conditions involving divisors and remainders. Such problems, now classified as linear indeterminate equations, can be expressed as follows: Given two specified integer divisors d_1 and d_2 and integers r_1 and r_2, find an integer N such that $N \equiv r_1 \pmod{d_1}$ and $N \equiv r_2 \pmod{d_2}$. The solution is accomplished by a process somewhat similar to the so-called

Euclidean algorithm for finding the greatest common divisor of two integers. That is, we use repeated divisions to produce pairs of smaller divisors and remainders, "grinding" the original numbers until they are small enough to allow an easy solution. I quote Āryabhaṭa's very abbreviated statement of the method and then illustrate its operation in detail.

> One should divide the divisor of the greater remainder by the divisor of the smaller remainder. The mutual division [of the previous divisor] by the remainder [is made continuously. The last remainder], having a clever [quantity] for multiplier, is increased [or decreased] by the difference of the [initial] remainders [and divided by the last divisor].
>
> The one above is multiplied by the one below, and increased [or decreased] by the last. When [the result of this procedure] is divided by the divisor of the smaller remainder, the remainder, having the divisor of the greater remainder for multiplier, and increased by the greater remainder, is the [quantity that has such] remainders for two divisors.

Consider the example of finding $N = 29a + 7 = 53b + 12$. We start out by dividing "the divisor of the greater remainder," 53, by the other divisor, 29, and get a new remainder, 24. Then we divide 29 by 24 and get a new remainder, 5, and divide 24 by 5 and get a new remainder, 4, and so forth. Each division essentially just represents substituting a new indeterminate equation with smaller coefficients for the previous one. This is illustrated in the following sequence of equations derived from our original equation $29a+7 = 53b+12$, with a new variable used to represent each new remainder:

$$a = \frac{53b + (12 - 7)}{29} = 1b + \frac{24b + (12 - 7)}{29} = 1b + c$$

$$b = \frac{29c - (12 - 7)}{24} = 1c + \frac{5c - (12 - 7)}{24} = 1c + d$$

$$c = \frac{24d + (12 - 7)}{5} = 4d + \frac{4d + (12 - 7)}{5} = 4d + e$$

$$d = \frac{5e - (12 - 7)}{4} = 1e + \frac{1e - (12 - 7)}{4} = 1e + f$$

$$e = \frac{4f + (12 - 7)}{1}$$

Now we have worked our way down to a very easy indeterminate equation, $e = 4f + (12 - 7)$, that we can solve by inspection. Let us take our "clever multiplier" f equal to 1 and add the product $4f = 4$ to the difference of the initial remainders: $4 + (12 - 7) = 9 = e$. Now we can just work our way back up from the bottom to the top of our sequence of divisors, substituting in the value of e to solve for d, the value of d to solve for c, and so forth, "multiplying the one above by the one below":

$$
\begin{array}{ll}
53 & \dfrac{53 \cdot 59 + (12 - 7)}{29} = 108 = a \\[2ex]
29 & \dfrac{29 \cdot 49 - (12 - 7)}{24} = 59 = b \\[2ex]
24 & \dfrac{24 \cdot 10 + (12 - 7)}{5} = 49 = c \\[2ex]
5 & \dfrac{5 \cdot 9 - (12 - 7)}{4} = 10 = d \\[2ex]
4 & \dfrac{4 \cdot 1 + (12 - 7)}{1} = 9 = e \\[2ex]
1 &
\end{array}
$$

Thus our solution is ultimately $N = 29 \cdot 108 + 7 = 53 \cdot 59 + 12 = 3139$.

5.1.2 Bhāskara's commentary on *Āryabhaṭīya* 2

Āryabhaṭa's treatment of mathematics is wide-ranging but extremely compressed, with no expository detail and no description of its structure. The earliest clues that we have as to how his formulas were actually presented and explained to students come from the first surviving commentary on the *Āryabhaṭīya* (indeed, the first surviving commentary on any Sanskrit mathematical work), which was composed by Bhāskara in 629, more than a century after Āryabhaṭa wrote. However, there was clearly no lack of mathematical activity in the intervening period, as is evident from Bhāskara's occasional quotes from mathematical sources now lost and references to mathematicians otherwise unknown.

Bhāskara's commentary is the first available Sanskrit source of detailed discussions on the nature of mathematics. This topic comes up with the very first verse of Āryabhaṭa's treatise, where he states his intention of expounding "the three [subjects of] calculation [*gaṇita*], time-reckoning, and the sphere." Bhāskara in his commentary on this verse attempts to untangle the various meanings, astronomical and nonastronomical, of the comprehensive term *gaṇita*:

> Calculation is figures, shadows, series, equations, the pulverizer, etc.... Here, this word "*gaṇita*" has the meaning "calculation as a whole." Because of the meaning "calculation as a whole," he is speaking about planet-calculation as much as figure-calculation. And because of the non-differentiation of planet-calculation from geometry [figures] etc., and the non-differentiation of time reck-

oning and the sphere from calculation, he just calls [it] "*gaṇita*."[12]

At the beginning of his commentary on the second chapter, he states further that mathematics consists of "increase" and "decrease." He quotes an anonymous source that describes multiplication and "going" (to a power, that is, exponentiation) as forms of increase, and division and root extraction as forms of decrease. Bhāskara notes the possible objection that division by a fraction makes the divided quantity bigger, not smaller, but counters it with a geometric example using a rectangular figure four by five units in area made up of twenty unit squares. The area of one such square is one-twentieth of the total and its sides are one-fourth and one-fifth of the total width and length, respectively. It is true that dividing the area 1/20 by the side 1/4 does appear to increase the divided quantity, but really it decreases it (presumably Bhāskara means that this is so because the division reduces an area to a length).[13]

Another definition of mathematics in this passage describes it as consisting of "quantities," relating to topics like proportion and the pulverizer, and "figures," comprising subjects like shadows and, unexpectedly, series. Although Bhāskara discusses series only in terms of sequences of numbers, presumably they were viewed as configurations of objects too, as the name "pile solid" in Āryabhaṭa's series rules suggests. Furthermore, mathematics is said to be established in four "seeds," or different varieties of unknowns (apparently a classification of types of equations). Bhāskara is evidently presenting here in his introduction a summary or sample of a rich variety of perspectives on how mathematics is to be defined and classified in seventh-century India.[14]

Most of Bhāskara's commentary consists of specific explanations and illustrations of Āryabhaṭa's cryptic verse rules. These explanations rely heavily on the core Sanskrit discipline of grammar: the individual words of the base text's verses are glossed with synonyms and their relationships in the sentences are explained in terms of their grammatical forms. Bhāskara sometimes cites standard grammatical texts to justify a particular construction or to excuse an anomalous verbal formation for a technical term. This focus on linguistic aspects not only reflects the centrality of grammar in Sanskrit learning, but also allows the commentator to unpack several different but related grammatically valid meanings from what at first seems to be a rule with a single meaning. For example, Āryabhaṭa's pulverizer algorithm from verses 2.32–33 is reinterpreted by Bhāskara to refer also to linear indeterminate equations where one of the given remainders equals zero (or in our previous notation, $N = ad_1 + r_1 = bd_2$). He achieves this by taking Ārya-

[12]Commentary on *Āryabhaṭīya* 1.1, [Shu1976], pp. 5–7.

[13]Commentary on *Āryabhaṭīya* 2, [Shu1976], p. 44. All the following citations of Bhāskara's comments, identified henceforth only by the corresponding verse number(s) of Āryabhaṭa's text, are from [Shu1976], pp. 43–171. For fuller details of the discussions and analyses of them, see [Kel2006].

[14]A deeper investigation of Bhāskara's presentation of these perspectives is given in [Kel2007].

bhaṭa's phrases for "divisor of the greater remainder" and "divisor of the smaller remainder" to mean now "divisor which is a larger number" and "divisor which is a smaller number," respectively.

Bhāskara also amplifies Āryabhaṭa's brief remarks into fuller descriptions of technical terminology. For instance, when Āryabhaṭa introduces triangles in verse 2.6, Bhāskara explains that triangles may be "equal" (equilateral), "bi-equal" (isosceles), or "unequal" (scalene). Similarly, he narrows down Āryabhaṭa's term for "square" (literally " quadrilateral") in verse 2.3, stipulating that it cannot mean a figure having four equal sides but unequal diagonals. Moreover, Bhāskara assigns his own topical structure to parts of the text by identifying particular rules with particular subjects, as when he speaks of the rule for the value of a "bead" in verse 2.30 as pertaining to equations (literally "equal-making"). Bhāskara as a commentator thus seems to have quite a lot of latitude in reading the base text to extract not only the literal sense of what Āryabhaṭa said but also what he thinks the student needs to know about mathematics in general. Or perhaps, on the other hand, his exposition follows an existing exegetical tradition, going back possibly as far as the personal teachings of Āryabhaṭa himself.

The commentary also reveals valuable information about the visual and physical presentation of mathematics. Unfortunately, since the oldest surviving manuscripts of it are more than a thousand years later than Bhāskara's own time, no firm conclusions can be drawn about details of Bhāskara's notation. But we can tell at least that he expected the reader to use a writing surface to lay out the digits of the given numbers for a particular problem and then perform calculations on them. This writing surface was apparently a board of some kind—probably a dust board or piece of smooth dusty ground, judging from some of Bhāskara's word problems where the student must reconstruct by means of the pulverizer some digits of a previously known answer that were erased by the wind. The vocabulary of "beads" and "seeds" or "grains" to mean unknown quantities suggests that small articles of that sort were placed in appropriate positions among the written numerals on the board to stand for numbers of unknown value.

Geometric figures would also have been drawn on the board with strings and a compass ("crab-[instrument]"). Bhāskara instructs the reader how to draw an isosceles triangle with a fish figure or perpendicular bisector made of two intersecting circular arcs (as described in the explanation of figure 4.11), and how to make straight lines with stretched strings. His long digression on the proper construction and use of gnomons for shadow measurement, and his occasional use of the term "calendar maker" for mathematician, emphasize the close connection of these subjects with astronomy.

Other graphical constructions are made in order to explain computations, as in the construction of Sines. Bhāskara's derivation of the standard twenty-four Sines, one for each $225'$ of arc, employs "triangles and quadrilaterals" as suggested in verse 2.11. First, a circle with imagined radius R is divided into twelve "zodiacal signs" or 30-degree arcs by inscribed triangles and rectangles (a simplified picture for a single quadrant is shown on the left

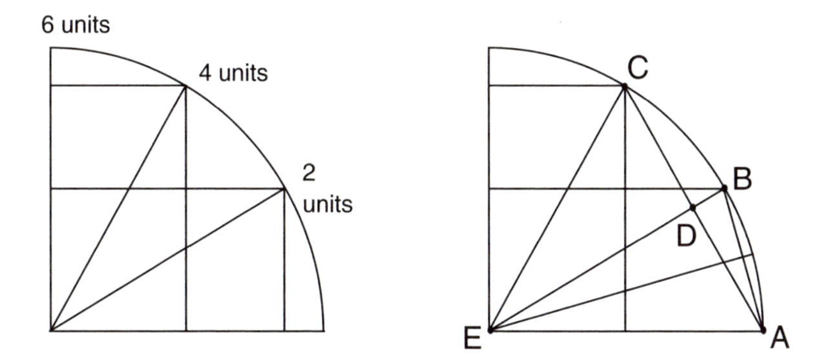

Figure 5.4 Constructing Sines in a circle.

side of figure 5.4). The unit of arc is considered to be half a sign. Then additional lines are drawn in the figure to create the right triangles on which the various Sine computations depend. We will use labeled points in the quadrant on the right side of figure 5.4 to keep track of the process, as follows:

The Chord AC of four arc-units (2 signs or 60°) is the Radius R, and half of it AD is the Sine of two arc-units or 30°. The Pythagorean theorem then gives the Cosine ED of two arc-units or Sine of four arc-units, and BD is its Versine. The known values of BD and AD in the right triangle ABD then give the Chord AB of two arc-units, half of which is the Sine of one arc-unit. After all the Sines in this configuration have been determined, the circle is then redivided into arcs equal to a quarter-sign or $7\frac{1}{2}^{\circ}$, and new right triangles are drawn to solve for other Sines. This process continues till all twenty-four Sines are found geometrically.

In parts of his commentary that take the form of a dialogue between Bhāskara's authorial voice and a questioner (presumably a hypothetical student), Bhāskara is concerned not just to explain how to apply the algorithms but also to convince the reader that they are true and useful. These persuasions do not consist of systematic detailed demonstrations but rather employ a variety of rhetorical devices, including citations from other works, analyses of semantic validity, verification by example, and geometric reasoning. However, too much exposition is evidently felt to be a bit insulting to the student's intelligence, as when Bhāskara speaks of drawing a geometric diagram "to convince the unintelligent" (verse 2.3) or explicitly listing all the Sines "for the slow-witted to perceive" (verse 2.12).

Nor does Bhāskara analyze critically the rules in the base text that we might consider open to challenge, such as the inaccurate formulas in verses 2.6–7 for the volumes of a triangular pyramid and a sphere. Although Bhāskara correctly determines the height of the pyramid with a geometric rationale, he accepts Āryabhaṭa's volume rule for it without comment. Similarly, he agrees with Āryabhaṭa that the prescribed computation for the volume of a sphere is "without remainder" or accurate, comparing it favorably to a "practical" rule $V = \frac{9}{2}r^3$ (which is actually less inaccurate than Āryabhaṭa's rule $V = \pi\sqrt{\pi}r^3$, which Bhāskara says is exact!). We cannot assume from this, though, that Bhāskara is ignorant of the subtleties of circular measurements. For example, in concurring that Āryabhaṭa's circumference/diameter ratio $\frac{62832}{20000}$ in verse 2.10 is indeed only "approximate," he cites an opinion that a nonapproximate value for this ratio is impossible—because "*karaṇīs* (surds, or square roots of non-square numbers) do not have a statable size." He dismisses the view that the *karaṇī* $\sqrt{10}$ is an accurate value for the circumference-to-diameter ratio with the remark that "here there is only a traditional precept, not a demonstration." That he should assert the inexactness of all circumference-to-diameter ratios (even the quite accurate value 3.1416) while defending the exactness of a very inaccurate volume rule is food for thought.[15]

5.1.3 "Arithmetic" and "algebra" in the *Brāhma-sphuṭa-siddhānta*

In 628 Bhāskara's contemporary Brahmagupta included in his *Brāhma-sphuṭa-siddhānta* two separate chapters on general mathematics. The first ten chapters of this long work are devoted to the standard *siddhānta* astronomy topics outlined in section 4.2; the eleventh presents critiques of earlier works, particularly Āryabhaṭa's. The twelfth chapter contains sixty-six verses dealing with what we popularly call arithmetic, or calculations with numbers. Then, after five more chapters containing additional material on various astronomical subjects, the eighteenth has 101 verses devoted to the pulverizer, along with other rules for finding unknown quantities, including methods for second-degree indeterminate equations.[16] This division of topics between the twelfth and eighteenth chapters reflects a deliberate distinction, seen here for the first time in a surviving Sanskrit text, between what later texts call "manifest computation" and "unmanifest computation," or mathematical operations on known and unknown quantities, respectively.

Chapter 12. Verse 12.1. The introductory verse of the twelfth chapter states:

> Whoever knows separately the twenty operations beginning with addition, and the eight procedures [or practices] ending with shadows, is a mathematician ["calculator"].

[15] A more detailed exploration of such enigmas and their implications for the role of a mathematical commentator is undertaken in section 6.4.

[16] It may be that Brahmagupta's choice of sixty-six verses for his *gaṇita* chapter was meant to pay homage to (or possibly to belittle) the thirty-three verses of Āryabhaṭa's smaller one. Verses 1–13, 20–47, and 62–66 of chapter 12, along with verses 1–6, 16, 30–45, 51, 60, 62–76, and 100 (mistakenly labelled 99) of chapter 18, are translated in [Plo2007b], pp. 421–434. All the citations from the *Brāhma-sphuṭa-siddhānta* in this section are taken from [DviS1901], with reliance on the above translations and an unpublished partial translation by David Pingree (2000).

Brahmagupta is the first known mathematical author to classify arithmetic topics in this way, but he apparently expects the reader already to be familar with them, as he does not specify each "operation" and "procedure" individually. His rules in the next twelve verses, supported by the remarks of the ninth-century commentator Pṛthūdakasvāmin as well as the mathematical categories in other texts, indicate that the "twenty operations" refer to the following: addition, subtraction, multiplication, division, squaring, square root, cubing, cube root, five methods for reduction of fractions (described in verses 12.8–9), the Rule of Three Quantities, the inverse Rule of Three Quantities, four other proportion rules from the Rule of Five to the Rule of Eleven, and barter or exchange. Similarly, the order of the remaining rules in verses 12.14–54 agrees with later standard usage in identifying the "eight procedures," as follows:

1. "Mixture," or analyzing combinations of quantities;
2. Series;
3. Figures (geometry);
4. "Excavations," or volumes;
5. "Piles," or computing the size of stacked objects;
6. "Sawing," or computations concerning lumber;
7. "Heaps," or computations concerning mounds of grain; and
8. Shadows of gnomons.

However, except for the above distinction between operations and procedures, Brahmagupta does not explicitly state a topical structure for this chapter: the following subject divisions, as in the case of the *Āryabhaṭīya*'s *gaṇita* chapter, are editorial.

Verses 12.2–5. Arithmetic of fractions. Brahmagupta does not bother explaining the first six operations for integer quantities; evidently the student is expected to have learned them in the course of earlier education. He does, however, describe their use in the case of fractions, with rules corresponding respectively to the following expressions involving fractional operands $\frac{a}{b}$ and $\frac{c}{d}$ and an integer n:

$$\frac{a}{b} \pm \frac{c}{d} = \frac{ad \pm bc}{bd}, \quad n + \frac{a}{b} = \frac{nb + a}{b}, \quad \frac{a}{b} \cdot \frac{c}{d} = \frac{ac}{bd}, \quad \frac{a}{b} \div \frac{c}{d} = \frac{ad}{bc},$$

$$\left(\frac{a}{b}\right)^2 = \frac{a^2}{b^2}, \quad \sqrt{\frac{a}{b}} = \frac{\sqrt{a}}{\sqrt{b}}.$$

Verses 12.6–7. Cubing and cube root. The process of cubing an integer is described as follows:[17]

> The cube of the last [digit] is to be set down, and the square of the last [digit] multiplied by three and by the preceding [digit], and the square of the preceding [digit] multiplied by the last [digit] and by three, and the cube of the preceding [digit], [with each result shifted] from the one before it. [Their sum is] the cube.

This method seems to include a description of how the computed quantities are to be spatially arranged to produce the final answer. For example, the cube of a two-digit number $10a+b$ is to be summed up from the products a^3, $3ba^2$, $3ab^2$, and b^3, each successively displaced "from the one before it"—in other words, aligned to reflect their actual place value in the computation. We will illustrate what this method might have been intended to look like with a sample calculation of the cube of 74:

$$\begin{array}{cccccc l} 3 & 4 & 3 & & & & = 7^3 \\ & 5 & 8 & 8 & & & = 3 \cdot 4 \cdot 7^2 \\ & & 3 & 3 & 6 & & = 3 \cdot 7 \cdot 4^2 \\ & & & 6 & 4 & & = 4^3 \\ \hline 4 & 0 & 5 & 2 & 2 & 4 & = (74)^3 \end{array}$$

(Note that here the "last digit" is the most significant one, as in the verbal number representation systems discussed in section 4.3.1.) Brahmagupta's cube root procedure is similar to that in *Āryabhaṭīya* 2.5.

Verses 12.8–9. Reduction of fractions. Here Brahmagupta gives rules for simplifying what are considered to be five typical kinds of fractional quantities. They are (1) the combination of two different fractions, $\frac{a}{b} + \frac{c}{d} = \dfrac{ad+bc}{bd}$; (2) a fraction of a fraction, $\frac{a}{b} \cdot \frac{c}{d} = \dfrac{ac}{bd}$; (3) a number added to a fraction, $n + \frac{a}{b} = \dfrac{nb+a}{b}$; (4) and (5) a fraction plus or minus a fraction of itself, $\frac{a}{b} \pm \frac{a}{b} \cdot \frac{c}{d} = \dfrac{a(d \pm c)}{bd}$. These reductions are very close to the previously stated rules for basic arithmetic operations on fractions.[18]

Verses 12.10–13. Rule of Three Quantities and other proportional rules. The remainder of Brahmagupta's twenty operations concern proportions. His *trai-rāśika* rule for solving for an unknown quantity x in the proportion $a : b = x : d$ is the same as Āryabhaṭa's, $x = ad/b$. He then considers the case when the same quantities are instead inversely proportional, so that an increase in the "measure" b produces a decrease in the "fruit" a.

[17] A different interpretation of this verse that doesn't depend on the alignment of the partial results is presented in [Plo2007b], p. 422.

[18] This presentation is a variant of four standard categories of fraction reduction, in which additive and subtractive forms of Brahmagupta's type (3) are merged with types (4) and (5); see [DatSi1962], vol. 1, pp. 190–191, and [Kus2004].

(To paraphrase an example from Pṛthūdakasvāmin's commentary, if a load measured in units of size b equals a units, then how much will it equal if measured in units of size d?) In this case the unknown x is given by $x = ab/d$.

For problems in compound proportions, Brahmagupta says that the "fruit" quantities are exchanged on both sides, and the answer is "the product of the greater [number of] quantities divided by the product of the fewer." Again, this seems to imply a particular notational layout, evidently standard, which we illustrate with a typical Rule of Five example: If a piece of cloth of length l_1 and width w_1 is bought for a price a_1, what is the width x of a piece of length l_2 that is bought for a price a_2? Brahmagupta's instructions suggest the following manipulations of the given quantities, which do in fact produce the correct answer:

$$\begin{array}{ll} l_1 & l_2 \\ w_1 & \\ a_1 & a_2 \end{array} \quad \rightarrow \quad \begin{array}{ll} l_1 & l_2 \\ w_1 & \\ a_2 & a_1 \end{array} \quad \rightarrow \quad x = \frac{l_1 \cdot w_1 \cdot a_2}{l_2 \cdot a_1}.$$

The final operation involves barter or direct exchange of goods with different prices. That is, if we can buy n_1 objects of a certain kind for a price p_1 and n_2 objects of another kind for a price p_2, how many objects x of the second kind would be a fair exchange for a quantity q_1 objects of the first kind? We would express this problem algebraically as $q_1 \frac{p_1}{n_1} = x \frac{p_2}{n_2}$. Brahmagupta's equivalent rule, based on his procedures for proportions, says "first exchange the prices; the rest is as stated," which we can illustrate in the same form as the previous rule, as follows:

$$\begin{array}{ll} p_2 & p_1 \\ n_2 & n_1 \\ q_1 & \end{array} \quad \rightarrow \quad \begin{array}{ll} p_1 & p_2 \\ n_2 & n_1 \\ q_1 & \end{array} \quad \rightarrow \quad x = \frac{q_1 \cdot n_2 \cdot p_1}{n_1 \cdot p_2}.$$

Verses 12.14–16. Mixtures. A few formulas for problems on mixed quantities are presented here, beginning with investment computations. Like Āryabhaṭa, Brahmagupta does not bother stating a specific rule for computing simple interest for a given principal, rate, and time, presumably considering that to be covered by the Rule of Three Quantities. But he does deal with some more complicated situations, such as finding an unknown initial principal from the given sum of principal plus interest at the end of the given investment period, or finding the time in which a given principal invested at a given rate will increase to a given multiple of itself. He also states a more general version of the quadratic rule for interest computation mentioned in verse 2.25 of the *Āryabhaṭīya*.[19] A general problem of mixture, where participants contributing different amounts to a commercial venture receive proportional amounts of the gain or loss, is treated here as well.

Verses 12.17–20. Series. Brahmagupta's rules for the sum, mean, and number of terms in an arithmetic progression are equivalent to those discussed

[19]See the detailed discussion of these problems in [DatSi1962], vol. 1, pp. 219–223.

above in *Āryabhaṭīya* 2.19–22. He also states rules equivalent to Āryabhaṭa's for the "pile solids" $\sum_{j=1}^{n}\sum_{i=1}^{j} i$, $\sum_{i=1}^{n} i^2$, and $\sum_{i=1}^{n} i^3$.

Verses 12.21–32. Geometry: Area, sides, and diagonals of triangles and quadrilaterals.[20] Brahmagupta begins his extensive collection of rules on geometry with the following verse for the "approximate" and "accurate" areas of triangles and quadrilaterals:

> The approximate area is the product of the halves of the sums of the sides and opposite sides of a triangle or a quadrilateral. The accurate [area] is the square root from the product of the halves of the sums of the sides diminished by [each of] the four sides. (verse 12.21)

He does not give any more details about these formulas or the figures to which they apply, but we can analyze them as follows. For any quadrilateral with sides a, b, c, d and perimeter $p = a+b+c+d$, as shown in the examples in figure 5.5A–C, its area A is to be found by the formulas

$$A \approx \frac{a+c}{2} \cdot \frac{b+d}{2} \qquad \text{approximately, and}$$

$$A = \sqrt{\left(\frac{p}{2} - a\right)\left(\frac{p}{2} - b\right)\left(\frac{p}{2} - c\right)\left(\frac{p}{2} - d\right)} \qquad \text{accurately.}$$

If we set one side, say d, equal to zero, the rules can also be used for triangles with sides a, b, c, as in figure 5.5D. The approximate formula is of course exact for rectangles but not for other quadrilaterals or triangles. The accurate rule is the so-called "Heron's formula" in the case of a triangle, and its quadrilateral version is sometimes called "Brahmagupta's generalization." It is exact only for figures that are cyclic, that is, meaning that they can be inscribed in a circle (which is true for all rectangles and isosceles trapezia, as well as for some scalene quadrilaterals). Although Brahmagupta does not say so, we will see from what follows that he appears to be assuming the cyclic condition for all his rules about quadrilaterals.[21]

After this verse appear various rules equivalent to the following formulas expressed in terms of the quantities in figure 5.5:

---

[20] See [Kus1981] for details on these geometry rules in the *Brāhma-sphuṭa-siddhānta*.

[21] See the discussion in [Sara1979], pp. 88–92. This source explains geometric demonstrations (in some cases supplied by Sanskrit commentators of the second millennium) for Brahmagupta's stated results about figures, which we will generally just cite instead of reproducing.

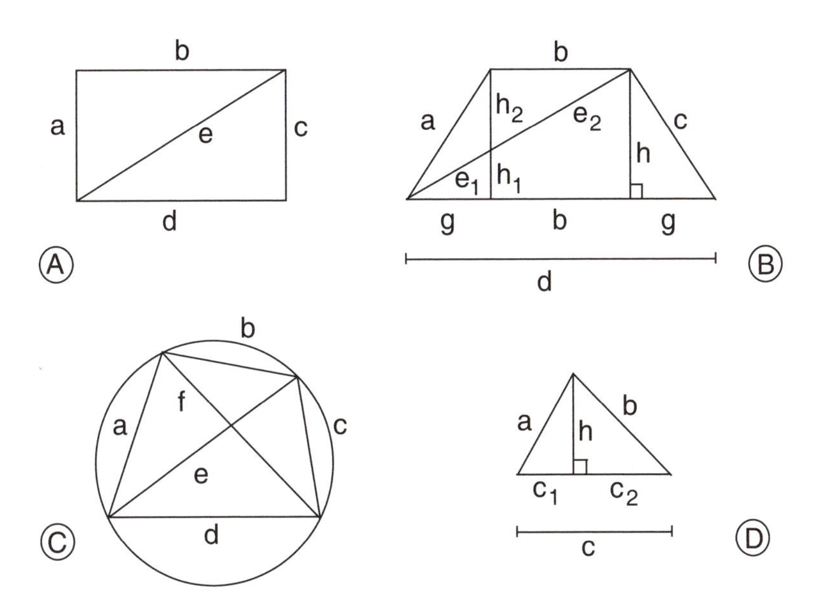

Figure 5.5 Rectilinear figures in the geometry rules of the *Brāhma-sphuṭa-siddhānta*: A. Rectangle. B. Isosceles trapezium. C. Scalene (cyclic) quadrilateral. D. Triangle.

Segments c_1 and c_2 in the base c of the triangle, and its altitude h:[22]

$$c_1 = \frac{1}{2}\left(c - \frac{b^2 - a^2}{c}\right), \quad c_2 = \frac{1}{2}\left(c + \frac{b^2 - a^2}{c}\right),$$
$$h = \sqrt{a^2 - c_1{}^2} = \sqrt{b^2 - c_2{}^2}.$$

The diagonal $e = e_1 + e_2$ and altitude h of the isosceles trapezium (literally "non-unequal quadrilateral," though the rule also holds for rectangles):[23]

$$e = \sqrt{ac + bd}, \quad h = \sqrt{e^2 - \left(\frac{b+d}{2}\right)^2}.$$

The sides of a right triangle (for example, the one with sides a, h, c_1 in figure 5.5D):

$$c_1 = \sqrt{a^2 - h^2}, \quad h = \sqrt{a^2 - c_1{}^2}, \quad a = \sqrt{h^2 + c_1{}^2}.$$

The segments h_1, h_2 of the altitude h of the trapezium and the segments e_1,

[22]These formulas can be proved using the juxtaposed right triangles with hypotenuses a and b, as shown in [Sara1979], pp. 119–120.

[23]Its applicability to squares and rectangles was noted by Brahmagupta's commentator Pṛthūdakasvāmin; see [Sara1979], pp. 71–72.

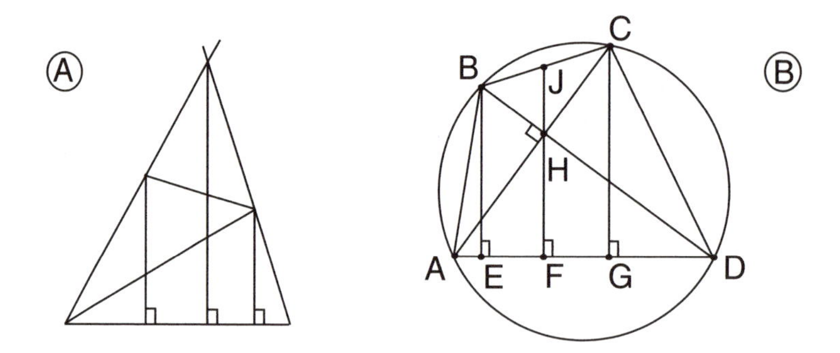

Figure 5.6 A. The "needle figure" of a quadrilateral. B. Altitude-segments of a quadrilateral in Brahmagupta's theorem.

e_2 of the diagonal e intersecting it:[24]

$$e_1 = \frac{g}{g+b} \cdot e, \quad e_2 = \frac{b}{g+b} \cdot e, \quad h_1 = \frac{g}{g+b} \cdot h, \quad h_2 = \frac{b}{g+b} \cdot h.$$

The radius r (not shown in the figure) of a circle circumscribing the trapezium, the scalene quadrilateral, and the triangle, respectively:[25]

$$r = \frac{ae}{2h} \qquad \text{(trapezium)},$$

$$r = \frac{\sqrt{a^2+c^2}}{2} = \frac{\sqrt{b^2+d^2}}{2} \qquad \text{(scalene quadrilateral)},$$

$$r = \frac{ab}{2h} \qquad \text{(triangle)}.$$

The diagonals e and f of the scalene (cyclic) quadrilateral:[26]

$$e = \sqrt{\frac{(ad+bc)(ac+bd)}{ab+cd}}, \quad f = \sqrt{\frac{(ab+cd)(ac+bd)}{ad+bc}}.$$

Brahmagupta then discusses the diagonal-segments and altitude-segments of a scalene quadrilateral and of a so-called "needle figure," a quadrilateral with two opposite sides produced until they intersect, as illustrated in figure 5.6A. In particular, he states the following rule for the segments of the altitude passing through the intersection of the diagonals of a quadrilateral, which evidently must be "Brahmaguptan" with perpendicular diagonals, as in figure 5.6B:

[24] The same verse can also be interpreted in a slightly different way to give a rule equivalent to the one for segments of the altitude in *Āryabhaṭīya* 2.8; it also applies to scalene quadrilaterals. See [Plo2007b], pp. 423–424, and [Sara1979], p. 72.

[25] See [Sara1979], p. 72, pp. 90–91, and p. 120 respectively. Note that the formula for the case of the scalene quadrilateral applies only to certain cyclic quadrilaterals, sometimes called "Brahmaguptan quadrilaterals," where the diagonals e and f are perpendicular.

[26] [Sara1979], p. 89 and pp. 111–113.

> The two lower [segments] of the two diagonals are two sides in a triangle; the base [of the quadrilateral is the base of the triangle]. Its perpendicular [altitude] is the lower portion of the [central] perpendicular; the upper portion of the [central] perpendicular is half of the sum of the [side] perpendiculars [each] diminished by the lower [portion of the central perpendicular]. (verse 12.31)

In the figure, the "lower segments of the two diagonals" AC and BD are AH and HD. The "central perpendicular" is JF and the "side perpendiculars" are BE and CG. Brahmagupta says that the upper portion JH of JF is given by

$$JH = \frac{(BE - HF) + (CG - HF)}{2}.$$

In other words, the height of J above the base AD is halfway between the heights of B and C. This implies that J bisects BC, a result now famous as "Brahmagupta's theorem" and usually stated in the following fashion: The perpendicular to any side of a Brahmaguptan quadrilateral passing through the intersection of the (orthogonal) diagonals will bisect the opposite side.

Verses 12.33–39. Geometry: Determining figures with non-surd sides from given numbers. These verses contain rules for constructing numerical examples involving rectilinear figures, based on arbitrarily chosen numbers. Although Brahmagupta does not explicitly mention it, all the formulas produce what we would call rational values for the sides, altitudes, and so forth of figures including triangles (isosceles, scalene, or right), trapezia (isosceles or "tri-equal," namely, with three equal sides), and scalene quadrilaterals. Evidently the goal is to allow the mathematician to avoid dealing with quantities in geometry problems that involve *karaṇīs*, that is, surd quantities or square roots of non-square numbers.[27]

For example, a scalene quadrilateral is said to be formed from two right triangles that share a common hypotenuse. Brahmagupta observes that such pairs of triangles can be produced by multiplying all the sides of one arbitrarily chosen right triangle by the hypotenuse of another, and vice versa. (This will produce two new triangles, each with hypotenuse equal to the product of the two original hypotenuses. Note that the resulting quadrilateral in every case will be cyclic, with one diagonal equal to the diameter of its circumscribed circle, because the juxtaposed right triangles with a shared hypotenuse can be inscribed in two equal semicircles.)

The non-*karaṇī* values for right-triangle sides to be used in such constructions are prescribed by rules equivalent to the following formulas, where m and n are arbitrarily given quantities (presumably either integer or fractional):

[27] See [Sara1979], pp. 135–141, for explanations and demonstrations of the individual rules. It is also noted there that all of them correspond to juxtaposing rational right triangles or rectangles to form the desired figures.

To determine a leg b and hypotenuse c for a right triangle with one given leg a, using an arbitrary quantity m:

$$b = \frac{1}{2}\left(\frac{a^2}{m} - m\right), \quad c = \frac{1}{2}\left(\frac{a^2}{m} + m\right).$$

To determine sides a, b, c for a right triangle, using arbitrary quantities m and n:

$$a = mn, \quad b = m + \frac{mn}{n+2}, \quad c = \sqrt{a^2 + b^2}.$$

Verses 12.40–43. Geometry: Circumference, area, and chords of a circle. Brahmagupta states rules for the "approximate" and "accurate" circumference and area of a circle equivalent to the following expressions (where C, A, and r are the circumference, area, and radius, respectively):

$$C \approx 3 \cdot 2r, \quad A \approx 3r^2 \qquad \text{approximately, and}$$
$$C = \sqrt{10(2r)^2}, \quad A = \sqrt{10r^2} \qquad \text{accurately.}$$

(Note that he uses $\sqrt{10}$ as his "accurate" value of π, although he was surely acquainted with the *Āryabhaṭīya*'s more accurate value of 62832/20000.) He then gives an expanded version of the rules for chords and "arrows" that we have already seen in *Āryabhaṭīya* 2.17–18.

Verses 12.44–46. Excavations. The fourth of Brahmagupta's eight procedures is treated very briefly. Most of it is given to his rules for the volume of an excavation with vertical sides (namely, depth times area of the base), the volume of a "needle" or pyramidal solid (one-third the depth times area of the base), and the volume of an excavation apparently corresponding to the frustrum of a square pyramid, where the top and bottom are squares of different sizes. If the depth of this last figure is h and the side of its smaller square face is s_1, while the side of the larger is s_2, then Brahmagupta's rule corresponds to the following expression for its volume V:[28]

$$V = h \cdot \left(\frac{s_1 + s_2}{2}\right)^2 + \frac{1}{3}\left(h \cdot \frac{{s_1}^2 + {s_2}^2}{2} - h \cdot \left(\frac{s_1 + s_2}{2}\right)^2\right).$$

Verse 12.47. Piles. The fifth procedure, on piles or stacks of bricks, is treated even more briefly in a single verse. The volume of a stack is said to be the area of the base times the height, or the mean of the widths at top and bottom times the height and the length.

Verses 12.48–49. Sawing. The sixth procedure lists standard labor rates for sawn rectangular planks of various kinds of timber. The rate depends on the total area of the cut surfaces and the type of wood.

[28]See [Sara1979], pp. 201–202, for the proof of its equivalence to the standard formula for the pyramidal frustrum.

Verses 12.50–51. Heaps. Similarly, the seventh procedure deals with metrology of different kinds of grain in heaps that are apparently roughly conical. The volume of a heap in all cases is said to be equal to the height times the square of one-sixth of the circumference. This is in fact equivalent to the standard formula for the volume V of a cone with height h and base of radius r, $V = h\pi r^2/3$, if π is considered to be equal to 3.

Verses 12.52–54. Shadows of gnomons. In addition to rules corresponding to the ones in *Āryabhaṭīya* 2.14–16 for finding heights and distances of light sources from their shadows, Brahmagupta includes here a rough algorithm for telling time by the sun. If the time between sunrise and sunset on a given day is d, the height of the gnomon is g, and the length of the shadow s, then the elapsed time t since sunrise (in the morning) or before sunset (in the afternoon) is given by

$$t = \frac{d}{2(s/g + 1)}.$$

This formula corresponds qualitatively although not accurately with the actual behavior of t—it will give small values when shadows are long at morning and evening, and approach $d/2$ most closely at noon. Of course, it is not exact even at noon except for the rare case when the shadow length s at noon is zero, that is, when the sun is directly overhead. On the other hand, this approximation is easier to use than the trigonometrically exact and complicated method for finding t that we examined in section 4.3.4.

Verses 12.55–65. Miscellaneous computations. Now that Brahmagupta has covered the twenty operations and eight procedures that he mentioned at the beginning of the chapter, he adds a final few verses that seem to be a collection of practical tricks and tips for facilitating basic computations of various kinds involving multiplication, division, squares, and square roots. For example, he instructs how to make the multiplication or division of two numbers, say m and n, easier by adding or subtracting some quantity p in one of them to make it round, and adjusting the computation accordingly, as follows:

$$m \cdot n = m \cdot (n \pm p) \mp mp, \qquad \frac{m}{n} = \frac{m}{n \pm p} \pm \frac{m}{n \pm p} \cdot \frac{p}{n}.$$

These manipulations also include approximate rules for squares and square roots of sexagesimal place value numbers.[29]

In the last verse of this twelfth chapter, Brahmagupta promises to "tell the rest in [the chapters on] Sine construction and the pulverizer"—namely, chapters 21 and 18, respectively. We turn first to chapter 18 on the pulverizer.

Chapter 18. Verses 18.1–2. Topics to be covered. The first two verses explain the chapter's purpose and content:

[29] See the description of these in [Plo2008a].

> Because generally [the solution of] problems cannot be known without the pulverizer, I will state the pulverizer together with problems.
>
> A master [*ācārya*] among those who know treatises [is characterized] by knowing the pulverizer, zero, negative [and] positive [quantities], unknowns, elimination of the middle [term, that is, solution of quadratics], single-color [equations, or equations in one unknown], and products of unknowns, as well as square nature [problems, that is, second-degree indeterminate equations].

Most of these methods for operating with unknown quantities, despite being classified here under the pulverizer, do not directly involve linear indeterminate equations. As promised in the first verse, their treatment includes not only solution rules but also numerous sample problems (without solutions) illustrating their application, a few of which we shall examine in the course of this section.

Verses 18.3–29. The pulverizer and its astronomical applications. The technique Brahmagupta gives for finding quantities by means of the pulverizer is essentially the same as that in *Āryabhaṭīya* 2.32–33. Most of his verses on the subject describe and illustrate how to apply it to astronomical problems. The heart of the matter is finding the time when a given planet or planets can have a particular given mean longitude. The background for Brahmagupta's rules is as follows.

Recall from section 4.3.2 that mean motions are determined by ratios of large integers representing revolutions about the earth in a period of many years, such as a *kalpa* or *yuga*. Such a period is said to contain a known number D of days, and a planet completes a known number R of revolutions in those D days, traversing in the process $360R$ degrees of longitude. (Let us assume for simplicity that longitudes are measured only to the nearest degree, ignoring arcminutes and so forth for the present.) So, in any number $d < D$ of days since the start of the period, the planet must traverse some number $k < R$ of integer revolutions plus $L < 360$ degrees of longitude, such that

$$\frac{360k + L}{d} = \frac{360R}{D} \qquad \text{or} \qquad d = (360k + L)\frac{D}{360R}.$$

A typical astronomical pulverizer arises when L is known but k and d are not: in particular, when the current weekday number $W = d \bmod 7$ is given instead of the desired day-count d. Then to find d we have to solve the linear indeterminate equation

$$d = 7x + W = (360k + L)\frac{D}{360R}$$

for k and x, where x is some integer number of weeks. Brahmagupta includes several variations on this theme, such as finding the day-count at which two or more planets have given longitudes, or finding the interval between two given weekdays on both of which a planet has the same given longitude.

Verses 18.30–35. Sign rules and arithmetic of zero. Brahmagupta gives here the first surviving explanation in Indian mathematics of the arithmetic of positive and negative numbers and zero, almost all of which are identical to their counterparts in modern algebra:

> [The sum] of two positives is positive, of two negatives negative; of a positive and a negative [the sum] is their difference; if they are equal it is zero. The sum of a negative and zero is negative, [that] of a positive and zero positive, [and that] of two zeros zero.
>
> [If] a smaller [positive] is to be subtracted from a larger positive, [the result] is positive; [if] a smaller negative from a larger negative, [the result] is negative; [if] a larger [negative or positive is to be subtracted] from a smaller [positive or negative, the algebraic sign of] their difference is reversed—negative [becomes] positive and positive negative.
>
> A negative minus zero is negative, a positive [minus zero] positive; zero [minus zero] is zero. When a positive is to be subtracted from a negative or a negative from a positive, then it is to be added.
>
> The product of a negative and a positive is negative, of two negatives positive, and of positives positive; the product of zero and a negative, of zero and a positive, or of two zeros is zero.
>
> A positive divided by a positive or a negative divided by a negative is positive; a zero divided by a zero is zero; a positive divided by a negative is negative; a negative divided by a positive is [also] negative.
>
> A negative or a positive divided by zero has that [zero] as its divisor, or zero divided by a negative or a positive [has that negative or positive as its divisor]. The square of a negative or of a positive is positive; [the square] of zero is zero. The square root of a square [is] what that [squared quantity is].

Note that a nonzero quantity divided by zero seems to remain somehow "zero-divided"; it is not clear whether the quantity is considered unchanged by the division by zero. The last sentence appears to indicate that the sign of a square's square root corresponds to that of the original quantity that was squared. Note also that the sequence of elementary operations now ends with the square root; cubing and cube root are omitted from algebra, presumably because no standard method exists for determining the cube root of an unknown quantity (i.e., there is no general solution of the cubic).

Verse 18.36. "Concurrence" and "the difference method." Here Brahmagupta defines two rules for finding quantities from their sums and products.[30]

[30]These two methods are described in [DatSi1962], vol. 2, pp. 43–44 and pp. 84–85, respectively.

If two unknown quantities a and b have known sum and difference $(a+b)$ and $(a-b)$, then the so-called "concurrence" operation gives

$$a = \frac{(a+b)+(a-b)}{2}, \qquad b = \frac{(a+b)-(a-b)}{2}.$$

Similarly, if their difference and the difference of their squares (a^2-b^2) are known, then the "difference method" prescribes

$$a = \frac{1}{2}\left(\frac{a^2-b^2}{a-b} + (a-b)\right), \qquad b = \frac{1}{2}\left(\frac{a^2-b^2}{a-b} - (a-b)\right).$$

Verses 18.37–40. Karaṇīs or surds. These are techniques for manipulating square roots of non-square numbers, or *karaṇīs*, in computations. Their purpose seems to be to minimize operating with fractional or approximate quantities by delaying as long as possible the actual extraction of the square root.[31] Brahmagupta's *karaṇī* rules may be expressed as follows in terms of non-square numbers K_1, K_2, K_3, K_4, and quantities a, b, m, where m is chosen such that K_1/m and K_2/m are both perfect squares:

$$\begin{aligned} a\sqrt{K_1}\cdot b\sqrt{K_1} &= abK_1, \\ \sqrt{K_1}\pm\sqrt{K_2} &= \sqrt{m\left(\sqrt{\frac{K_1}{m}}\pm\sqrt{\frac{K_2}{m}}\right)^2}, \\ \sqrt{K_1}\cdot(\sqrt{K_2}+K_3) &= \sqrt{K_1K_2}+\sqrt{K_1K_3}, \\ \frac{\sqrt{K_1}+\sqrt{K_2}}{\sqrt{K_3}+\sqrt{K_4}} &= \frac{(\sqrt{K_1}+\sqrt{K_2})(\sqrt{K_3}-\sqrt{K_4})}{(\sqrt{K_3}+\sqrt{K_4})(\sqrt{K_3}-\sqrt{K_4})} \\ &= \frac{(\sqrt{K_1}+\sqrt{K_2})(\sqrt{K_3}-\sqrt{K_4})}{K_3-K_4}. \end{aligned}$$

Verses 18.41–42. Combining terms. About to discuss the process of determining unknown quantities—that is, solving equations—Brahmagupta lays the groundwork by stating some useful facts about combining their terms. Like powers of an unknown (Brahmagupta mentions powers up to the sixth) may be added or subtracted, while unlike ones must be taken separately. Products in one unknown produce a higher power of it, while the product of different unknowns or "colors" is a separate term with its own name, the "[color]-product."

Verses 18.43–59. Techniques and examples for solving equations in one unknown. The text now turns to the topics described in the second verse of the chapter as "single-color" and "elimination of the middle [term]": that

[31] The nature and use of *karaṇī* or surd quantities in Indian mathematics is discussed in detail in, for example, [ChKe2002], [Jai2001], ch. 5, pp. 118–154, and [PatF2005].

is, rules for finding the value of an unknown quantity or "color" when it is combined with known quantities in various ways. Brahmagupta, like Bhāskara in his commentary on the *Āryabhaṭīya*, uses Sanskrit words meaning "equal" or "equal-making" to refer to such combinations (for instance, in verses 18.43 and 18.63), which we will just translate as "equation."

The first method is equivalent to the technique for finding the value of a "bead" in *Āryabhaṭīya* 2.30: essentially, when b_1 unknowns plus c_1 units equals b_2 unknowns plus c_2 units, then the value x of the unknown is given by $x = \frac{c_2 - c_1}{b_1 - b_2}$. Next, when some multiple b of the unknown added to some multiple a of the square of the unknown equals a known number c, then the value x of the unknown is found by algorithms equivalent to the following:

$$x = \frac{\sqrt{4ac + b^2} - b}{2a} \quad \text{or} \quad x = \frac{\sqrt{ac + (b/2)^2} - b/2}{a}.$$

What we call b here is named by Brahmagupta the "middle [term]," from which the procedure "elimination of the middle [term]" takes its name. This nomenclature implies that there was a standard notation for laying out equations, with the multiplier of the unknown always placed in the middle between the multiplier of the square of the unknown and the known number, somewhat analogous to our $ax^2 + bx = c$. The solution process removed the middle term, leaving one unknown equal to an expression in terms only of known quantities.[32]

The possibility of having two solutions according to the sign of the square root in the quadratic algorithms is not explicitly mentioned.[33] If there are different unknowns in separate terms of an equation, Brahmagupta prescribes (verse 18.51) solving for the first in terms of the others. If there are too many unknowns—that is, if the problem is indeterminate—the pulverizer is necessary.

A typical instance of his sample problems is the following:

> The square root of the remainder of the revolutions of the sun minus two [is] decreased by one, multiplied by ten, [and] increased by two. When does [the result] become [equal to] the remainder of the revolutions minus one on a Wednesday? (verse 18.49)

In other words, we first have to state and solve a quadratic equation for the sun's "remainder of revolutions," that is, its longitude L. A later commentator works the problem by setting the longitude equal to the square of what we might call a "dummy" unknown plus two, or, as we would write it, $L = x^2 + 2$. The resulting quadratic equation is equivalent to

$$0x^2 + 10x - 8 = x^2 + 0x + 1.$$

---

[32] See [DatSi1962], vol. 2, pp. 69–70.

[33] But it has been argued that some of Brahmagupta's sample problems indicate that he knew that either the positive or the negative square root could be used; see [DatSi1962], vol. 2, pp. 74–75.

By an "elimination of the middle" rule we find $x = 9$ (ignoring the negative square root), and hence $L = 83$.[34] Then we would have to use the pulverizer to find a day-count d that produces this value of L on a Wednesday, when $d = 3 \bmod 7$. Astronomically speaking, the problem seems quite artificial: one might want to know when the sun would have a particular longitude on a particular weekday, perhaps for astrological purposes, but there is no reason that the longitude would need to be extracted from a quadratic equation. The rest of the sample problems combine the use of the pulverizer with these equation-solving techniques in similar ways; several of them are embellished with the encouraging comment, "[Whoever] computes [the answer] within a year is a mathematician."

Verses 18.60–63. Solving an equation in two unknowns and their product. The previous "single-color" methods will not work for equations that combine more than one unknown or "color" in the same term. If a problem contains two unknowns and their product, say $axy = bx + cy + d$, then the rule for solution can be paraphrased by the following: Let an arbitrary number m equal $ax - c$, and let $n = \frac{ad + bc}{m}$. Then also $n = ay - b$, which can be shown by substituting $d = axy - bx - cy$. So $y = \frac{n + b}{a}$ and $x = \frac{m + c}{a}$. Or else, Brahmagupta notes, one can simply assume all but one of the unknowns equal to arbitrary numbers, and then solve for the remaining unknown.

Verses 18.64–71. "Square-nature," or second-degree indeterminate equations. The Sanskrit name of this technique, *varga-prakṛti*, means "nature (or principle) of the square"; it may also involve a pun on *prakṛti* "nature" and *kṛti* "square." These square-nature problems (of which Brahmagupta's text preserves the first record) require finding two (rational) quantities whose squares satisfy certain conditions. In particular, for a given multiplier N, it is desired to find (rational) numbers x and y such that $Nx^2 + 1 = y^2$ (a form of the so-called Pell's equation). As in the case of the linear pulverizer, the solution is found essentially by changing the given equation into another of the same form that we can solve trivially, and then manipulating the values in the latter equation to fit the former.

To find the required square roots x and y, Brahmagupta tells us to choose an arbitrary "desired square," say a^2, such that when it is multiplied by the given number N and increased or decreased by an arbitrary "desired number," say k, it produces another perfect square, say b^2. In other words, we construct an auxiliary second-degree indeterminate equation $Na^2 + k = b^2$, where we can find a solution a, k, and b just by inspection. Then we form a second auxiliary equation as follows:

$$N(2ab)^2 + k^2 = (Na^2 + b^2)^2,$$

and the unknown x and y are derived by an algorithm corresponding to the

[34] See [Col1817], pp. 346–347, and [DatSi1962], vol. 2, p. 75.

following:

$$x = \frac{2ab}{k}, \qquad y = \frac{Na^2 + b^2}{k}.$$

We can demonstrate the validity of Brahmagupta's method algebraically by manipulating our first auxiliary equation, $Na^2 + k = b^2$:

$$\begin{aligned} k &= Na^2 - b^2, \\ k^2 &= N^2a^4 - 2Na^2b^2 + b^4, \\ 4Na^2b^2 + k^2 &= N^2a^4 + 2Na^2b^2 + b^4, \\ N\left(\frac{2ab}{k}\right)^2 + 1 &= \left(\frac{Na^2 + b^2}{k}\right)^2. \end{aligned}$$

So these expressions for x and y are indeed a solution to the given equation $Nx^2 + 1 = y^2$.

The same rule can also be read in a more general sense as a method for "composing" new square-nature equations out of two different existing ones. For if $N{a_1}^2 + k_1 = {b_1}^2$, and $N{a_2}^2 + k_2 = {b_2}^2$, then[35]

$$N(a_1b_2 + a_2b_1)^2 + k_1k_2 = (Na_1a_2 + b_1b_2)^2.$$

The rest of Brahmagupta's rules for manipulating square-nature equations can be briefly expressed as follows:[36]

If $Na^2 + 4 = b^2$, then

$$N\left(\frac{a(b^2-1)}{2}\right)^2 + 1 = \left(\frac{b(b^2-3)}{2}\right)^2.$$

If $Na^2 - 4 = b^2$, then

$$N\left(\frac{ab(b^2+3)(b^2+1)}{2}\right)^2 + 1 = \left((b^2+2)\left(\frac{(b^2+3)(b^2+1)}{2} - 1\right)\right)^2.$$

If $N = n^2$ for some n, k is given, and m is an arbitrarily chosen number, then

$$N\left(\frac{1}{2n}\left(\frac{k}{m} - m\right)\right)^2 + k = \left(\frac{1}{2}\left(\frac{k}{m} + m\right)\right)^2.$$

If $Na^2 + k = b^2$ and N is divisible by some square n^2, then

$$\frac{N}{n^2} \cdot (na)^2 + k = b^2.$$

[35]For proofs of this and Brahmagupta's other square-nature rules, along with a discussion of the terminology, see [DatSi1962], vol. 2, pp. 141–161.

[36]The later development of the so-called "cyclic method" for solving arbitrary square-nature problems is discussed in section 6.2.2.

If $Na^2 + k = b^2$ and $h = kn^2$ for some square n^2, then

$$N(na)^2 + h = (nb)^2.$$

If for given N_1 and N_2 a solution x, y, m is sought such that $N_1 m + 1 = x^2$, $N_2 m + 1 = y^2$, then

$$m = \frac{8(N_1 + N_2)}{(N_1 - N_2)^2}, \quad x = \frac{3N_1 + N_2}{N_1 - N_2}, \quad y = \frac{3N_2 + N_1}{N_1 - N_2}.$$

Verses 18.72–74. Other methods of producing squares. Now Brahmagupta turns to techniques for finding square numbers that satisfy other conditions, which we can paraphrase as follows:

If x and y are sought such that $(x+y)$, $(x-y)$, and $(xy+1)$ are all squares, then if a and b are two arbitrarily chosen quantities, we can put

$$x = (a^2 + b^2) \cdot \frac{2a^2}{b^4}, \quad y = (a^2 - b^2) \cdot \frac{2a^2}{b^4}.$$

If for given quantities a and b some x is sought such that $(x + a)$ and $(x - b)$ are both squares, and if m is an arbitrarily chosen quantity, then we can put

$$x = \left(\frac{1}{2}\left(\frac{a+b}{m} - m\right)\right)^2 + b.$$

Verses 18.75–98. Sample problems and applications. Brahmagupta closes the technical part of the chapter with a variety of sample problems for applying indeterminate methods to astronomical situations, such as the following:

> [Whoever] computes within a year a given remainder of revolutions diminished by ninety-two, multiplied by eighty-three, and increased by one [so that it is] a square, is a mathematician. (verse 18.79)

Again, these generally seem unrealistic as instances of practical problems, especially those that require the computation of longitudes or parts of longitudes that are perfect squares.

In the closing verses of this chapter, however, Brahmagupta stresses the importance of mastering such test questions: "With what is thus stated, [a mathematician] may solve the questions given in other works. In the assemblies of the people he will destroy the brilliance of [other] astronomers as the sun [destroys that] of the stars" (18.100–101). Evidently these puzzles were useful at least for professional prestige, as a badge of mastery in a difficult subject.

The section on Sine constructions. Brahmagupta mentioned at the close of the twelfth chapter, on arithmetic, that the rest of calculation would be explained not only in the discussion of the pulverizer but also in that of Sine constructions. The latter topic is dealt with in verses 17–23 of the twenty-first chapter, which treats the sphere. Verse 16 of the same chapter refers to the *Brāhma-sphuṭa-siddhānta*'s unique trigonometric radius $R = 3270$, whose mysterious value was noted above in section 4.3.3:

> Because the Radius of the minutes of a rotation is [expressed] in minutes but not with seconds, the Sines are also not accurate. Therefore another Radius was made [by me].[37]

This suggests that Brahmagupta chose $R = 3270$ because he thought it produced more accurate Sines than those arising from other values of R known to him, which presumably were all approximations derived from $360 \cdot 60/2\pi$ for some value of π. The implication seems to be that an independently chosen integer R value is expected to be more precise. However, Brahmagupta does not explain why he selected the particular value 3270, nor are his (integer) Sine values significantly more accurate than those of Āryabhaṭa.

The Sine constructions in verses 21.17–23 are concise directions for determining Sine values geometrically. Like Bhāskara in his commentary on the same topic in *Āryabhaṭīya* 2.11, Brahmagupta solves various right triangles in the circle by the Pythagorean theorem to get new Sine quantities from known ones.

5.2 THE BAKHSHĀLĪ MANUSCRIPT

The surviving independent treatises on *gaṇita* as a topic in its own right, rather than a subdivision of astronomy, start to make their appearance (probably) shortly after the time of Brahmagupta. The first of these treatises, one of the most famous medieval Sanskrit mathematical works, is also by many centuries the earliest surviving Sanskrit mathematical manuscript. This is the text known as the Bakhshālī Manuscript, which had been buried in a field about eighty kilometers from modern Peshawar in Pakistan, a major trading center in ancient times. It was found by a farmer in 1881 and acquired by the Indologist A.F.R. Hoernle, who presented it to the Bodleian Library at Oxford in 1902, where it resides to this day.[38]

Although the Bakhshālī Manuscript is undated, its approximate age (within a few centuries, at least) can be tentatively deduced from a number of factors. In the first place, it is written on birch bark rather than paper, which came into wide use in northern India in the second millennium. Second, the script it employs, called Śāradā, is a descendant of a North Indian script used under the Gupta emperors,[39] and appears to have been in use in northwest India since at least the eighth century. The version of Śāradā script appearing in the Bakhshālī Manuscript most closely corresponds to those seen in inscriptions from the eleventh and twelfth centuries; since epigraphic script styles tend to be conservative, it seems reasonable to say that the Bakhshālī Manuscript itself might have been written as early as the eighth century

---

[37] *Brāhma-sphuṭa-siddhānta* 21.16; see [Ike2002], pp. 203–204.

[38] The Bakhshālī Manuscript is listed in the Bodleian collection as MS. Sansk.d.14. The description of it in these paragraphs is based on the study in [Hay1995], which contains an edition, translation, and analysis of the available part of the manuscript. A few of the translated rules and explanations have been reproduced in [Plo2007b], pp. 436–441.

[39] See [DanA1986], pp. 110–111.

but no later than the twelfth. The language of the Bakhshālī Manuscript fits in with this hypothesis, being a form of Classical Sanskrit somewhat modified by dialect features corresponding to medieval north-west Indian vernaculars.[40]

Because of the deterioration and disorder of the birch bark leaves of the manuscript, it is impossible to reconstruct with certainty exactly what the text originally looked like. It was evidently a collection of several dozen algorithms and sample problems in verse, with a commentary explaining them in a combination of prose and numerical notation. It may have been a single author's treatise or a compilation of rules and problems from multiple sources. In either case, at least some part of the content is doubtless older than the manuscript itself, although it is difficult to estimate how much older. The date of the commentary, based on similarities to the style and vocabulary of other medieval mathematical works, has been estimated around the seventh century.[41]

The colophon of the Bakhshālī Manuscript states that it was written by a Brāhmaṇa identified only as the son of Chajaka, a "king of calculators," for the use of Vasiṣṭha's son Hasika and his descendants. Nothing further is known about any of these people. Chajaka's son was perhaps the author of the commentary, or some part of it, as well as the scribe of the manuscript.

5.2.1 Mathematical notation

The Bakhshālī Manuscript is unique and extremely precious as a source of direct information about how medieval Sanskrit mathematics was actually written. Unsurprisingly, numerals are invariably in decimal place value form, with zero represented by a round dot. The following paragraphs describe the major notational and graphical features of the text, most of which are also used in mathematical manuscripts from later centuries. Several of these characteristics can be seen below in the sample page reproduced and transcribed in figure 5.7, which is translated and analyzed in section 5.2.2.

Displayed notation. Quantities expressed in numerals are generally set off from text in the manuscript by boxes that may extend vertically through two or more lines of text. The scribe sometimes starts a new line under the bottom of a box in the previous line, or sometimes starts level with the dangling box and just skips over it, frequently breaking a word between syllables in the process. The result is a pretty good compromise between maintaining readability and saving space.

Fractions. Fractional quantities are written vertically with the numerator above the denominator, but no horizontal line between them; the numerator

[40] [Hay1995], pp. 23–55.

[41] [Hay1995], pp. 148–150. Other historians (for instance, [Rao2004], p. 5) follow the estimate of Hoernle published in the late nineteenth century, dating the work as early as the third century CE. Hoernle's estimate, however, was based on chronological assumptions about the text's language and verse meter that have since been challenged; see [Hay1995], pp. 53, 56.

may be larger than the denominator. An integer to be combined with a fraction, as in $3\,\frac{3}{8}$, is written above the fraction's numerator.

Negative numbers. A negative quantity is indicated by a small cross figure following it, somewhat resembling the modern plus sign. It was probably originally an abbreviation of the Sanskrit word *ṛṇa*, or "negative," whose initial letter *ṛ* in some early script forms is roughly cross-shaped.[42] In later manuscripts, however, and in modern printed editions, a negative number is usually indicated instead by a dot above it (see section 6.2.2).

Abbreviations for operations and quantities. Words meaning "added," "subtracted," "result," "remainder," and so forth are sometimes abbreviated to their initial syllables, especially when they accompany numerical quantities in boxes. The names or units of unknown quantities are also routinely abbreviated in this way.

"Proto-symbolism," or problems in tabular form. The combination of a tabular format for displaying numerical quantities with accompanying abbreviations for operations and unknowns creates what might be called a sort of "syncopated algebra" or "proto-symbolism" for the layout of problems. An example of this is seen in a problem about finding a number that when increased and decreased by given numbers produces perfect squares (a topic we saw earlier in *Brāhma-sphuṭa-siddhānta* 18.72–74). The problem states:[43]

> What quantity increased by five gives a square root? That quantity decreased by seven gives a square root; what is that quantity?

In its graphical layout, zeros or dots are used to represent unknown quantities and abbreviations (*yu* for *yuta*, "added," and *mū* for *mūla*, "square root") are used for the operations. In roman transcription it looks like this:

0	5	yu	mū	0	sā	0	7	+	mū	0
1	1			1		1	1			

We can translate this notation both literally and mathematically as follows:

$$\frac{0}{1}\ \frac{5}{1}\ \text{plus}\ \ \text{root}\ \frac{0}{1}\ \ \|\ \ \text{that}\ \frac{0}{1}\ \frac{7}{1}\ \text{minus}\ \ \text{root}\ \ 0$$

$$x + 5 = y^2, \qquad\qquad x - 7 = z^2.$$

(The prescribed solution procedure is a variant of the one given by Brahmagupta and yields $x = 11$, $y = 4$, $z = 2$.)

Figures. The surviving part of the Bakhshālī Manuscript contains little geometry and hence hardly any figures; fragments of simply sketched shapes, labeled with their numerical dimensions, can be seen on two leaves.[44]

---

[42] [Hay1995], pp. 88–89.
[43] Bakhshālī Manuscript f.59r; see [Hay1995], pp. 235, 333, 576.
[44] See [Hay1995], pp. 319, 322.

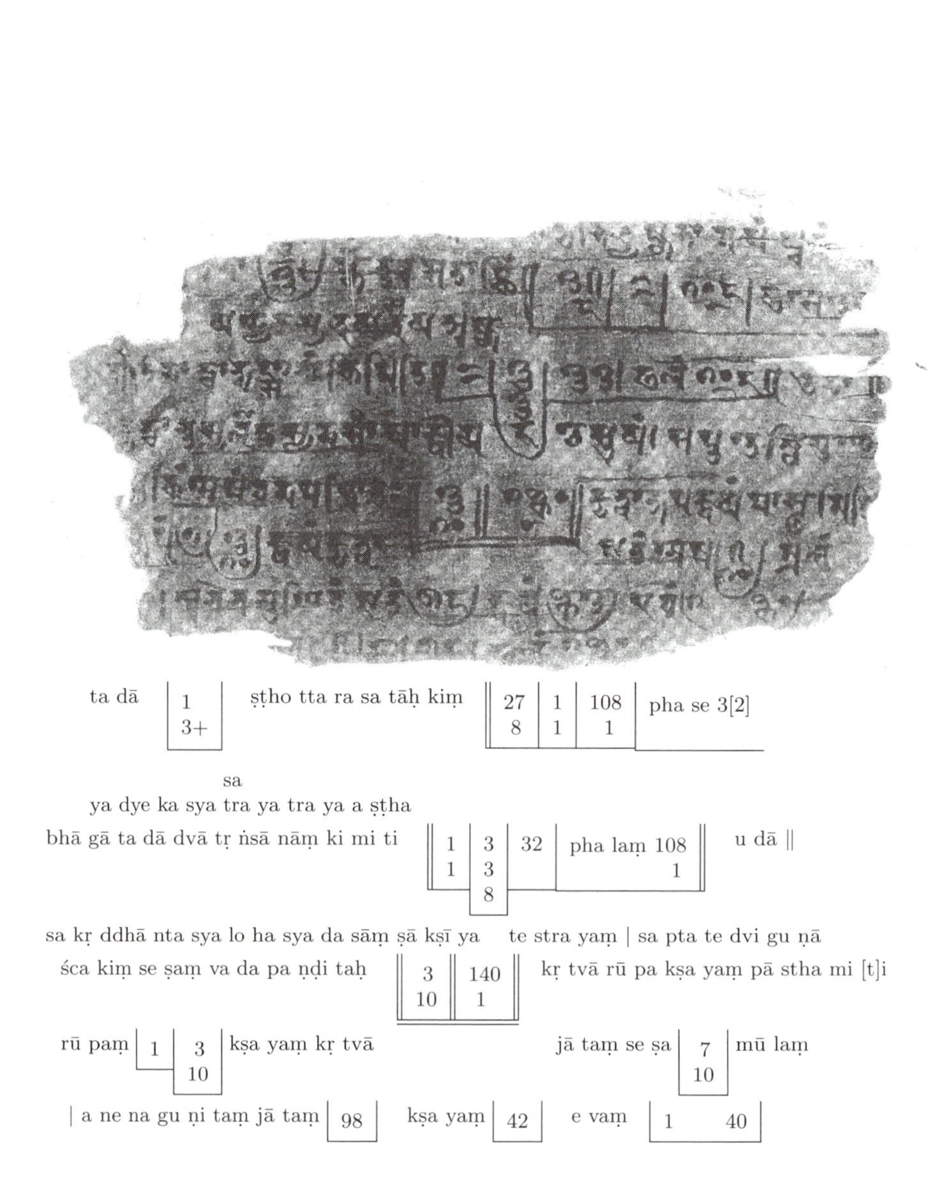

Figure 5.7 A page from the Bakhshālī Manuscript and transcription of the legible parts of its central seven lines, with reconstructed text in square brackets.

5.2.2 Mathematical content

The computational rules in the Bakhshālī Manuscript are not arranged in any topical order, at least as far as we can tell from the manuscript as it survives, but rather consist of an unsystematic collection of problem-solving procedures. Many of the verse rules are somewhat specialized applications of arithmetic to particular problems such as barter, composition of alloys, and the like, rather than prescriptions for fundamental operations. (A few verses explaining elementary procedures such as multiplication and division of fractions and addition and subtraction of negative numbers are cited in the commentary as it works through problems that require their use.) Various forms of the Rule of Three Quantities are used to perform checks on the results of the computations.

The problems contained in the manuscript page reproduced in figure 5.7 concern a typical rule for computing the loss in weight of a quantity of impure metal during the process of refining it. If the impure metal loses some fraction $\frac{a}{b}$ of its total weight during each stage of the refining, then the final amount of pure metal after the nth stage will be the original weight times $\left(1-\frac{a}{b}\right)^n$. In the words of the text on the previous page of the manuscript, "Having subtracted the lost [fraction] from unity at [every] smelting, [and found] the product [of the remainders from subtraction], then one should indicate the result of the original quantity times [that] product [as the desired] residue."[45]

A sample problem is then stated where the original weight is 108 units and one-third of the metal is lost as slag each time it is melted; after three repetitions of the process, the metal is pure. The commentary correctly calculates the final weight as 32 units and verifies it in various ways by the rule of three quantities. Since the ratio of the initial weight to the final weight is $\frac{27}{8}$: 1, the text goes on to say (translating from the start of the above transcription of the reproduced page):

> ... then [the fraction $-\frac{1}{3}$ intrudes from the broken-off line above] for a hundred and eight, what [is obtained]? $\frac{27}{8}$, $\frac{1}{1}$, $\frac{108}{1}$, result: remainder 3[2] ... If for one, three [and] three-eighths, then for thirty-two, what? $\frac{1}{1}$, $3\,\frac{3}{8}$, 32, result: $\frac{108}{1}$.
>
> [Another] example: Of a smelted [amount of] iron, three tenths are lost. Seventy times two [units produces] what remainder? Say, oh learned one! $\frac{3}{10}$, $\frac{140}{1}$. "Having subtracted the lost [fraction] from unity":... Unity 1. $\frac{3}{10}$, having subtracted, the

[45] [Hay1995], pp. 189 (text), 299 (translation).

resulting remainder $\frac{7}{10}$ [is] produced.... Multiplied by that, 98. Subtracted, 42. Thus, 140.

The second example assumes a loss of three-tenths of the initial weight of 140 units in the smelting process: $\left(1 - \frac{3}{10}\right) \cdot 140 = 98$.

5.3 THE *GAṆITA-SĀRA-SAṄGRAHA*

This treatise, the "Epitome of the essence of calculation," is the work of a medieval namesake of Mahāvīra, the founder of Jainism, who was evidently himself a Jain, judging from the invocations in many of his verses. The *Gaṇita-sāra-saṅgraha*'s lavish praise of the ninth-century king Amoghavarṣa, a member of the Rashtrakuta dynasty in southern India, suggests that Mahāvīra may have worked at Amoghavarṣa's court, so his work is generally dated to about the middle of the ninth century. The *Gaṇita-sāra-saṅgraha* is the first known independent mathematical treatise in Sanskrit to survive in its entirety, and it contains more than 1100 verses; thus it supplies an unprecedented amount of explicit detail about the organization of mathematical topics. However, some of this structure appears rather idiosyncratic, varying in several respects from the classifications of mathematics hinted at by the content and arrangement of earlier works.

After an unnumbered introductory chapter, Mahāvīra divides the rest of his book into eight numbered chapters on individual mathematical subjects called "procedures," although most of them do not correspond in content to what Brahmagupta identified as the "eight procedures" in the *Brāhma-sphuṭa-siddhānta*. The chapters consist of verses containing both rules and sample problems (without solutions), with occasional prose sentences linking them. Their content is briefly outlined in the remainder of this section, although space does not permit working through their verses or groups of verses individually, as we tried to do with some of the earlier and shorter works.[46]

[*Chapter 0*]: *"Terminology Chapter." Preface and technical terms.* This introductory part begins with an invocation of the author's renowned namesake, the first Mahāvīra, "the lamp of the knowledge of numbers by whom the whole universe is enlightened." Next come the verses praising Amoghavarṣa and then Mahāvīra's frequently quoted eulogy on *gaṇita*, acclaiming it as applicable in all fields of practical or religious relevance, including love, commerce, music, theater, cooking, medicine, building, prosody, rhetorical orna-

[46] All the following citations of the *Gaṇita-sāra-saṅgraha* are taken from the edition in [Ran1912]. Verses 1.1–2, 1.9–28, 1.46–52, 2.61, 2.65, 2.69, 2.72, 2.93, 2.99, 4.4, 4.23–27, 4.29–30, 4.40–45, and 6.216–221 of the accompanying English translation in [Ran1912] have been reproduced in [Plo2007b], pp. 441–447. Note that the chapter numbers in this translation, which begin with chapter 1, are off by one from Mahāvīra's own arrangement, which begins with an unnumbered section that we call here "chapter 0."

mentation, poetry, logical argument, grammar, astronomy, and cosmology. The knowledge of numbers is likened in an elaborate metaphor to an ocean whose water, shores, fish, shark, waves, jewels, bottom, sands, and tide represent the subjects treated in the remaining eight chapters, respectively.

Moving on to technical terms and basic metrology, since "it is not possible to understand the meaning of anything without its name," Mahāvīra identifies various units of measure for amount and area, for time, for grain, for gold, for silver, and for iron or base metals in general; the last four categories are all various kinds of weights. The smallest of the given units in space and time are evidently infinitesimals, and bring to mind the interest in quantification of the very small and very large expressed in earlier Jain works mentioned in section 3.4. The minimum possible length is considered to be the size of an indestructible quantity of substance, or "ultimate atom." The smallest finite quantity, the "atom," is defined as an infinite number of those "ultimate atoms," and $8^7 = 2{,}097{,}152$ atoms make a digit. The smallest possible amount of time, or instant, is that required for one atom to move past another, and the smallest finite time unit is defined as innumerable instants. All the weight units, on the other hand, are finite amounts in ordinary metrology.

Mahāvīra then names the eight fundamental arithmetic operations, which surprisingly do not include addition and subtraction; the first six cover multiplication through cube root, while the last two involve adding and subtracting quantities in progression, or series. We have seen in earlier works that a mathematical author generally would not bother to explain the details of elementary addition and subtraction, but it seems quite daring thus to cast them out of the classificatory structure of mathematics altogether. After that, the arithmetic of positive and negative quantities and zero is explained, beginning with the following statement:

> A quantity multiplied by zero is zero; it is unchanged [if] divided, increased, or decreased by zero. A product etc. with zero is zero. In addition, zero [takes] the form of the addend. (verse 0.49)

The rules for adding, subtracting and multiplying positive and negative quantities are similar to Brahmagupta's in chapter 18 of the *Brāhma-sphuṭa-siddhānta*. Squares and square roots are described as follows:

> The square of a positive or negative is positive; the square roots of those two [squares] are positive and negative respectively. Because a negative quantity, on account of its own form, is non-square, therefore [there is] no square root of that. (verse 0.52)

The last sentence seems to mean that because no two equal quantities can have a negative product, negative quantities can have no square root.

Mahāvīra ends this preface with names for twenty-four successive decimal places (up to 10^{23}) and a description of the desired attributes of a mathematician:

> A mathematician ["calculator"] is to be known by eight qualities: swift work, deliberation, refutation, non-idleness, comprehension, concentration [or memory], inventiveness, and having the answers. (verse 0.69)

Chapter 1: "Procedure of Operations." Mahāvīra begins this chapter with a description of multiplication. He uses distinct Sanskrit terms for the "multiplier" and the "to-be-multiplied" (multiplicand), which are to be set down with the so-called "door-hinge method"—evidently, lining up their corresponding digits vertically. The ultimate result comes out as a horizontal string of digits, as illustrated by this sample problem:

> Setting down [a concrete number expression representing] 142,857,143 and multiplying it by seven, call that the "king's necklace." (verse 1.13)

(The reader will find the reason for the name immediately obvious from the calculation.)

The subsequent rules for division, squaring, square root, cubing, and cube root include the fundamental methods we have seen in the *Āryabhaṭīya* and *Brāhma-sphuṭa-siddhānta.* They also encompass a variety of other techniques involving spatial manipulation, factoring, and so forth, of the operands. These techniques are generally just referred to by name without explanation, as in "the rule of division of equals" for removing common factors in a dividend and divisor, or "value-part" for a factor that may be transferred from a multiplier to a multiplicand or vice versa. Evidently such concepts would have been familiar to students before they tackled the higher mathematics in the *Gaṇita-sāra-saṅgraha.*

Mahāvīra's operations on series likewise are rooted in the same basic rules given in the texts we investigate earlier, but arranged and amplified in a novel manner. For example, the rules for quantities such as sums and means of arithmetic progressions in *Āryabhaṭīya* 2.19–22 could be applied either to an entire sequence starting with its first term, or to the final part of it ending with its last term. Mahāvīra, on the other hand, breaks up these techniques into two distinct operations, identified as addition and subtraction with respect to series. Summing the terms of a sequence beginning with the first is considered to be addition. But computing the "remainder-series"—summing up a subset of terms ending with the last—is called subtraction, as it consists of finding the difference between the sum of the entire series and the sum of its first part.

Not only arithmetic progressions but also so-called "multiplication summations," or series in geometric progression, are dealt with here. The sum s of a geometric series with n terms, a constant ratio r, and first term a_1 is defined by a rule equivalent to the following:

$$s = \frac{a_1 r^n - a_1}{r - 1}.$$

Another method is stated that apparently originates in the ancient prosody rule we saw in discussing the work of Piṅgala in section 3.3, where a succession of squaring and doubling operations was used to calculate 2^n, the number of possible poetic meters that can be constructed from n syllables that can be of two different kinds. Mahāvīra presents a more general version of this technique to compute r^n in the above rule, which we will illustrate in working one of Mahāvīra's sample problems:

> What is the wealth of a merchant [represented by a geometric series] whose first term is seven, [constant] multiplier three, and number of terms the square of three? (verse 1.100)

Given $n = 9$ as stated, we find $r^n = 3^9$ by the squaring-and-multiplying method, as follows: Since 9 is odd, subtract 1 and write 1; since $9 - 1 = 8$ is even, halve it and write 0; since $8 \div 2 = 4$ is even, halve again and write 0; for $4 \div 2 = 2$ write 0, for $2 \div 2 = 1$ write 1, and stop. Then work back through the sequence from the end to the beginning, multiplying by $r = 3$ at every step marked with a 1 and squaring at every step marked with a 0, as illustrated in the worked procedure below.[47]

$$\begin{array}{ccccccccc} 9 & \rightarrow & 8 & \rightarrow & 4 & \rightarrow & 2 & \rightarrow & 1 \\ 1 & & 0 & & 0 & & 0 & & 1 \\ 3 \cdot (81)^2 & \leftarrow & (81)^2 & \leftarrow & 9^2 & \leftarrow & 3^2 & \leftarrow & 3 \cdot 1 \end{array}$$

We end up with $3^9 = 3 \cdot (81)^2 = 19{,}683$, and from the above formula the sum s is 68,887.

Chapter 2: "Procedure of Fractions." Mahāvīra's first six operations on fractions, from multiplication up to cube root, are treated essentially as in earlier texts. To make his seventh and eighth operations of series summation and series subtraction apply to fractions, he discusses series involving not only fractional terms and fractional constant differences but even fractional values of n for their numbers of terms. Exactly what is meant by the idea of a fractional number of terms is not explained; Mahāvīra's sample problems in this section just state the given quantities numerically instead of setting them up in word problems that might give clues about their conceptual meaning. Most of the same rules apply to these fractional series as in the case of integer series, and in fact some of the same verses are reused.[48] Geometric series with fractional terms and constant ratios are also dealt with, but they use integer numbers of terms rather than fractional ones.

The remainder of this chapter discusses six kinds of manipulations or

---

[47]Note that this layout of the successively computed numbers is an editorial invention; Mahāvīra's verse (1.94) gives no explicit instructions on the details of writing down the steps, although he does specify that even quantities are marked with a 0 and odd ones with a 1.

[48]Compare, for example, 1.69 and 2.33, or 1.93 and 2.40. See [Sara1979], p. 239, for a reference to such series in other works.

simplifications of fractions.[49] The first four of them are equivalent to the methods enumerated by Brahmagupta in *Brāhma-sphuṭa-siddhānta* 12.8–9, namely: adding fractions with different denominators, taking fractions of fractions (multiplying fractions), and increasing or decreasing an integer or fraction by a fraction of itself. The fifth deals with division by a fraction and the sixth, called "mother of fractions," involves the combination of any or all of the first five, as in the following sample problem:

> A third; a fourth; a half of a half; a sixth of a fifth; one divided by three fourths; one divided by five halves; one and a sixth; one and a fifth; a half plus its own third; two sevenths plus its own sixth; one minus a ninth; one minus a tenth; an eighth minus its own ninth; a fourth minus its own fifth. Adding [all] these in a rule [like] a lotus-garland of types of fractions, dear one, say [the answer]. (verses 2.139–140)

That is, the student is asked to compute

$$\frac{1}{3}+\frac{1}{4}+\frac{1}{2}\cdot\frac{1}{2}+\frac{1}{5}\cdot\frac{1}{6}+\frac{1}{3/4}+\frac{1}{5/2}+\left(1+\frac{1}{6}\right)+\left(1+\frac{1}{5}\right)+\left(\frac{1}{2}+\frac{1}{2}\cdot\frac{1}{3}\right)$$
$$+\left(\frac{2}{7}+\frac{2}{7}\cdot\frac{1}{6}\right)+\left(1-\frac{1}{9}\right)+\left(1-\frac{1}{10}\right)+\left(\frac{1}{8}-\frac{1}{8}\cdot\frac{1}{9}\right)+\left(\frac{1}{4}-\frac{1}{4}\cdot\frac{1}{5}\right)$$
$$=\frac{121}{15}.$$

Chapter 3: "Procedure of Miscellaneous [Methods]." This is a collection of methods that we would classify as solving equations in one unknown. Mahāvīra divides the solution rules into ten categories, depending on the characteristics of the given quantities. For example, if the unknown amount is composed of given fractional parts plus a given known quantity, that is called the "fraction" class of problem. If the unknown consists of a given known quantity plus given fractional parts of successive "remainders," that is the "remainder" class, which we can describe more clearly with a sample problem, as follows:

> A king took a sixth of a quantity of mangoes, the queen a fifth of the remainder, three princes a fourth, a third [and] a half [of successive remainders]. The youngest son took the three remaining mangoes. [You], skilled in the miscellaneous [methods], tell [me] the amount of them. (verses 3.29–30)

In general, if x is the unknown, c the known quantity, and n the number of successive fractional parts, then such a problem may be written as

$$x=\frac{x}{n+1}+\frac{1}{n}\cdot\frac{nx}{n+1}+\ldots+\frac{1}{2}\cdot\frac{2x}{n+1}+c,$$

[49]The simplification or reduction techniques used for fractions by Mahāvīra and by a later mathematician, Nārāyaṇa Paṇḍita, are discussed in detail in [Kus2004].

and the solution rule prescribes the equivalent of

$$x = \frac{c}{\left(\frac{n}{n+1}\right)\left(\frac{n-1}{n}\right)\cdots\left(\frac{1}{2}\right)}.$$

Thus in the problem at hand, $x = 18$ mangoes.

Similarly, an unknown x expressed in terms of its square root, or what we would call a quadratic in $\sqrt{x}$, falls into the "square root" class of problems and is solved by a version of the quadratic formula (again, only one of the two square roots in the quadratic solution is mentioned). Other classes arise in the case of terms involving the square root of part of x, or part of the square root of x, and so on. One more sample problem relating to one of the more complicated classes, the so-called "square root with two known quantities," will serve as an illustration:

> One bee is seen in the sky, a fifth and a fourth of the (successive) remainders [and] a third of the [next] remainder [and] the square root [of the whole] in a lotus-flower, [and] two in a mango-tree. How many are they? (verse 3.48)

That is, the desired number x of bees is expressed by

$$x = 1 + \frac{x-1}{5} + \frac{1}{4}\cdot\frac{4(x-1)}{5} + \frac{1}{3}\cdot\frac{3}{4}\cdot\frac{4(x-1)}{5} + \sqrt{x} + 2.$$

Chapter 4: "Procedure of Rule of Three Quantities." Here Mahāvīra gives solutions for direct and inverse proportions from Rule of Three up to Rule of Nine, as well as barter. His algorithms and examples are not substantially different from those of previous texts, and we will not examine any of them closely here.

Chapter 5: "Procedure of Mixtures." The "mixture" problems cover every other kind of algorithmic procedure that does not involve geometry. These subjects include problems on interest and investments, with simple or mixed capitals; barter; and various types of solution techniques called "pulverizers," which may deal with mixtures of determinate or indeterminate quantities. The class of pulverizers familiar from earlier texts, that is, those that treat linear indeterminate equations, is not applied in this work to astronomical problems; instead, the examples involve dividing heaps of fruits among given numbers of travellers with a given number left over, and so forth.

The last of these sections on "pulverizer" methods is called the "variegated" or "diverse" pulverizer, and, as its name suggests, it comprises a miscellany of techniques. These include the following rule and sample problem for combinations, or ways to choose among n things a subset of size r:

> When [one has] set down a sequence of [numbers] beginning with and increasing by one, above and below, in order and in reverse

> order [respectively], a reverse-product [in the numerator] divided by a reverse-product [in the denominator] is [the size of] the extension [of the different varieties of combination].
>
> Oh mathematician, tell [me] now the characters and varieties of the combinations of the flavors: astringent, bitter, sour, pungent, and salty, together with the sweet flavor. (verses 5.218–219)

That is, make a fraction consisting of products of the first n integers, in order (in the numerator) and in reverse order (in the denominator), as follows: $\frac{1 \cdot 2 \cdot 3 \ldots n}{n(n-1)(n-2)\ldots 1}$. Then the "reverse" product, or the product of the $(n-r)$ rightmost factors, is selected in both numerator and denominator. The two reverse products form a new fraction, $\frac{(r+1)(r+2)\ldots n}{(n-r)(n-r-1)\ldots 1}$, which is the number of possible combinations and is easily seen to be equivalent to the modern $\frac{n!}{r!(n-r)!}$.

The remainder of the chapter is devoted to series computations, some of which are equivalent to those in the second chapter. General formulas for sums of squares and cubes are also dealt with here. The section closes with the combinatorial mathematics of prosody similar to what we saw in Piṅgala's text in section 3.3: how to find the number of poetic meters with n syllables (i.e., 2^n, as mentioned in the survey of Mahāvīra's chapter 1); how to order the different meters in a standard sequence or "extension" according to the arrangement of their light and heavy syllables; how to identify a particular meter from its number in the sequence, and vice versa; and how to compute the number of meters containing a given number of light or heavy syllables.

Chapter 6: "Procedure of Figures." In introducing his chapter on geometry, Mahāvīra categorizes different types of results as approximate or accurate, and classifies figures as triangular, quadrilateral, or circular—but here "circular" refers also to other curvilinear figures and surfaces (discussed in more detail below). According to Mahāvīra, "[all] remaining figures are just different [forms] of the varieties of these" (verse 6.3).

"Approximate" rules are first given for computing the characteristics of all these figures, followed by a set of corresponding "accurate" ones. Mahāvīra often uses the same criteria as Brahmagupta in distinguishing approximate rules from exact ones. For example, for triangles and quadrilaterals the approximate area is the product of the averages of pairs of opposite sides, while the accurate area is found by Heron's/Brahmagupta's formula. And for circular figures, the distinction generally depends on whether the rule uses 3 or $\sqrt{10}$ for the value of the circumference/diameter ratio. We will examine these rules in the case of a few of Mahāvīra's nonregular "circular" or "circuloid" figures.

Besides the circle and semicircle, Mahāvīra places in this category a "long circle," a "conch-circle," a "depressed [or] raised" circle, and an "outer

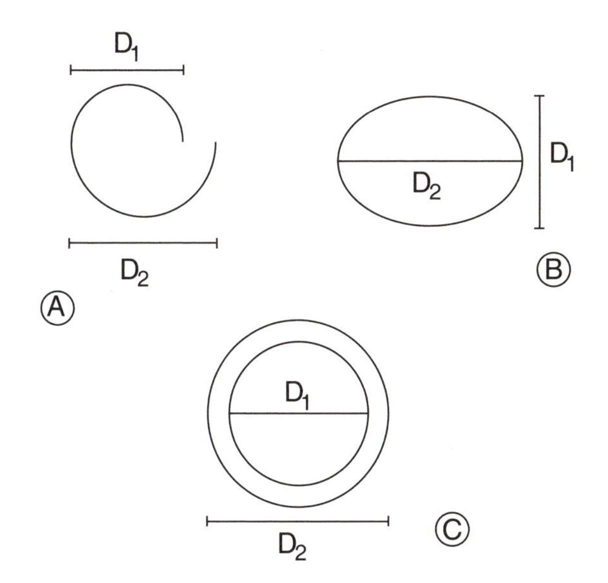

Figure 5.8 Curvilinear figures in the *Gaṇita-sāra-saṅgraha*: A. The "conch-circle." B. The "long circle." C. The "wheel-circle."

[or] inner wheel-circle" (verse 6.6). These are evidently specific technical terms, just as names meaning "oblong quadrilateral" were recognized technical terms for rectangles; but as the figures they refer to are not described by a detailed construction, their exact shape has to be deduced from the formulas applied to them. The "conch-circle," the "long circle," and the "wheel-circle" are imagined here as the figures shown in figure 5.8A–C respectively. (The "depressed or raised circle," not shown, would seem to be a concave or convex spherical cap.) Their areas A and circumferences or perimeters C are said to be computed exactly by procedures equivalent to the following expressions:

"Conch-circle" (apparently a spiral formed of two semicircles with diameters D_1, D_2; we can consider its opening $D_2 - D_1$ to be sealed off by a line segment to make it a closed figure):

$$C = \sqrt{10}\left(D_2 - \frac{D_2 - D_1}{2}\right),$$

$$A = \sqrt{10}\left(\left(\frac{1}{2}\left(D_2 - \frac{D_2 - D_1}{2}\right)\right)^2 + \left(\frac{D_2 - D_1}{4}\right)^2\right).$$

"Long circle" (an ellipse with major and minor axes D_2, D_1?):

$$C = \sqrt{6{D_1}^2 + 4{D_2}^2}, \qquad A = C \cdot \frac{D_1}{4}.$$

"Wheel-circle" (an annulus with outer circumference and diameter C_2, D_2, and inner ones C_1, D_1:

$$A \approx \frac{C_1 + C_2}{2} \cdot \frac{D_2 - D_1}{2} \qquad (\text{"approximate"}),$$

$$A = \frac{C_1 + C_2}{6} \cdot \frac{D_2 - D_1}{2} \cdot \sqrt{10} \qquad (\text{"accurate"}).$$

The formulas for the spiral and annulus agree with those of modern geometry up to the value of π (except that the so-called "accurate" rule for the annulus area has an extraneous factor of $\sqrt{10}/3$ in it, which was probably intended to correct for having computed the circumferences C_1, C_2 with 3 instead of $\sqrt{10}$).[50] The "long circle" rules are only approximate, unless this name is meant to represent some other, non-elliptical oval shape for which they would be exact.[51]

The geometry chapter also includes rules for computing other figures, for operations with *karaṇīs* in such computations, and for creating figures with non-*karaṇī* sides, as discussed in *Brāhma-sphuṭa-siddhānta* 12.33–39. Its final section is the mysteriously named "Demonic Procedure" containing rules for various involved geometry problems, such as determining the sides of a rectangle when some given multiple of the area is stated to be equal to the sum of given multiples of the diagonal, the length, and the width.

Chapter 7: "Procedure of Excavations." Mahāvīra's chapter on "excavations" also covers the earlier solid geometry categories of "piles" and "sawing" (see the discussion under *Brāhma-sphuṭa-siddhānta* 12.1 in section 5.1.3). The category "heaps" of grain, however, is omitted. A sample of the pile computations is illustrated in figure 5.9, where a wall built in the form of a trapezoidal prism has length d, height h, and width w_b and w_t at its bottom and top respectively. If it is broken along the slanting surface shown, then the volumes V_t and V_b of its top and bottom portions are given correctly by

$$V_t = \frac{dh}{6}(2w_t + w_b), \qquad V_b = \frac{dh}{6}(2w_b + w_t).$$

Chapter 8: "Procedure of Shadows." This final topic is said in the introductory chapter to be the "tide" of the ocean of mathematics, "bound up with the [astronomical] branch of *karaṇas*" (verse 0.22). Indeed, Mahāvīra's shadow rules are simple timekeeping approximations and right triangle proportions of the sort we saw in the *Āryabhaṭīya*'s and *Brāhma-sphuṭa-siddhānta*'s *gaṇita* chapters, not the trigonometrically exact rules used in theoretical *siddhānta* astronomy.

[50] But in [Ran1912], p. 205, the translator opines that the insertion of $\sqrt{10}/3$ is simply an error in the formula: "the mistake seems to have arisen from a wrong notion that in the determination of the value of this area, π is involved even otherwise than in the values of [the circumferences]."

[51] These rules are analyzed in detail in [GupAJM], pp. 173–177.

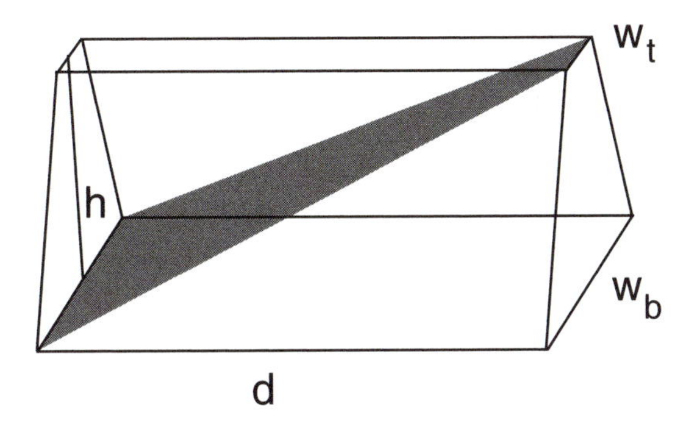

Figure 5.9 Volume computation for a broken pile.

To sum up, Mahāvīra has indeed covered in this work an "ocean" of *gaṇita*, including some topics such as sign rules and zero, the pulverizer and *karaṇī* computations, that were elsewhere classified as algebra. Thus the foregoing overview of a few Sanskrit texts composed within a time span of (probably) less than four centuries makes it clear that although they share many results and methods, they also diverge widely in their ideas even about the fundamental organization and presentation of mathematics. (Contrast, for example, the firmly fixed standard didactic structure of much of Hellenistic Greek mathematics prevailing over similar time scales.) The next chapter attempts to draw some general conclusions about this eclectic subject and its cultural and intellectual environment in medieval India.

Chapter Six

The Development of "Canonical" Mathematics

Mathematics and mathematical astronomy in the Indian tradition were constantly changing and growing throughout their history. But they also developed over time a certain degree of standardization in their textual format and content. The various classifications of mathematical topics apparently influencing early texts such as the *Āryabhaṭīya*, the *Brāhma-sphuṭa-siddhānta*, and the Bakhshālī Manuscript were mostly replaced by a more uniform structure for organizing mathematical knowledge.

The closest thing to a universal canon of mathematical texts that emerged in Sanskrit was the output of the twelfth-century astronomer Bhāskara (II). His *Siddhānta-śiromaṇi*, *Līlāvatī*, and *Bīja-gaṇita* became by the middle of the second millennium the most widespread Indian expository works on mathematical astronomy, arithmetic, and algebra, respectively. Even today they are generally considered the "standard" Sanskrit treatises on their subjects and the chief exemplars of the "standard" structure for mathematical content, although they never decisively displaced earlier authorities in the way that the works of, say, Euclid and Ptolemy did.

This chapter attempts to describe some of the ways in which mathematics and its practitioners fit into medieval Indian society, and the institutional and intellectual frameworks that supported them. It also considers the content and context of Bhāskara II's writings, and some possible factors in their emergence and persistence as canonical mathematical science. The mathematical works of Nārāyaṇa Paṇḍita in the fourteenth century, closely modeled on those of Bhāskara, illustrate how new topics and expansions of old ones were integrated into the now largely standardized structure of *gaṇita*.

6.1 MATHEMATICIANS AND SOCIETY

The role of mathematics and mathematicians in premodern Indian society of any period is a potentially vast subject that is still poorly understood. How many people were engaged in mathematical pursuits? How did they earn their living? What were the social goals that fostered or retarded their work, and what was their social status? How did they perpetuate their professional occupations and mathematical knowledge from one generation to the next? What was the intellectual position of mathematical knowledge within the universe of Sanskrit learning? How did all these aspects change during South Asia's historical development over the course of centuries?

The answers to most of these questions can only be guessed at, owing to the scarcity of detailed sources and the difficulty of building up accurate explanations from the few documentary traces of a very complex and varied history. We will start with a brief description of the available historical sources, followed by an overview of the most relevant social structures, and proceed from there to constructing our (very speculative) picture of medieval Indian mathematical professions.

6.1.1 Biographical information in textual sources

Such information as we have about the lives of Sanskrit authors is usually derived from autobiographical details in the opening or closing verses of their texts. Sometimes a commentator or scribe will elaborate on this information with additional details presumably derived from pedagogical tradition. But the data thus supplied seldom go beyond some (usually small) subset of the following names:

- The author himself;
- The hereditary social groups to which he belonged;
- His father and perhaps other relatives such as paternal grandfather, uncles, or brothers (occasionally mother);
- His teacher;
- His place of residence, birth, or family origin;
- His royal or noble patron;
- His favorite deity.

Sometimes an author states the date when he was born or when he completed the work in question, but in many cases only an approximate date can be inferred from context. Astronomical texts frequently also contain indirect chronological information in their parameters or worked examples, which can be used to date their composition within at least a few decades.

When specific dates are supplied, the calendar eras most often used and most securely established are the Śaka era (78 CE) and the Vikrama era (57 BCE), the second of which is generally called just "*saṃvatsara*," "year."[1] The conversion of months and days in Indian dates into their Julian calendar equivalents is sometimes uncertain, because we don't know whether or when in a particular year a local calendar might have included an intercalary month. If the author includes the name of the corresponding weekday, that can be used as a check on the conversion.

Several authors, on the other hand, have left a textual record lacking almost all identifying details. Even their dates can be estimated only within

[1]See [Sal1998], pp. 180–198, for a survey of Indian calendar eras.

a century or so, by comparing the dates of earlier authors whom they mention and later authors who mention them. An example of such minimal identification is provided by the following invocatory verse from Śrīdhara's *Tri-śatikā* ("Three hundred [verses]"):

> Paying homage to Śiva, having extracted the epitome of *gaṇita* by means of the *Pāṭī-gaṇita* ("Arithmetic") composed by himself, the teacher Śrīdhara will now state [that] for the sake of worldly activities.[2]

This tells us no more than the bare identity of the author, his preferred deity, and the name of an earlier composition of his. These facts constitute almost all we know about Śrīdhara as a historical figure. Intertextual references indicate only that he wrote after Brahmagupta and before Govindasvāmin, presumably sometime in the eighth century.[3]

The physical copies of texts can add at least something to the contextual information derived from their content. As we have seen in the case of the Bakhshālī Manuscript, Sanskrit manuscripts often contain colophons where the scribe lists information about the work, such as its title and author, and post-colophons, where he supplies information about the individual copy, such as the date and place of its completion. (Many scribes, however, omit some or all of these useful data.) He may also include his own name, that of his patron, or that of the designated recipient of the copy. The age of the earliest dated manuscript of a work thus puts a lower bound on the age of the work itself. And the number of surviving manuscript copies, their geographical distribution, and the information about their owners and users contained in their colophons and notes tell us something about the impact of the work and its audience. (This is how we know, for example, about the regional preferences for different astronomical *pakṣas* described in section 4.2.) Unfortunately, almost none of this information has been systematically studied in the case of individual works.

Moreover, the known manuscript corpus is not an entirely reliable historical guide. There are still a great many undocumented Sanskrit manuscripts about which we know nothing, and those that are known were mostly copied in the past few hundred years. Very few copies of mathematical or astronomical texts are older than the sixteenth century or thereabouts. Consequently, Indian mathematicians of the early modern period were the gatekeepers of the science of earlier times. What we have of medieval texts, for the most part, is what the scholars of the seventeenth century and afterward thought it necessary to copy and study.[4]

---

[2] [DviS1899], p. 1.

[3] [Pin1981a], p. 58. Even these straightforward inferences may not be entirely reliable; see [GupAJM], pp. 251–253, for arguments that Śrīdhara was actually a Jain.

[4] [Pin1981a], p. 118.

6.1.2 Social and professional groups in Indian society

What little biographical information there is about Indian mathematicians often involves social categories that are unfamiliar to non-Indian readers, or are understood only vaguely as part of something known as "the caste system." The most significant of these categories for our purposes, as they operated in medieval India, are briefly described in the following list.

The *varṇa*. Literally meaning "color," although not in the sense of skin color, the hereditary *varṇas* are the four fundamental categories of the Hindu social structure. They constitute a hierarchy that has been recognized since Vedic times, extending from the Brāhmaṇas (priests and scholars) through Kṣatriyas (rulers and warriors) and Vaiśyas (merchants and farmers) down to Śūdras (artisans and servants); their associated symbolic colors are white, red, yellow, and black, respectively. Almost all Hindu authors of Sanskrit literary or didactic works are Brāhmaṇas. The three higher *varṇas*, especially Brāhmaṇas, are called "twice-born," referring to a ritual for adolescent males considered a spiritual second birth. Śūdras are traditionally ineligible to undergo this ceremony or to study the Sanskrit scriptures, or even to hear them recited. The lowest-status groups, nowadays known as "dalits" or "untouchables," are outside the caste system entirely and have no recognized *varṇa*.

The *jāti*. Within each *varṇa* (and sometimes crossing the boundaries between them) are innumerable instances of the hereditary occupational groups called *jātis*, commonly translated as "castes." Individuals are born into their parents' *jāti* and are expected to marry within it (the technical term for this characteristic is endogamy). A *jāti* is defined not only by kinship but also by livelihood and locality; each type of occupation has one or more *jātis* associated with it, which may differ somewhat from region to region in their level of social or economic status. The social status of a *jāti* generally reflects that of a particular *varṇa*, but this is not an iron-clad rule: a high-status Vaiśya *jāti* might be more socially powerful and more wealthy than a low-status Brāhmaṇa one, for example.

The *gotra*. A *gotra*, unlike a *jāti*, is an exogamous group, whose members are not allowed to marry one another. Most often, *gotras* are associated with the Brāhmaṇa *varṇa*, where they are considered to be lineages representing descent from the legendary Vedic sages.

The *jñāti*. This is an endogamous group often roughly equivalent to *jāti*, but frequently a sub-lineage within a *jāti*.

In addition to their various caste designations, Hindus are sometimes grouped in accordance with their choice of a primary deity from the Hindu pantheon. This choice may be individual but is frequently a sort of sectarian identification shared by members of a family or a *jāti*, with the major sects being Shaivism (devotees of Śiva) and Vaishnavism (devotees of Viṣṇu).

The caste system in the Classical Sanskrit period became much more complex and rigid than the original *varṇa* division in Vedic India. Its inflexibility was stressed by many medieval Hindu writings on *dharma* (religious law or right conduct), although other texts sometimes disagreed, describing spiritual qualifications for high status as more important than genealogical ones. Moreover, in practice there were ways to get around the rules of caste: for one thing, cross-*jāti* and even cross-*varṇa* matings inevitably occurred from time to time. Although *dharma* texts officially disapproved of such matings, they acknowledged them by defining new hereditary categories for their various offspring. Furthermore, members of a low-status group who moved to a new location among strangers might upgrade their status by identifying themselves as members of a higher *jāti*, or even a higher *varṇa*.

Buddhists and Jains as well as some heterodox Hindu groups originally rejected the premise of hereditary caste distinctions as well as the Vedic deities. In medieval times, though, a number of them (particularly Jains) assimilated more or less into the predominant social structure of caste, and some of the Hindu gods were recognized in their worship. Sectarian division among the Jains produced the two main groups named after the dress customs of the ascetics belonging to them, the Digambaras ("space-clothed" or naked) and the Śvetāmbaras ("white-clothed"). Within the Śvetāmbara sect a number of groups called *gacchas* emerged around the beginning of the second millennium; a *gaccha* is a devotional or pedagogical lineage passing down both religious instruction and secular learning to successive generations of young monks.

Among the vast majority of Hindus, Buddhists, and Jains, the family structure was patrilineal and patriarchal.[5] Family identity was transmitted through the males, with females joining their husbands' families, to which they generally brought dowries, and family authority was at least theoretically vested in the husband. Moreover, male descendants were required for the performance of Hindu religious rituals honoring deceased ancestors. Thus a lack of sons was seen as a dangerous deprivation, and an excess of daughters as a financial burden. As *dharma* rules about caste purity became more strict in post-Vedic times, the autonomy of women became more limited. Although women were responsible for running the household and required a wide range of skills, including handling money, *dharma* texts frowned on letting them earn their own livelihood or live without the supervision and control of a male relative. Girls of the educated classes might study a variety of learned subjects, including Sanskrit and arithmetic, as evidenced by the occasional use of feminine vocatives (e.g., "girl" or "beautiful one") in sample problems posed to students in arithmetic texts.[6] But women make up only a very small minority of the authors represented in medieval Sanskrit

---

[5]Exceptions included the matrilineal societies of parts of southwest India, briefly discussed in section 7.2.

[6]See [Hay1995], pp. 462–463.

scholarship and literature.

Most authors are known by a single Sanskrit given name, which is generally a name or epithet of a deity, seer, or saint. Although these personal names sometimes reflect kinship ties (for example, when a son was named after his grandfather or other male ancestor), there was no exact equivalent in premodern India of the modern practice of assigning a hereditary family surname to male descendants. What sometimes appear to be surnames in the modern designations of medieval authors are generally instead honorifics (e.g., "Ācārya," "teacher"; "Rāja," "king"; "Sūri," "teacher," frequently used among Jains; "Dvivedī," "one who has memorized two Vedas"; and so forth) or else the names of their *varṇas*, *jātis*, *gotras*, or other subgroups (such as "Gārgya" and "Somayājin").

Historians draw upon all of the above social categories and practices in the effort to determine which ancient or medieval Indian mathematician is which—not always successfully. Personal names are seldom unique to any one scholar, so if reliable dates and biographical details are lacking, it can be impossible to distinguish the works of one author from those of another bearing the same name. To further complicate matters, the titles of treatises too are sometimes shared by several different works, while on the other hand some treatises are known by more than one title.

6.1.3 Mathematics as a field of study and profession

Bearing in mind these various caveats, we can nonetheless make the following generalizations about the teachers and learners of Sanskrit mathematics in medieval India. Among Hindus, astronomer-mathematicians were almost exclusively Brāhmaṇas pursuing the subject as part of their *jāti* occupation.[7] Court astronomers serving noble or royal patrons tended to have a higher *jāti* status than village astrologers casting horoscopes and making annual calendars for humbler customers. Jain and Buddhist scholars might also have court patronage or might study and teach in sectarian monasteries. Some Hindu scholars too received support from endowed institutions, often attached to temples, that taught various subjects to novice priests. Some authors of mathematical textbooks were presumably teachers supported by the community in return for the practical instruction they provided. Moreover, the ancient occupation of court *gaṇaka*, "calculator" or accountant, must have required the services of many mathematically literate people throughout the medieval period.[8]

Probably all children who received any literate education at all were taught the fundamentals of numeracy.[9] Whatever the details of this most elemen-

[7]A handful of known exceptions, mostly from the late second millennium, are named in [PinVJEI]. An eleventh-century king ([Pin1970a], vol. 4, pp. 336–339), a fifteenth-century merchant ([Pin1970a], vol. 5, p. 426), and a goldsmith ([Pin1970a], vol. 4, p. 445, [PinVJEI]) have also been mentioned as authorities on astronomy or mathematics.

[8]See [PinEMEI], [PinVJEI], [PinSSEI], and (particularly for references to professional "calculators" in ancient texts such as the *Mahābhārata* and the *Arthaśāstra*), [PinGEI].

[9][DatSi1962], vol. 1, p. 6.

tary mathematical training may have been, they were beneath the scope of all the Sanskrit treatises considered here. Even before learning the first verse of the *Gaṇita-sāra-saṅgraha* or the *Līlāvatī*, the pupil was evidently already expected to understand how to form the shapes of the decimal digits, how to add or multiply single-digit numbers, and so on. The emphasis on various types of commercial problems in arithmetic texts suggests that all students, even if they were not members of mercantile castes, were expected to understand the workings of basic financial transactions such as interest on loans and shares in an investment. Those students who completed the study of Sanskrit arithmetic might go on to learn algebra and astronomy, depending on their opportunities and the demands of their hereditary occupation. However, the details of the schooling process for pupils of various ages, *varṇas*, *jātis*, periods, regions, and social classes are all pretty vague.[10]

We do know that education for most Brāhmaṇas involved individual study with a teacher in the same *jāti*, frequently a relative such as a father, uncle, or older brother. Over the years, many learned families accumulated large libraries of manuscripts, which enriched the studies of future generations. According to the testimony of post-colophons of surviving copies (although these, as noted in section 6.1.1, generally belong to a considerably later period), some of these manuscripts were copied or paid for by a father or other relative for the use of a younger pupil. Some are described as copied by the pupil himself for his own use; in some such manuscripts, the rather crude script seems to suggest that the young copyist had not been many years literate before beginning his task.

There also existed some educational institutions larger than the family tutorial and the local school that maintained collections of manuscripts of religious and other works,[11] but we know little of their role in generating and disseminating mathematical learning. Perhaps the most famous of these institutions are medieval India's great Buddhist universities, or endowed temple complexes with resident teachers and students, which had an international reputation; however, they are not known to be associated with any Sanskrit mathematical works. Although the important eleventh-century Buddhist treatise *Kāla-cakra-tantra* ("Doctrine of the Wheel of Time") describes aspects of mathematical astronomy apparently derived from one or more of the medieval *pakṣas*,[12] there is only one known author of independent mathematical astronomy Sanskrit texts who may have been a Buddhist.[13]

---

[10]Indian regional traditions of mathematical practice, including vernacular primer texts teaching basic numeracy, are currently being studied by the author of [BabuCT]. However, most of the available textual sources for this study are very recent, although they may well represent much older pedagogical traditions. Some details about those traditions in medieval and early modern times are discussed in [PinEMEI] and [PinSSEI].

[11]See [PinILEI].

[12]See [Oha2000], pp. 349–360. Moreover, a seventh-century Chinese pilgrim records that astrology was part of the curriculum at the Indian Buddhist monastery he visited; see [PinEMEI].

[13]This is the eleventh-century astronomer Daśabala, whose name is an epithet of the Buddha and who is described in manuscript colophons of one of his works as a "bod-

We have mentioned above the monasteries and temple institutions where many Jains and Hindus studied, but mathematical authors are generally silent about them.[14] Some remarks in a couple of astronomical commentaries that seem to be addressed to a classroom rather than to an individual reader may hint at the existence of special schools for advanced mathematics instruction; mostly, though, the possible existence of such schools has to be tentatively inferred from the appearance of groups of (apparently unrelated) mathematicians in about the same geographical area at about the same time.[15]

Evidence about other types of scientific or scholarly institutions is even more vague and uncertain. Āryabhaṭa I speaks of astronomical knowledge "honored in Kusumapura," which may be Pāṭaliputra (the modern Patna)[16] in the medieval Buddhist heartland of the Bihar region in the northeast of the subcontinent, but we are not told who honored that knowledge or what institutions they may have been connected with. Similarly, the city of Ujjain in Madhya Pradesh, which in astronomical texts is typically said to be on the line of zero longitude or prime meridian, is sometimes alleged to have had an ancient observatory. But we have no detailed record or description of such an institution and no reliable evidence supporting the modern legends that one or another famous medieval mathematician was the chief astronomer there. Occasional references are made to mathematicians discussing their opinions in "assemblies of the eminent" (see, e.g., section 6.2.3), but we don't know what form these conferences took or who was eligible to participate in them, or what role they played in a tradition of professional training that seems to have been focused on individual instruction in family dwellings. In short, there is still a great deal of uncertainty about all but the broadest generalizations concerning the ways mathematical practitioners were trained, evaluated, and supported.

Nor is it easy to estimate the total number of such practitioners in the medieval period. The survey in [Pin1970a] of all known authors on the Sanskrit mathematical sciences, including astrology and divination, will include when complete about five to ten thousand names. (Some five or ten of them, or about one-tenth of one percent, are female, all writing on nonmathematical aspects of *jyotiṣa*.) Perhaps a third or a half of these authors may be dated to the past five or six centuries or may represent duplicate references to an author of the same name. And probably no more than ten or twenty percent of the remainder wrote on the technical aspects of mathematics or mathematical astronomy. We can put a lower bound on this estimate by considering that there are perhaps six or eight premodern Indian mathematicians who are (sort of) universally known, such as Ārya-

---

hisattva"; the work itself, however, is dedicated to the Hindu goddess Parvati. See [Pin1988], p. 2.

[14] However, the *gacchas* of many Jain mathematical authors are known; see [PinVJEI].

[15] See [PinSSEI].

[16] So says Bhāskara I in his commentary on this statement in *Āryabhaṭīya* 2.1; see [ShuSa1976], pp. xvii–xix.

bhaṭa and Brahmagupta. These famous names and about forty more have been (somewhat arbitrarily) selected for inclusion in the biographical listing in appendix B, owing to their association with the mathematical works or discoveries discussed in this book. Again, about a third of that total lived within the past five hundred years or so. So we can deduce that there were probably somewhere between fifty and a thousand medieval Indian mathematicians prominent enough to have handed down their works or fame as far as the present day. How many others there might have been who left no record of their mathematical activities it is impossible to say.

6.1.4 Sanskrit culture and science in Southeast Asia and the Himalayan region

We know much less about the mathematical sciences on the northern and southeastern fringes of the medieval Sanskrit world—that is, in Tibet and Southeast Asia—than we do about their counterparts in India itself, so a very brief mention of them will have to suffice. As noted in section 1.2, around the middle of the first millennium CE, Indian trade and culture began to spread overseas to the south and east. Sanskrit mathematical astronomy of the medieval *siddhānta* period is attested by technical terms appearing in dates of Cambodian inscriptions as early as the seventh century, and in Javanese ones from the eighth century to the fifteenth.[17]

Calendrics, astrology and divination seem to have been the chief applications of this knowledge, as indeed they also were in India. The decimal place value numerals along with some sexagesimal metrology accompanied this transmission; in fact, as discussed in section 3.1, some of the earliest surviving inscriptions showing decimal place value and the zero are found in Southeast Asia rather than India itself. Toward the end of the Old Javanese period in the early second millennium, this Indo-Javanese mathematical astronomy spread to Bali as well.[18]

Turning to the northern periphery, we see a somewhat similar development. Chinese Buddhist pilgrims had been journeying to the Buddha's homeland of India since the early Classical period and had brought back many Sanskrit texts. A few astrological texts from the late first millennium indicate that Indian mathematics and astronomy were not unknown in China itself, although the extent to which they influenced the indigenous Chinese tradition is unclear.[19] But the spread of medieval Sanskrit mathematical astronomy into the northern Himalayas apparently occurred around the time of the so-called "second transmission" or revitalization of Buddhism in Tibet in the eleventh century, largely influenced by Esoteric Buddhism in India. Its chief vehicle seems to have been the abovementioned Sanskrit Buddhist work *Kāla-cakra-tantra*, which was translated into Tibetan and Mongolian. The Indian mathematical astronomy therein described became the foundation of

---

[17][Gom1998]; [Gom2001], pp. 122–124.

[18][Hid2000], p. 371.

[19]See [Mar2000], pp. 383–385.

traditional calendars in Tibet, Mongolia, and Bhutan, which coexisted with a different form of mathematical astronomy derived from China.[20]

6.2 THE "STANDARD" TEXTS OF BHĀSKARA (II)

The illustrious Bhāskara (born 1114), commonly known as Bhāskara II to distinguish him from his seventh-century predecessor, is possibly the first famous Indian mathematician to have been immortalized on a monument. An inscription in a temple in Maharashtra, claiming to have been erected by Bhāskara's grandson Caṅgadeva, records Bhāskara's lineage for several generations before him (as well as two generations after him). The inscription commemorates a grant from a local ruler, on a date corresponding to 9 August 1207, to support Caṅgadeva's founding of a school for the study of Bhāskara's works: a *siddhānta*, a *karaṇa*, and one treatise each on arithmetic and algebra. This unusually abundant biographical information, supplemented by some details in Bhāskara's own verses, reveals that Bhāskara came from a long line of court scholars, mostly astronomer-astrologers in the central-western parts of India. His father, Maheśvara, and apparently Bhāskara himself were natives of the city called Vijjaḍaviḍa in the Western Ghat mountains between the Tapti and Godavari rivers. According to Bhāskara, it was Maheśvara who taught him astronomy; Bhāskara modestly describes himself in the last verse of his *Bīja-gaṇita* as having attained only a fraction of his father's knowledge.[21]

Bhāskara thus possessed the advantages of a renowned scholarly lineage and a tradition of royal and noble patronage, extending even to the sponsorship of a school after his death for the study of his writings. These advantages were doubtless partly responsible for the successful dissemination and lasting popularity of his works, but other factors were involved as well. Unlike many earlier mathematical authors, Bhāskara wrote his own concise prose commentary on each of his treatises (except the *karaṇa*), including sample problems for all his rules in the arithmetic and algebra texts—and supplying solutions for all the sample problems. These pedagogical aids, along with the careful organization of the works and their clear exposition, were doubtless highly appreciated by students and teachers.

6.2.1 The *Līlāvatī*

The most famous Sanskrit mathematical work ever composed is probably Bhāskara's *Līlāvatī*, literally "Beautiful" or "Playful." It has been speculated that the work was named after Bhāskara's own daughter, although there is

---

[20]See the overview of Indo-Tibetan astronomy in [Oha2000].

[21]See [Jai2001], pp. 19–23, and [Pin1970a] 4, p. 299. The temple of Caṅgadeva's inscription was identified by Bhau Daji in the nineteenth century in the village of Patna near Chalisgaon; see the Gazetteer of the Bombay Presidency, www.maharashtra.gov.in/pdf/gazeetter__reprint/Khandesh/appendix__p.html#6, and [PatK2001], p. xviii.

no direct mention of any such person in any of Bhāskara's writings or in most of their commentaries.[22] This speculation has been strengthened by the presence in the text of several verses addressed to a feminine gender "beautiful one" or "fawn-eyed one." And it could be imagined that the following invocatory verse was crafted to appeal to someone with a fondness for the letter "l":[23]

> *līlāgalalulallolakālavyālavilāsine* |
> *gaṇeśāya namo nīlakamalāmalakāntaye* ||
>
> Homage to [the deity] Gaṇeśa, delighting in the writhing black snake playfully twining about his beautiful neck, bright as a blue and shining lotus. (*Līlāvatī* 9)

The punning final verse of the text similarly might hint at a tender attitude:

> Those who keep in their throats the *Līlāvatī* having entirely accurate [arithmetic] procedures, illustrating elegant sentences, [whose] sections are adorned with excellent [rules for] reduction of fractions and multiplication and squaring [etc.]...
>
> [Alternative translation:] Those who clasp to their necks the beautiful one completely perfect in behavior, enticing through the delight of [her] beautiful speech, [whose] limbs are adorned by the host of good qualities [associated with] good birth ...
>
> attain ever-increasing happiness and success.[24] (*Līlāvatī* 272)

On the other hand, numerous other verses in the same work are addressed to imaginary hearers in the masculine gender called, for instance, "best of merchants" or "intelligent mathematician." So it cannot be said for certain how much of this fond language represents personal feeling and how much is just literary device in praise of the text.

The following brief overview of the *Līlāvatī* focuses on a few of its techniques and explanations that expand or illuminate the ones of earlier works discussed in chapter 5. The text is presented as a stream of continuously numbered verses, as in Brahmagupta's chapter 12 on *gaṇita* in the *Brāhma-sphuṭa-siddhānta*, but Bhāskara's own prose commentary identifies each topic separately. The sectioning of the text employed here lumps some of these topics together for convenience, but mentions Bhāskara's identifications of

---

[22]See section 8.2.3 for a discussion of a later legend concerning Bhāskara's putative daughter.

[23]All verse numbers in the *Līlāvatī* are given in accordance with the edition in [Apt1937]. Verses 1–2, 9–33, 45–52, 60–64, 73–74, 77–79, 82–83, 132–143, 151–152, 156–157, 163–167, 169–173, 183–190, 199–204, 206–208, 239–241, 261–262, and 271–272, with their commentary, have been translated in [Plo2007b], pp. 448–467.

[24]My thanks to Gary Tubb for checking and improving my attempt at rendering the poetic version of the above verse. Professor Tubb observes that the last line of the verse also could be interpreted in accordance with the poetic version, taking the words for "increase" and "attainment of happiness" in their explicitly sexual meanings.

them.[25] Quoted excerpts show the verses indented within Bhāskara's own prose commentary. Note that the commentary, unlike the verses, frequently expresses numbers as numerals rather than in verbal form. Verse 1 is an invocation of the deity Gaṇeśa (who is traditionally praised at the start of Sanskrit texts as the god of beginnings and remover of obstacles), and verses 9 and 272 have been discussed above.

Verses 2–11. Metrology and decimal places. Bhāskara does not go into elaborate detail as Mahāvīra did in the *Gaṇita-sāra-saṅgraha* on different types of units of measure. He simply states what were apparently some of the most common units of money, weight, and linear, square, and cubic measure, adding, "The remaining definitions concerning time and so forth are to be understood [as they are] generally known from popular usage" (commentary on verse 8). He names the decimal places up to 10^{17}.

Verses 12–47. Eight operations on integers, fractions, and zero. The eight fundamental operations are, as usual, addition, subtraction, multiplication, and so on, up to cube root. The section on fractions begins with rules for four types of simplification of fractions: namely, reduction to a common denominator, fractions of fractions, and integers increased or decreased by a fraction. (Compare these with the varieties given by Brahmagupta in verses 12.8–9 of the *Brāhma-sphuṭa-siddhānta* and by Mahāvīra in chapter 2 of the *Gaṇita-sāra-saṅgraha.*) Then the eight operations are explained all over again for fractions, although rather sketchily in some cases, as illustrated by this rule and example for the last four operations:

> For squaring, the two squares of numerator and denominator—and for cubing, [their] two cubes—are to be given; [and] for determining the root, the two roots [of numerator and denominator]. (verse 43)

Here is an example:

> Friend, tell [me] quickly the square of three and one-half, and then the square root of the square, and the cube [of it], and then the root of the cube, if you know fractional squares and cubes. (verse 44)

Statement: $\begin{matrix} 3 \\ \frac{1}{2} \end{matrix}$. After the integer is multiplied by the denominator, the result is $\frac{7}{2}$. The square of that is $\frac{49}{4}$. Then the root is $\frac{7}{2}$. The cube is $\frac{343}{8}$; its root is $\frac{7}{2}$.

Those are the eight operations for fractions.

The rules for operations on zero are standard. Like Brahmagupta, Bhāskara considers a quantity divided by zero to remain "zero-divided":

[25] I have relied on [Hay2000], p. 2, for much of this topical structure.

> In addition, zero [produces a result] equal to the added [quantity], in squaring and so forth [it produces] zero. A quantity divided by zero has zero as a denominator; [a quantity] multiplied by zero is zero, and [that] latter [result] is [considered] "[that] times zero" in subsequent operations.
>
> A [finite] quantity is is understood to be unchanged when zero is [its] multiplier if zero is subsequently [its] divisor, and similarly [if it is] diminished or increased by zero. (verses 45–46)

Here is an example:

> Tell [me], what is zero plus five, [and] zero's square, square root, cube, and cube-root, and five multiplied by zero, and ten divided by zero? And what [number], multiplied by zero, added to its own one-half, multiplied by three, and divided by zero, [gives] sixty-three? (verse 47)

Statement: 0. That, added to five, [gives] the result 5. The square of zero is 0; the square root, 0, the cube, 0; the cube-root, 0.

Statement: 5. That, multiplied by zero, [gives] the result 0.

Statement: 10. That, divided by zero, is $\frac{10}{0}$.

[There is] an unknown number whose multiplier is 0. Its own half is added: $\frac{1}{2}$. [Its] multiplier is 3, [its] divisor 0. The given [number] is 63. Then, by means of the method of inversion or assumption [of some arbitrary quantity], [which] will be explained [later], the [desired] number is obtained: 14. This calculation is very useful in astronomy.[26]

Those are the eight operations involving zero.

Verses 48–55. Inversion and method of assumed quantity. Bhāskara's rules for working backward and for "false position," for finding an unknown quantity from a given result, are typical. One of his sample problems is the following, incorporating the "remainder class" of miscellaneous methods that we saw in Mahāvīra's chapter 3:[27]

---

[26]See section 6.2.3 for a discussion of what this remark may mean.

[27]In fact, [Hay2000], p. 2, notes that verses 48–89 constitute an entire section on "miscellaneous methods" corresponding to Mahāvīra's chapter 3. Bhāskara evidently acknowledges this model at the end of verse 89 when he says "These are the miscellaneous methods in the *Līlāvatī* on arithmetic," using the same term as Mahāvīra.

> A traveler on a pilgrimage gave one-half [of his money] at Prayāga [modern Allahabad], two-ninths of the rest at Kāśī, [Varanasi/Benares], one-fourth of the remainder in toll fees, and six-tenths of the remainder at Gayā [near Patna]. Sixty-three gold coins [were] left over, [and he] returned with that to his own home. Tell [me] the [initial] amount of his money, if [the method for] the remainder-class is clear [to you]. (verse 54)

Statement: $\frac{1}{1}, \frac{1}{2}, \frac{2}{9}, \frac{1}{4}, \frac{6}{10}$. The given [answer] is 63. Taking the assumed quantity here [to be] unity (1), subtracting the remaining fractions from the remainder by the rule for subtraction of fractions, when simplified, the resulting remainder is $\frac{7}{60}$. Dividing this into the given [answer] 63, multiplied by the assumed [quantity], the resulting amount of money is 540. This may also be done by the rule for inversion.

Verses 56–72. "Pseudo-algebraic" algorithms. These include the basic rules for what we called "concurrence" and the "difference method" in *Brāhma-sphuṭa-siddhānta* 18.36, and methods for finding quantities that satisfy certain conditions involving perfect squares, as in *Brāhma-sphuṭa-siddhānta* 18.72–74, along with some quadratic formula rules like those in Mahāvīra's "miscellaneous methods." Brahmagupta, of course, classed these topics as part of algebra, and Bhāskara's presentation of them suggests that he considered them a sort of "pseudo-algebra" for finding certain types of unknowns before the methods of algebra were formally studied, as in the following excerpt:

> Tell [me], my friend, the numbers whose squares, subtracted and [separately] added [to each other] and [then] diminished by one, produce square roots [i.e., are perfect squares]; [a problem] with which those skilled in algebra, [who] have mastered the algebra [techniques] called "six-fold," torment the dull-witted.... The square of the square of an assumed number, and the cube of that number, each multiplied by eight and the first product increased by one, are such quantities, in the manifest [arithmetic] just as in the unmanifest [algebra]. (verses 62–63)

The assumed quantity is $\frac{1}{2}$. The square of its square, $\frac{1}{16}$, multiplied by eight $\left(\frac{1}{2}\right)$, is added to 1; the resulting first quantity is $\frac{3}{2}$. Again, the assumed quantity is $\frac{1}{2}$; its cube, $\frac{1}{8}$, is multiplied by eight. The resulting second quantity is $\frac{1}{1}$. Thus the two quantities are $\frac{3}{2}, \frac{1}{1}$.

Now with one as the assumed quantity, [they are] 9, 8; with two, 129, 64; with three, 649, 216. And [computation can be done] to an unlimited extent in this way in all procedures, by means of assumed quantities.

> Algebra, [which is] equivalent to the rules of arithmetic, appears obscure, but it is not obscure to the intelligent; and it is [done] not in six ways, but in many. Concerning the [rule of] three quantities, [with all of] arithmetic and algebra, the wise [have] a clear idea [even] about the unknown. Therefore, it is explained for the sake of the slow[-witted]. (verse 64)

"Six-fold" refers to the six elementary operations in algebra (no cube or cube root), as discussed under *Brāhma-sphuṭa-siddhānta* 18.35. The task here is to find (rational) quantities x and y such that $x^2 \pm y^2 - 1 = z^2$ (for z an integer or a fraction). This rule manipulates an assumed quantity a to give $x = 8a^4 + 1$, $y = 8a^3$. It is easy to show that then $x^2 \pm y^2 - 1 = 16a^4(4a^4 \pm 4a^2 + 1)$, which is a perfect square.

Verses 73–89. Rule of Three Quantities and variations. Bhāskara treats the direct and inverse Rule of Three Quantities and gives a general procedure for rules of other odd quantities, as well as the standard rule for barter.

Verses 90–116. Mixed quantities. Here Bhāskara begins his treatment of several of the classic "eight procedures" in their classic order: namely, mixtures, series, figures, excavations, piles, sawing, and heaps. The procedure on mixtures deals with standard topics such as interest, investments, and the purity of gold. At the end of this section, as at the end of Mahāvīra's chapter 5, appear combinatorial rules dealing with prosody and other contexts where permutations and combinations occur:[28]

> This is considered to be general. Its application in prosody among [those] who know it [is given] in the chapter on metrics [in the *Brāhma-sphuṭa-siddhānta*; in architecture in variations of window frames ... and in medicine in the variation of tastes. These are not described for fear of over-extension [of this book]. (verses 113–114)

Verses 118–134. Series. Formulas for the sums of the first n integers, squares, and cubes are given, as are rules for the summation of series in arithmetic and geometric progression. More prosody rules appear at the

[28] This translation is taken from [Kus1993], p. 89.

end: they treat not only the number of distinct poetic meters with n syllables in a quarter-verse, as before (i.e., 2^n), but also the number of distinct half-verse patterns that can be made by combining two different n-syllable quarter-verses and the number of distinct verse patterns that can be made by combining four different n-syllable quarter-verses. Those numbers are equal to $(2^{2n} - 2^n)$ and $(2^{4n} - 2^{2n})$ respectively (because we must subtract out the excluded cases where two or more of the combined quarter-verse patterns are identical).

Verses 135–198. Geometry of triangles and quadrilaterals. Weighing in at seventy-nine verses, Bhāskara's "procedure of figures" or plane geometry in general receives the most ample treatment of any of his subjects. He begins with the Pythagorean theorem and other rules for solving right triangles, with an interesting digression on figures that are geometrically impossible:

> A straight-sided figure described in [over]-confidence where the sum of [all] the sides except [one] is smaller than or equal to [that] one side is to be known as a "non-figure." (verse 163)

Example:

> When the sides are given by an [over]-confident one [as] three, six, two, and twelve in a quadrilateral, or [as] three, six, and nine in a triangle, one must consider that a "non-figure."

These two figures are impossible to demonstrate. [Put] in the places of the sides straight sticks equal to the [proposed] sides; the impossibility of laying out [such an example] will be obvious.

On the other hand, interestingly enough, a quantity in a geometric figure may have an absolute negative value. Bhāskara explains such a concept in the case of solving for a triangle's two "base-segments" produced by the intersection of the base and altitude. The general formula for the base-segments is equivalent to that in *Brāhma-sphuṭa-siddhānta* 12.22, but Bhāskara explains how to apply it when the altitude falls outside the triangle:

> The sum of the two sides in a triangle is multiplied by their difference and divided by the base. The quotient, standing in two [places], is [separately] diminished and increased by the base, [and] halved. [The results] are the segments of the base [adjacent] to those two [sides]. (verse 165)...

Example with a negative base-segment:

> In a triangle, when the two sides are ten and seventeen and the base is nine, then tell me quickly, mathematician, the two base-segments, the altitude, and also the computed [area]. (verse 168)

> Statement: The two sides are 10, 17, [and] the base 9. Here, by [the rule] beginning "The sum of the two sides in a triangle ..." the quotient is 21. The base [can] not be diminished by that, so the base is subtracted from it. Half the remainder is the negative base-segment, that is, in the opposite direction [outside the triangle]. In this way the two resulting base-segments are 6, 15. Here in either case [i.e., with or without a negative base-segment], the resulting altitude is 8, the area 36.

Bhāskara repeats the Heron/Brahmagupta formula for the area of a triangle or quadrilateral but calls it "approximate" in the latter case (verse 169). He criticizes it as follows:

> Now, a rule for determining the inaccuracy....
>
>> Since the two diagonals of the quadrilateral are not determined, then how can the area in this [figure] be determined? Its two diagonals computed according to [the method of] the earlier [authorities] are not appropriate in other cases; and [since] with the same sides [there may be] different diagonals, therefore the area is not unique. (verse 171)
>
> In a [given] quadrilateral, when one and another [opposite] corner have approached [each other ...] they compress the hypotenuse attached to them. But the other two, stretching outward, expand their own hypotenuse. Hence it is said "with the same sides [there may be] different diagonals."
>
>> Without specifying either one or the other of the altitudes or diagonals, how can one ask for the definite area of that [quadrilateral] even though it is indefinite? He who asks for or states [that area] is a demon, or completely and permanently ignorant of the determination of quadrilateral figures.

Indeed, an arbitrary quadrilateral can be distorted just as Bhāskara describes, producing different figures with the same four sides but different areas, so the lengths of the sides alone do not determine the area. This accusation echoes an earlier criticism by Āryabhaṭa II in the *Mahā-siddhānta*.[29] It is not entirely clear whether Āryabhaṭa II and Bhāskara simply failed to recognize that Brahmagupta's rules applied only to cyclic quadrilaterals or whether they were criticizing him for not having stated that condition explicitly. Bhāskara mentions a quadrilateral with four equal sides and unequal diagonals (verse 173), that is, a rhombus, so it is clear that he at least is not restricting the geometry of quadrilaterals only to those that can be inscribed in a circle.

---

[29]See [Sara1979], p. 87.

Verses 199–213. Geometry of circles, spheres, and chords. According to Bhāskara, the "accurate" ratio of circumference to diameter in a circle is $\frac{3927}{1250}$, and the "practical" ratio is $\frac{22}{7}$.[30] He gives rules for finding the area A of a circle and the surface area S and volume V of a sphere that are equivalent to the following:

$$A = \frac{D \cdot C}{4}, \quad S = \frac{4D \cdot C}{4}, \quad V = \frac{S \cdot D}{6},$$

where D is the diameter of the circle or sphere and C its circumference. This is apparently the first statement in a Sanskrit mathematical text of exact expressions for S and V.[31] Bhāskara also gives rules for sides of inscribed regular polygons in a circle and for chords and "arrows" of arbitrary arcs.

Verses 214–240. Excavations, piles, sawing, heaps, shadows. The *Līlāvatī* deals rather summarily with the last five traditional procedures, allotting on average only about five verses to each, including sample problems. At the end of the shadows section he includes an interesting digression on the foundational role of the Rule of Three Quantities in mathematics.

Verse 241. On the Rule of Three Quantities.

> Thus the [rule of] five or more quantities is completely explained by considering the Rule of Three Quantities. Just as this universe is pervaded by Lord Nārāyaṇa [Viṣṇu] (who removes the sufferings of those who worship him and is the sole generator of this universe), with his many forms—worlds and heavens and mountains and rivers and gods and men and demons and so on—in the same way, this whole type of computation is pervaded by the Rule of Three Quantities.
>
> If so, then what is [the purpose of] the many [different rules]? He responds:
>
> > Anything that is calculated in algebra or here [in arithmetic] by means of a multiplier and a divisor is understood by those of clear intelligence as just a Rule of Three Quantities. But for increasing the intelligence of dull-witted ones like us, it has been explained by the wise in many different and easy rules. (verse 241)

(The author of the base text, namely, Bhāskara himself, is referred to in the third person in accordance with commentarial etiquette.)

Verses 242–260. The pulverizer. Here Bhāskara gives rules for the operation of the pulverizer in its accustomed sense, that is, the "Euclidean division" solution method for linear indeterminate equations. Among other things, he

---

[30]The π value 22/7 was first mentioned by Āryabhaṭa II as an "accurate" value, although there are indications that it was known earlier: see [HayKY1989], pp. 8–11.

[31][Hay1997b]; [Sara1979], pp. 208–210.

points out that the solution can sometimes be simplified by first removing common factors from the remainder and one of the divisors, and adjusting the quantities that satisfy the new equation to give a solution for the original one. In other words, let the given indeterminate equation be $d_1 x = d_2 y + r$, where d_1, d_2, and r are known, and suppose, for example, that d_1 and r have a common factor a. Then if we find a solution to the new equation $(d_1/a)x = d_2 y + (r/a)$, say, $x = w$ and $y = z$, we will have $x = w$ and $y = az$ as a solution to the original equation.[32]

Verses 261–271. The "net of digits." Bhāskara seems to have introduced this name for problems in permutations and combinations. We have seen that certain combinatorial rules had been known for a very long time in the field of prosody and that some medieval mathematicians, including Bhāskara himself, discussed some more general rules on combinations as part of series or mixtures. At the end of the *Līlāvatī*, Bhāskara now treats them as a separate subject also embracing, for example, permutations of given digits in a number (hence the term "net of digits" to describe the grid-like appearance created when one writes down the "extension" or ordered list of all the possible permutations). Here is one of his sample problems illustrating the rule that the number of possible permutations of n things is $1 \cdot 2 \cdots n$, applied to the distribution of n iconographic attributes among the n hands of a statue of a deity:

> How many statues of Śiva [can] there be, with [his attributes] the rope, the elephant-hook, the serpent, the drum, the skull, the trident, the corpse-bier, the dagger, the arrow, and the bow held in his different hands? And how many of Viṣṇu with the club, the discus, the lotus, and the conch-shell? (verse 263)

Statement: The [number of] places is 10. The resulting [number of] different statues [of Śiva] is 3,628,800, and in the same way [the number of statues] of Viṣṇu is 24.

> [Only] a summary is stated [here] for fear of [going on too] long, since the ocean of calculation is endless. (verse 269)

6.2.2 The *Bīja-gaṇita*

The *Bīja-gaṇita*, or "Seed-computation," was evidently intended as a more advanced work than the *Līlāvatī* and consequently has accumulated fewer extant manuscript copies, fewer commentaries, and no doubt fewer readers. It also seems to have had fewer rivals, being the earliest extant independent treatise on algebra in the Indian tradition. Like the *Līlāvatī*, it appears to have been written as a continuously numbered stream of verses divided

---

[32] An overview of the major medieval developments in the pulverizer is provided in [Jai2001], pp. 202–204.

into topics or sections. This division is followed in the overview below, where we have space to examine only a few of the work's many interesting points. (Again, quoted verses are doubly indented within Bhāskara's own prose commentary.)[33]

Section 1. The six operations on positive and negative quantities. After an invocatory verse, Bhāskara gives a brief introduction to the current work:

> Previously mentioned [in the *Līlāvatī*] was the manifest whose source is the unmanifest. Since generally questions cannot be very well understood by the slow-witted without the application of the unmanifest [algebra], therefore I tell also the operation of algebra. (verse 2)

In the process of explaining the basic sign laws and operations involving them, he makes some brief remarks about mathematical notation. Most of the abbreviations he prescribes for algebraic symbols are familar from the Bakhshālī Manuscript (see section 5.2.1), but Bhāskara's negative sign is a dot over the negative number rather than a cross following it:

> > Three units and four units are together [both] negative or [both] positive or separately positive and negative or separately negative and positive. Tell me quickly [if] you know the addition of the two positive and negative [quantities]. (verse 4)
>
> Here the first letters of units and unknowns are written in order to represent [them], and quantities that are negative [have] a dot above [them].

Like Mahāvīra, Bhāskara holds that "there is no square root of a negative [number] due to its non-squareness" (verse 7).

Section 2. The six operations with zero. In the rules for the arithmetic of zero, there is an explicit association of division by zero with infinity, which was absent in the *Līlāvatī*:

> In this quantity also, which has zero as its divisor, there is no change even when many [quantities] have entered into it or come out [of it] just as at the time of destruction and of creation, when throngs of creatures enter into and come out of [him, there is no change] in the infinite and unchanging one [i.e., Viṣṇu]. (verse 11)

Does this mean that a "zero-divided" quantity is acknowledged as infinite in algebra but assigned some sort of indeterminate status in arithmetic? Or is

---

[33]Translations of verses 1–11, 23–26 (accompanied by selections from the commentary of Sūryadāsa), 70–76, 89–94, 113, 120, 125, 128, and 129 (the last few accompanied by Bhāskara's commentary) have been published in [Plo2007b], pp. 468–477.

it simply that the challenging concept of actual numerical infinity is deferred to the more advanced stage of instruction? The situation is further obscured by the fact that in sample problems in later sections of the *Bīja-gaṇita*, Bhāskara seems to expect that quantities both multiplied and divided by zero will be treated as finite, just as they were in the sections of the *Līlāvatī* covering operations with zero.

Section 3. The six operations with unknowns. Unknown quantities are represented by the names of colors, which are manipulated arithmetically in accordance with the following rules. The two sides of equations are explicitly mentioned in one of the sample problems, where *yā* stands for the unknown or *yāvattāvat* and *rū* for *rūpa* or unit:

> [One] unknown (*yāvattāvat*) is the color black, another blue, yellow, and red. [Colors] beginning with these have been imagined by the best of teachers to be the designations of the measures of the unknowns in order to accomplish their calculation.
>
> Among these [unknown quantities] the sum and difference of two having the same character [is as usual], but [for the sum and difference] of two having different characters, putting them separately [is required].
>
> [Example:] One positive unknown together with one unit and a pair of positive unknowns diminished by eight units.
>
> Oh friend! Tell [me] quickly, what is [the result] in the summing of these two sides?... (verses 1–14)

Statement: *yā* 1 *rū* 1. *yā* 2 *rū* $\dot{8}$. In adding them the result is *yā* 3 *rū* $\dot{7}$ ...

> ... There is, however, in the multiplication of units and a color, a color [as the result].
>
> But, in the multiplication of two, three, and so on [unknowns] which have the same character, there are their squares, cubes, and so on [as the results]. In the multiplication of [unknowns] which have different characters, [the result is] their product. (verses 15–16)

Section 4. The six operations with karaṇīs or surds. Here Bhāskara, rather than just giving a few techniques for manipulating *karaṇīs* or square roots of non-square quantities, attempts to set out a complete treatment for them in terms of the six operations of algebra. This requires some care, as not all of the procedures are obvious from their counterparts using ordinary numbers. In particular, he wants to have a meaningful square root operation for *karaṇī* expressions that will express the result in terms of *karaṇīs*, such that, for example,

$$\sqrt{n - \sqrt{k_1}} = \pm(\sqrt{k_2} - \sqrt{k_3}),$$

where n and k_1 are given non-square integers and k_2 and k_3 are integers. Thus, Bhāskara requires *karaṇī* expressions to have both positive and negative square roots, just as integers and fractions do. His square root method in this case gives $\frac{n \pm \sqrt{n^2 - k_1}}{2}$ for k_2 and k_3. We can see why this works by noting that

$$n - \sqrt{k_1} = k_2 + k_3 - 2\sqrt{k_2 k_3},$$

and then setting $k_2 + k_3 = n$ and $2\sqrt{k_2 k_3} = \sqrt{k_1}$; these two equations in two unknowns yield a quadratic in k_2 or k_3 that gives Bhāskara's result. Notice, however, that if $n^2 - k_1$ is not a perfect square, more *karaṇī* manipulations will be required to get the desired k_2 and k_3.[34]

Section 5. The pulverizer. Most of Bhāskara's rules in this section on linear indeterminate equations, and most of his sample problems as well, are repeated from the corresponding section in the *Līlāvatī*. It seems to be not entirely determined whether the pulverizer is properly part of arithmetic or algebra.

Section 6. Square-nature. This short section of seven verses gives some of the same basic rules for second-order indeterminate equations—that is, to find integers x, y such that $Nx^2 + k = y^2$ for given integers N and k—that we saw in chapter 18 of the *Brāhma-sphuṭa-siddhānta*.

Section 7. The cyclic method (for square-nature problems). The *Bīja-gaṇita* is apparently the first surviving text to present the famous general solution for $Nx^2 + 1 = y^2$ that is known as the "circle" or "cyclic" method. However, the method itself did not originate with Bhāskara, according to a citation that attributes it to (or at least attests to it in) the now lost work of an earlier mathematician.[35] The method relies on the fact that if a solution a, b is known for any one of the particular cases of the "auxiliary" second-order indeterminate equation $Na^2 + k = b^2$, where the given constant k equals -1, ± 2, or ± 4, then a solution x, y can be derived from it for $Nx^2 + 1 = y^2$. (The rules for deriving the desired solution in most of these cases were stated in the square-nature section in *Brāhma-sphuṭa-siddhānta* 18.64–71.) The cyclic method is an algorithm for reducing an auxiliary equation with arbitrary k to one of those particular cases.

The procedure is as follows. When a solution a, b is known for an arbitrary auxiliary equation $Na^2 + k = b^2$ for a given N and some integer k, then the pulverizer is used to choose an integer m such that $\frac{am + b}{k}$ is also an

[34] A full explanation of Bhāskara's *karaṇī* rules is given in [Jai2001], pp. 118–154.

[35] This is the author Jayadeva, quoted in a commentary dated to the late eleventh century; see [Shu1954].

integer and $|m^2 - N|$ is as small as possible. Now, because $Na^2 + k = b^2$ by assumption and $N \cdot 1^2 + (m^2 - N) = m^2$ trivially, we can "compose" these two simple auxiliary equations according to the method described in *Brāhma-sphuṭa-siddhānta* 18.64–71 to form a new one, namely

$$N(am + b)^2 + k(m^2 - N) = (bm + Na)^2.$$

And since $k(m^2 - N)$ is obviously divisible by k, and m was originally chosen so that $(am + b)$ would be divisible by k, it follows that $(bm + Na)$ is also divisible by k. Therefore we can produce yet another new auxiliary equation, $Na_1{}^2 + k_1 = b_1{}^2$, where

$$a_1 = \frac{am + b}{k}, \quad k_1 = \frac{m^2 - N}{k}, \quad b_1 = \frac{bm + Na}{k},$$

and a_1, k_1, and b_1 are all integers.

Now this method has come full circle: if k_1 is not equal to one of the desired values, namely, -1, ± 2, or ± 4, we simply start over, employing the pulverizer again to choose an integer m_1 such that $\frac{a_1 m_1 + b_1}{k_1}$ is also an integer and the difference $|m_1^2 - N|$ is as small as possible. Then we use that m_1 to create still another auxiliary equation $Na_2{}^2 + k_2 = b_2{}^2$ in the same way as before, and so on until eventually some k_i attains one of the desired values, when a solution to the original equation $Nx^2 + 1 = y^2$ can be derived from it.[36]

Section 8. Equations in one unknown. Most of this section on solving linear equations is taken up by sample problems, most of which are identical to ones illustrating variou arithmetic topics in the *Līlāvatī.* Perhaps Bhāskara is driving home his point from *Bīja-gaṇita*, verse 2, that "the unmanifest (algebra) is the source of the manifest (arithmetic)," and they are just different approaches to the same problems.

Section 9. "Elimination of the middle" (quadratics). Here Bhāskara mentions, among other things, the two square roots that may give two distinct solutions to a quadratic equation, as in the following sample problem:

> A fifth part of a troop of monkeys, minus three, squared,
> has gone to a cave; one is seen [having] climbed to
> the branch of a tree. Tell [me], how many are they?
> (verse 130)

> ... And here as before the result is a twofold amount, 50, 5. In this case the second is not to be taken due to its inapplicability. People have no confidence in [or "comprehension of"] a manifest [quantity] becoming negative.

[36]Bhāskara does not offer a demonstration for the claim that the cyclic method will always eventually produce some k_i equal to one of the desired values; in fact, this was not proved until the work of H. Hankel in the nineteenth century. See [DatSi1962], vol. 2, p. 172.

In this case, what we would write as the quadratic equation in x, where x is the desired number of monkeys in the troop, that is, $\left(\frac{1}{5}x - 3\right)^2 + 1 = x$, has been solved by Bhāskara as the equation *yā va* 1 *yā* 5̇5 *rū* 0 [=] *yā va* 0 *yā* 0 *rū* 2̇50, where *yā va* stands for *yāvattāvat varga*, "the square of the unknown." The two resulting values for x, from the two square roots in the quadratic, are 50 and 5. Only the first of these values, Bhāskara says, is "applicable" to this particular problem, because the second would make the "fifth part of the troop, minus three" equal to a negative number, which makes no sense in this context.

Section 10. Equations in more than one unknown. The remaining three sections build on the basic rules of Brahmagupta in the *Brāhma-sphuṭa-siddhānta.*[37] Linear equations in more than one unknown are solved by finding one unknown in terms of the others. If there are more unknowns than equations, so that the problem is indeterminate, the pulverizer is used.

Section 11. Elimination of the middle with more than one unknown. A single quadratic equation with more than one unknown is to be transformed into a square-nature problem if possible, and solved by indeterminate methods.

Section 12. Products of unknowns. Equations containing products of two or more unknowns are solved with arbitrarily chosen numbers, as directed by Brahmagupta.

Section 13. Conclusion. The last few verses contain information about Bhāskara's background and his work, surveyed at the beginning of this section.

6.2.3 The *Siddhānta-śiromaṇi*

This work (literally "Crest-jewel of *siddhāntas*") was composed when Bhāskara was 36, that is, in 1150. It lived up to the boast in its title by gaining a high place among astronomical treatises, although as an orthodox Brāhma-pakṣa work it could not supplant the canonical texts of the other *pakṣas*. The *Siddhānta-śiromaṇi* is sometimes described as containing the *Līlāvatī* and *Bīja-gaṇita* in addition to its two sections on astronomy proper, and it is evident that Bhāskara considered the subject matter of all these compositions to be very closely linked. But since the arithmetic and algebra texts have individual titles and have usually been copied as individual manuscripts, we will follow the tradition of considering the *Siddhānta-śiromaṇi* a separate work devoted to astronomy.

The treatise is divided into two sections: the first, on planetary calculations, presents standard computational algorithms like discussed in section 4.3 for calculating mean motions, true motions, the Three Questions, lunar and solar eclipses, and so forth. The second, on *gola*, contains chapters on the following subjects: praise of the sphere, the form of the sphere, the

[37] They are described more fully in [DatSi1962], vol. 2, pp. 57–59, 181–193, and 199–201, respectively.

sphere of the earth, explanations of various subjects from the *gaṇita* section, instruments, a poetic description of the seasons, and questions to test the student's knowledge. As it is not possible to do justice here to even a substantial portion of this comprehensive work, we will content ourselves with pointing out a few examples of Bhāskara's ingenious manipulations of small quantities and his explanations in his own commentary *Vāsanā-bhāṣya*, or "Commentary of rationales."

The first example, from the first section's chapter on true motions, involves calculating the speed of a planet's motion through the sky. We have seen in section 4.3.2 how to calculate a planet's mean speed R/D, where R is the number of the planet's integer revolutions in a given time period and D is the number of days in that period. But usually a planet's apparent motion will be slower or faster than that mean motion. Roughly, the speed will be least when the planet's anomaly κ (see section 4.3.3) is zero (i.e., when the planet is at apogee), greatest when the anomaly is $180°$, and close to the mean speed when the anomaly is about $90°$. If we want to know how fast a planet appears to be moving at some given time, how should we compute that? This problem of *tātkālika* or "at-that-time" motion was tackled in various ways by Indian astronomers; one typical strategy involved calculating the difference between the true and mean speeds as approximately proportional to the Sine-difference corresponding to the value of the anomaly. In following this approach, Bhāskara made use of a concept he called an "instantaneous Sine-difference," computed from the Cosine by a Rule of Three Quantities:

> The difference between today's and tomorrow's true [positions of a] planet ... is the true [daily] speed.... [At some point] within that time [or, on average in that time] the planet is required to move with that speed. Yet this is the approximate speed. Now the accurate [speed] for that time [or, instantaneous (*tātkālika*) speed] is described....
>
> If a Sine-difference equal to five-two-two [i.e., 225] is obtained with a Cosine equal to the Radius, then what [is obtained] with a desired [Cosine]? Here, five-two-two is the multiplier and the Radius is the divisor of the Cosine. The result is the accurate Sine-difference at that time.[38]

In other words, the "at-that-time" Sine-difference $\Delta \operatorname{Sin}$ for a given arc α is considered simply proportional to the Cosine of α:

$$\Delta \operatorname{Sin} = \operatorname{Cos} \alpha \cdot \frac{225}{R}.$$

It has been noted[39] that this and related statements reveal similarities between Bhāskara's ideas of motion and concepts in differential calculus. (In fact, perhaps these ratios of small quantities are what he was referring to in

---

[38] *Vāsanā-bhāṣya* on *Siddhānta-śiromaṇi* Ga.2.36–38, [SasB1989], pp. 52–53. The method is explained in detail in [Ike2004].

[39] For example, in [Rao2004], pp. 162–163.

his commentary on *Līlāvatī* 47 when he spoke of calculations with factors of 0/0 being "useful in astronomy.") This analogy should not be stretched too far: for one thing, Bhāskara is dealing with particular increments of particular trigonometric quantities, not with general functions or rates of change in the abstract. But it does bring out the conceptual boldness of the idea of an instantaneous speed, and of its derivation by means of ratios of small increments.

The *gola* section of the *Siddhānta-śiromaṇi* begins with an exhortation on the importance of understanding astronomy's geometric models that is somewhat reminiscent of Lalla's remarks quoted in section 4.5, but also emphasizes the need for their mathematical demonstration:

> A mathematician [knowing only] the calculation of the planets [stated] here [in the chapters on] mean motions and so forth, without the demonstration of that, will not attain greatness in the assemblies of the eminent, [and] will himself not be free from doubt. In the sphere, that clear [demonstration] is perceived directly like a fruit in the hand. Therefore I am undertaking the subject of the sphere as a means to understanding demonstrations.
>
> Like flavorful food without ghee and a kingdom deprived of [its] king, like an assembly without a good speaker, so is a mathematician ignorant of the sphere.[40]

An example of what Bhāskara means by a demonstration can be seen in his chapter on the terrestrial globe, where he criticizes the value for the size of the earth given by Lalla in the *Śiṣya-dhī-vṛddhida-tantra*, arguing that it is erroneous because of Lalla's erroneous rule for the surface area of a sphere. This rule, which Lalla allegedly stated in a now lost work on *gaṇita*, says that the surface area is equal to the area of a great circle times its circumference (or $2\pi^2 r^3$). Bhāskara remarks, "Because of the error in the computation stated by Lalla, the surface area of the [spherical] earth is wrong too."[41] He justifies his criticism by explaining his own rules from *Līlāvatī* 199–201 for the sphere's surface area and volume, which are equivalent to $4\pi r^2$ and $\frac{4}{3}\pi r^3$, respectively, for a sphere of radius r. He starts out by imagining equidistant great circles like longitude circles on the sphere's surface, and approximating the area of one spherical lune (a portion of the surface cut off between two adjacent great semicircles, like the skin of a segment of an orange):

> The circumference of a sphere is to be considered [as having] measure equal to the amount of Sines, times four [i.e., $24 \times 4 = 96$].

[40] *Siddhānta-śiromaṇi* Go.1.2–3, [SasB1989], pp. 175–176. See also [Srin2005], pp. 228–229.

[41] *Vāsanā-bhāṣya* on *Siddhānta-śiromaṇi* Go.3.54–57, [SasB1989], p. 187. See also [Cha1981] 2, pp. xx–xxi, 250–251, and [Hay1997b], pp. 198–199.

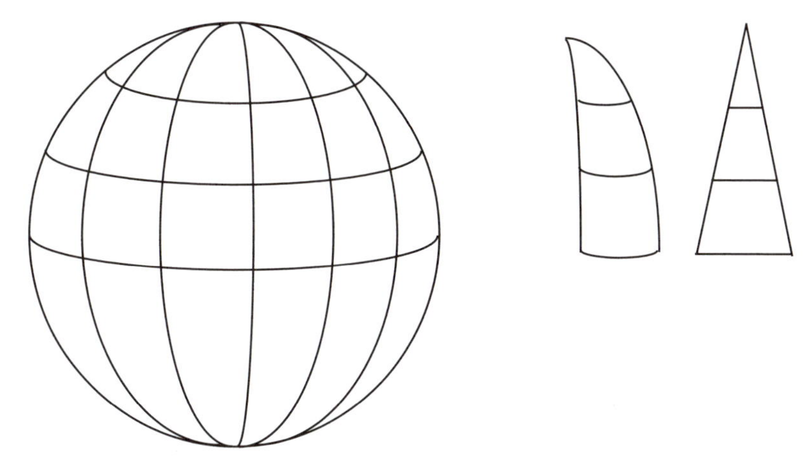

Figure 6.1 Dividing the surface of the sphere to compute its area.

> Spherical lunes [literally "mounds" or "circumferences"] are perceived [when the surface is] divided by multiple lines going from the top to the bottom as on the surface of the ball of an *amla* [Indian gooseberry] fruit. When one has set out spherical lunes [equal in] number to the prescribed [divisions of the] circumference ... the area in one lune is to be determined thus:...
>
> The circumference in the sphere is assumed [to be] equal to ninety-six cubits, and that many lunes are made with vertical lines at each cubit. Then when one has made horizontal lines at each cubit-interval on half of one lune, twenty-four lune-portions are imagined, [equal in] number to the Sines. Then the Sines separately divided by the Radius are the measures of the horizontal lines. In that case the lowest line is equal to [one] cubit, while [the lines] above are successively somewhat less, in accordance with [the sequence of] the Sines. The altitude is always just equal to [one] cubit. When one has found the areas of the portions equal to the sum of the base and the top multiplied by the altitude, [they are all] added. That is the area in half a lune; that times two is the area in one lune.[42]

The following explanation and amplification of Bhāskara's arguments rely on the sphere shown in figure 6.1, with circumference $4n$, where n is the number of Sines in a quadrant. (Bhāskara takes $n = 24$, but our figure for

[42] *Vāsanā-bhāṣya* on *Siddhānta-śiromaṇi* Go.3.58–61, [SasB1989], p. 188. This rationale is preceded in Bhāskara's discussion by a similar one that imagines a hemisphere divided into zones by small circles parallel to the equator, with a spherical cap on top. The areas of the zones spread out into long trapezoidal strips, plus the circular area of the spherical cap, add up to the total area of the hemisphere (*Vāsanā-bhāṣya* on *Siddhānta-śiromaṇi* Go.3.54–57, [SasB1989], pp. 187–188). See the description of both methods in [Sara1979], pp. 211–213, and the translation and explanation in [Hay1997b], pp. 202–217.

simplicity shows $n = 3$.) The sphere's surface is divided vertically into $4n$ equal lunes of unit width at the equator, and the top half of each lune is divided horizontally into n segments by small circles parallel to the equator at unit intervals.

The radius of each ith parallel circle (taking $i = 0$ at the equator) is proportional to the Cosine of its elevation above the equator. (If this is not immediately obvious, think—as Bhāskara certainly would have—of parallel day-circles on the celestial sphere whose radii are equal to the Cosines of their declinations, as explained in section 4.3.4.) If we let A_L be the area of one lune, we can consider that the area $A_L/2$ of the individual half-lune shown on the right side of figure 6.1 is approximately equal to that of the triangular figure corresponding to it. That triangle is a stack of $(n-1)$ trapezoids with a triangle on top, all of which are considered to have unit altitude. The equally spaced horizontal line segments dividing them are the chords of the corresponding unit arcs of the parallel circles in the half-lune. Assuming that the lowest of these line segments s_0 has unit length, and that the unit arc on the sphere contains u degrees, we can express the length s_i of each ith horizontal line segment by

$$s_i = 1 \cdot \frac{\mathrm{Cos}(iu)}{R} = 1 \cdot \frac{\mathrm{Sin}(90 - iu)}{R}.$$

Hence, as Bhāskara says, the measures of the line segments s_i are the Sines separately divided by the Radius. Then the area of the lowest trapezoid in the half-lune will be $\frac{s_0 + s_1}{2} \cdot 1$, and so on up to the top triangle, whose area will be $\frac{s_{n-1} + 0}{2} \cdot 1$. The sum of all of them will be the total area $A_L/2$ of the half-lune, which we may express in terms of the n Sines as follows:

$$\frac{A_L}{2} = \left(\frac{\mathrm{Sin}_n + \mathrm{Sin}_{n-1}}{2} + \frac{\mathrm{Sin}_{n-1} + \mathrm{Sin}_{n-2}}{2} + \ldots + \frac{\mathrm{Sin}_2 + \mathrm{Sin}_1}{2} + \frac{\mathrm{Sin}_1 + 0}{2}\right)\frac{1}{R}$$

$$= \left(\frac{\mathrm{Sin}_n}{2} + \sum_{i=1}^{n-1} \mathrm{Sin}_i\right)\frac{1}{R} = \left(\sum_{i=1}^{n-1} \mathrm{Sin}_i - \frac{\mathrm{Sin}_n}{2}\right)\frac{1}{R}.$$

The area A_L of the whole lune must be twice that amount. Since $\mathrm{Sin}_n = R$, area A_L will indeed be the sum of all the Sines minus half the Radius and divided by half the Radius, just as Bhāskara says.

To find a simpler expression for that sum of all the Sines, he then switches from a geometrical demonstration to a numerical illustration—that is, just adding up their known values:

> For the sake of determining that [lune area], this rule [was stated]: "The sum of all the Sines is decreased by half the Radius [and divided by half the Radius]" [verse Go.3.60cd]. Here the sum of all the sines beginning with 225 is 54,233. [When] that is decreased by half the Radius, the result is 52,514. [When] that is

> divided by half the Radius, the result is the area of one lune [and] equal to the diameter, 30;33. Because the diameter of a sphere with circumference ninety-six is just that much, 30;33, and the lunes are equal [in number] to [the divisions of] the circumference, therefore the area of the surface of the sphere is equal to the product of the circumference and the diameter; thus it is demonstrated.[43]

The sum of all the Sines in the standard Sine table, from $\text{Sin}_1 = 225$ to $\text{Sin}_{24} = 3438 = R$, is 54232 (Bhāskara says 54,233), which diminished by $R/2$ equals 52,513 (Bhāskara says 52,514). Dividing by $R/2$, we get a little over $30\frac{1}{2}$, or $30;33$ to the nearest sixtieth, which is in fact the diameter of a circle with circumference ninety-six units. From this Bhāskara infers the general result that the area A_L of the lune is equal to the diameter of the sphere. Since there are as many lunes as there are units in the sphere's circumference, the total area A of the sphere's surface therefore is just its circumference times its diameter.

Now that the formula for the surface area is demonstrated, Bhāskara uses it to explain the formula for the volume:

> And in the same way, that area produced from the surface of a sphere, multiplied by the diameter [and] divided by six, is called the accurate solid [volume] within the sphere.... Here is the demonstration: Square pyramidal holes [literally "needle-excavations"] with unit [base]-sides [and] depth equal to the half-diameter, [equal in] number to [the divisions of] the area, are imagined in the surface of the sphere. The meeting-point of the tips of the pyramids is inside the sphere. Thus the sum of the pyramid amounts is the solid amount; thus it is demonstrated.[44]

Using the same imagined unit grid, this time covering the whole of the sphere's surface, Bhāskara now considers it as made up of unit squares which are the bases of square pyramidal holes bored into the sphere, with depth equal to the sphere's radius r. The sum of the volumes of all the pyramidal holes is the total volume of the sphere. Bhāskara leaves it to the reader to recall that the volume of each pyramid will be one-third the product of its depth and the area of its base. So the sum of the volumes will be one-third the product of the total surface area times the depth, or $\frac{1}{3}A \cdot r = \frac{1}{6}A \cdot 2r$.

Elsewhere in the *Siddhānta-śiromaṇi*, in his chapters on eclipses, Bhāskara again criticizes Lalla's mathematics, this time concerning the geometry underlying computations for eclipse diagrams. As we saw in section 4.3.5, these computations involve a quantity called the "deflection" or deviation of the path of the ecliptic away from the east-west direction on the disk of the

[43] *Vāsanā-bhāṣya* on *Siddhānta-śiromaṇi* Go.3.58–61, [SasB1989], pp. 188–189.

[44] *Vāsanā-bhāṣya* on *Siddhānta-śiromaṇi* Go.3.58–61, [SasB1989], p. 189. See also [Sara1979], p. 213.

moon. Lalla computes a component of this deflection with the Versine of a particular quantity, but Bhāskara is adamant that the Sine should be used instead:[45]

> Some make the deflection computation with the Versine; for the sake of rejecting that, the regular [i.e., non-versed] Sine is [explicitly] specified here.... Those who state these two deflection [components] by means of a rule [using] a Versine do not state the motion of the sphere correctly.

We can explain this debate by considering the spherical geometry of the deflection and its two components. If (a) the celestial equator were always perpendicular to the horizon and (b) the moon always moved parallel to the celestial equator, there would be no deflection: the moon in a solar or lunar eclipse would always be moving due east (while the entire sky turned due west) without any drifting northward or southward. Condition (a) holds at locations with zero terrestrial latitude, where the celestial equator passes directly overhead through the local zenith. Condition (b) holds when the moon is at a solstice point, where the instantaneous direction of motion along the ecliptic is parallel to that at the corresponding point of the equator. In most cases, though, the celestial equator leans southward from the zenith (for observers in the northern hemisphere), and the direction of motion on the ecliptic is not parallel to that on the equator; both these effects contribute to the deflection of motion on the ecliptic away from due east-west motion. The two effects are computed separately in Indian astronomy as "deflection due to terrestrial latitude" and "deflection due to tropical longitude."

It is the deflection due to tropical longitude that sparks Bhāskara's dispute with Lalla, so we will treat it in isolation by imagining ourselves at a locality on the earth's equator, where the other deflection component is zero. Figure 6.2 shows the celestial sphere for a location at zero terrestrial latitude, where the celestial equator passes through the zenith Z and the west point W of the horizon and the north celestial pole P coincides with the north point N of the horizon. The ecliptic pole is at P', and the moon is on the ecliptic (or quite close to it) at M. The vernal equinox or intersection of the ecliptic and equator, where tropical longitude is zero, falls at W; the summer solstice, where tropical longitude equals $90°$, falls at C. The arc WM is the moon's present tropical longitude λ_M, MH is its declination δ_M, and PP' is the obliquity of the ecliptic, approximately $24°$.

In the spherical triangle $CP'M$, $\angle CP'M = \angle PP'M$ is the complement of the longitude WM, or $90° - \lambda_M$, and the arc PH is $90°$. The deflection due to tropical longitude at M, or the angle between the direction of the ecliptic at M and the equator turning due east-west, is considered in Sanskrit astronomical texts to be what we would call the angle PMP' between the arcs $P'M$ and PMH perpendicular to the ecliptic and equator, respectively. That is, the deviation from the north-south direction of arc $P'M$ perpendic-

[45] *Vāsanā-bhāṣya* on *Siddhānta-śiromaṇi* Ga.5.20–23, [SasB1989], pp. 119–120.

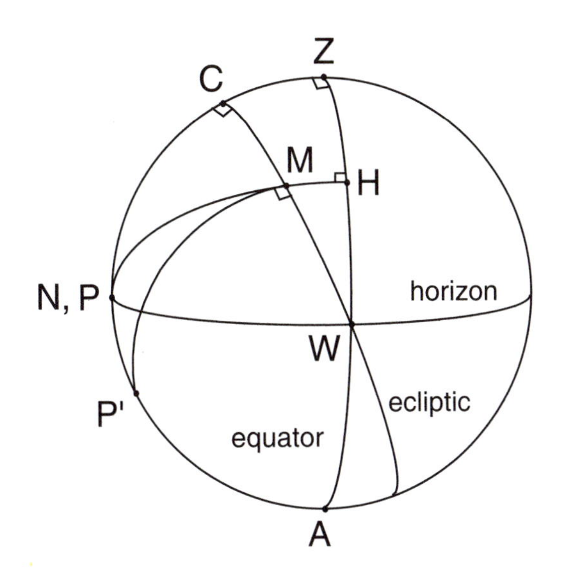

Figure 6.2 The longitude component of the deflection of the ecliptic.

ular to the ecliptic is equated to the deviation from the east-west direction of the moon's motion at M.

Lalla computes this angle with a rule equivalent to the following formula:[46]

$$\operatorname{Sin} PMP' \approx \operatorname{Sin} 24^\circ \cdot \frac{\operatorname{Vers}(90^\circ - \lambda_M)}{R},$$

while Bhāskara uses instead one equivalent to

$$\operatorname{Sin} PMP' \approx \operatorname{Sin} 24^\circ \cdot \frac{\operatorname{Sin}(90^\circ - \lambda_M)}{R}.$$

Both formulas will give identical results at the two extremes, when the longitude λ_M is 0° or 90°. But which is more correct in between?

Appealing to the law of sines for spherical triangles, as applied to triangle $PP'M$, we find the true expression for the desired angle to be:

$$\frac{\operatorname{Sin} PMP'}{\operatorname{Sin} PP'} = \frac{\operatorname{Sin} PP'M}{\operatorname{Sin} PM} = \frac{\operatorname{Sin} PP'M}{\operatorname{Cos} MH}$$

or

$$\operatorname{Sin} PMP' = \operatorname{Sin} 24^\circ \cdot \frac{\operatorname{Sin}(90^\circ - \lambda_M)}{\operatorname{Cos} \delta_M}.$$

Evidently, both Lalla's and Bhāskara's formulas are somewhat approximate in that they use R in place of $\operatorname{Cos} \delta_M$. If that approximation is overlooked,

[46]For Lalla's deflection rule in *Śiṣya-dhī-vṛddhida-tantra* 5.25, see [Cha1981], vol. 2, pp. 123–127. Bhāskara's is given in *Siddhānta-śiromaṇi* Ga.5.20–21, [SasB1989], p. 119.

then Bhāskara is right that the use of the Sine is mathematically preferable to that of the Versine. Moreover, he provides in the eclipse chapter of his *gola* section a physical rationale for his claim somewhat similar to the one explained above, where he isolates the desired component of deflection by considering a location at zero latitude, just as we did (although he does not invoke spherical trigonometry as such).[47]

It is all the more mystifying, then, that when Bhāskara wrote a commentary on Lalla's *Śiṣya-dhī-vṛddhida-tantra* and discussed its deflection rules, he supported Lalla's use of the Versine in this formula. His justification of it there seems somewhat halfhearted, merely pointing out that the Versine in question "increases in sequence" with the deflection—that is, the two are approximately proportional.[48] But he does not explicitly contradict Lalla's position while expounding Lalla's work. Clearly, the concept of "demonstration" in Sanskrit mathematics and the role it plays in determining right answers are much more complex than they may at first appear, as we discuss further at the end of this chapter.

6.2.4 The *jyotpatti*

Evidently intended as a part of the *Siddhānta-śiromaṇi*'s *gola* section, but often copied separately, is Bhāskara's composition called the *jyotpatti* or "construction of Sines," the same name used by Bhāskara I in his explanation of Sine-construction discussed in section 5.1.2.[49] This is a group of twenty-five verses stating formulas for determining a variety of Sines, not just the twenty-four values of the standard Sine table. Bhāskara introduces it in his first verse, as follows:

> Bhāskara, since he follows the path of the teachers in knowing the construction of Sines, discusses that in its many forms for the pleasure of clever mathematicians.

His first "form" of constructing Sines is purely mensurational: make a circle of the desired size, mark it with degrees, and measure the half-Chords in it with a string. He then states some rules for finding "by means of computation half-Chords which are very accurate," which in order of appearance are equivalent to the following basic trigonometric identities (where α is some given arc):

$$\operatorname{Cos}\alpha = \sqrt{R^2 - \operatorname{Sin}^2\alpha},$$
$$\operatorname{Vers}\alpha = R - \operatorname{Cos}\alpha,$$

[47] *Vāsanā-bhāṣya* on *Siddhānta-śiromaṇi* Go.8.30–74, [SasB1989], pp. 224–228.

[48] Commentary on *Śiṣya-dhī-vṛddhida-tantra* 5.25, [PandeC1981], p. 112.

[49] An earlier chapter of the *gola* section also treats the same subject, after which Bhāskara says, "Hence we will state the remaining Sine constructions subsequently" (*Vāsanā-bhāṣya* on *Siddhānta-śiromaṇi* Go.5.2–6, [SasB1989], p. 199.) All the cites from this "subsequent" *jyotpatti* chapter are taken from the version of it in *Siddhānta-śiromaṇi* Go.14, [SasB1989], pp. 281–285.

$$\operatorname{Sin} 30^\circ = \frac{R}{2},$$

$$\operatorname{Sin} 45^\circ = \sqrt{\frac{R^2}{2}},$$

$$\operatorname{Sin} 36^\circ = \sqrt{\frac{5R^2 - \sqrt{5R^2}}{8}} \approx R \cdot \frac{5878}{10000},$$

$$\operatorname{Sin} 18^\circ = \frac{\sqrt{5R^2} - R}{4},$$

$$\operatorname{Sin}\left(\frac{\alpha}{2}\right) = \frac{\sqrt{\operatorname{Sin}^2\alpha + \operatorname{Vers}^2\alpha}}{2} = \sqrt{\frac{R \operatorname{Vers}\alpha}{2}}.$$

Bhāskara says that the above methods were "thus stated by previous authorities; hence I will state the rest." This appears to imply that the remaining rules, equivalent to the following identities where α and β are given arcs, are Bhāskara's own discoveries:

$$\operatorname{Sin}\left(\frac{90^\circ \pm \alpha}{2}\right) = \sqrt{\frac{R^2 \pm R \operatorname{Sin}\alpha}{2}},$$

$$\operatorname{Sin}\left(\frac{\alpha - \beta}{2}\right) = \frac{\sqrt{(\operatorname{Sin}\alpha - \operatorname{Sin}\beta)^2 + (\operatorname{Cos}\beta - \operatorname{Cos}\alpha)^2}}{2},$$

$$\operatorname{Sin}\left(\frac{\alpha - (90^\circ - \alpha)}{2}\right) = \sqrt{\frac{(\operatorname{Sin}\alpha - \operatorname{Cos}\alpha)^2}{2}},$$

$$\operatorname{Sin}(\alpha - (90^\circ - \alpha)) = R - \frac{2\operatorname{Sin}^2\alpha}{R},$$

$$\operatorname{Sin}(\alpha \pm 1^\circ) = \operatorname{Sin}\alpha \cdot \frac{6568}{6569} \pm \operatorname{Cos}\alpha \cdot \frac{10}{573},$$

$$\operatorname{Sin}(\alpha \pm 3;45^\circ) = \operatorname{Sin}\alpha \cdot \frac{466}{467} \pm \operatorname{Cos}\alpha \cdot \frac{100}{1529},$$

$$\operatorname{Sin}(\alpha \pm \beta) = \frac{\operatorname{Sin}\alpha \operatorname{Cos}\beta \pm \operatorname{Sin}\beta \operatorname{Cos}\alpha}{R}.$$

Of this *pièce de résistance*, the general rule for $\operatorname{Sin}(\alpha \pm \beta)$, Bhāskara says that "this accurate Sine-creation is called in one case the sum-creation and in the other the difference-creation." Indeed, most of his other rules appear to be derived from applications of this formula or the previous half-angle formula. Intriguingly, his own short commentary on these verses merely explains how to use the rules, without providing demonstrations or derivations for them.

6.2.5 The *Karaṇa-kutūhala*

This text, whose name means loosely "the wonder of *karaṇas*," has an epoch date corresponding to 1183; it condenses and approximates the techniques of Bhāskara's earlier *Siddhānta-śiromaṇi* in typical *karaṇa* form. Its ingeniously devised rules recall the remarks of Vaṭeśvara and Sūryadeva that we saw in section 4.4.1, to the effect that *karaṇa* techniques are supposed to be somewhat obscure and markedly original. A characteristic example is Bhāskara's formula for computing the terrestrial latitude:

> A sixtieth part of the [noon] equinoctial shadow increased by 410 is increased by the equinoctial hypotenuse. The equinoctial shadow is divided by this [and] multiplied by 90. The degrees of latitude south [of the location, i.e., between the location and the equator] should be [the result].[50]

Considering the same situation and using the same notation as in our earlier discussion of figure 4.5, we let ϕ be the desired latitude, and take the noon equinoctial shadow of the standard twelve-digit gnomon to be $s = 12 \tan \phi$ and the shadow hypotenuse to be $h = \sqrt{s^2 + 144}$. Thus Bhāskara's rule can be rewritten as the following approximation:

$$\phi \approx \frac{90s}{h + \dfrac{s + 410}{60}}.$$

Although the text of the *karaṇa* itself does not derive or demonstrate this peculiar rule, we can suggest the following possible reconstruction as an attempt to understand the process of creating such approximations. Let us speculate that Bhāskara started with a much simpler and cruder version of this approximation, namely,

$$\frac{\phi}{90} \approx \frac{s}{h}.$$

and looked at a couple of easily computed solutions to it (noting that it is already exactly accurate for $\phi = 0$). For example, let us consider the results for $\phi_1 = 30°$ and $\phi_2 = 45°$, respectively. We see that the denominator of the approximating ratio s/h is far too small:

$$\frac{s_1}{h_1} = \frac{6;56}{13;52} >> \frac{30}{90} \quad \text{when} \quad \frac{\phi_1}{90} = \frac{30}{90} = \frac{1}{3} = \frac{6;56}{13;52 + 6;56},$$

and

$$\frac{s_2}{h_2} = \frac{12}{16;58} >> \frac{45}{90} \quad \text{when} \quad \frac{\phi_2}{90} = \frac{45}{90} = \frac{1}{2} = \frac{12}{16;58 + 7;2}.$$

(We express the shadow lengths and their hypotenuses sexagesimally, as is standard in Sanskrit astronomical texts.)

---

[50] *Karaṇa-kutūhala* 3.16, [Mish1991], p. 45. For a fuller discussion of this rule (somewhat disfigured by misprints), see [Plo1996b], pp. 60–61.

The accuracy of the approximation in both cases could be improved by adding something to h_1 or h_2 in the denominator. We see from the above examples that the added term should increase slightly as ϕ increases. The shadow s itself of course is something that increases with increasing ϕ, and it would be just the right size for the added term in the first case but too big in the second case. Instead, let us add in one sexagesimal part or sixtieth of the shadow plus some extra unknown term:

$$\frac{30}{90} = \frac{6;56}{13;52 + 6;56} \approx \frac{6;56}{13;52 + s_1/60 + x_1} = \frac{6;56}{13;52 + 0;6,56 + x_1},$$

and

$$\frac{45}{90} = \frac{12}{16;58 + 7;2} \approx \frac{12}{16;58 + s_2/60 + x_2} = \frac{12}{16;58 + 0;12 + x_2}.$$

Solving both these equations for their unknown terms x_1 and x_2, we see that the value of x_1 in the first case is $6;49,8$, while in the second case x_2 is $6;50$—almost identical! So if we take $6;50$ for our desired constant term in the denominator, we can write our final version of the approximation generally as

$$\frac{\phi}{90} \approx \frac{s}{h + \frac{s}{60} + 6;50} = \frac{s}{h + \frac{s + 410}{60}},$$

which is exactly Bhāskara's rule.

Such a technique of tweaking a simple and crude approximation to agree with a few known values, and thus deriving a much more accurate approximation, has been used plausibly to account for the form of other Indian approximation rules as well.[51] Nonetheless, the technique is not explicitly described in any Sanskrit text and thus remains a conjectural reconstruction of how the secretive *karaṇa* makers actually formed their algorithms.

6.3 THE WORKS OF NĀRĀYAṆA PAṆḌITA

The most significant Sanskrit mathematics treatises after those of Bhāskara II (if we pass over the works of the Kerala school discussed in the following chapter, which were brilliantly original but apparently not widely read) are those of an obscure fourteenth-century scholar named Nārāyaṇa, usually titled Paṇḍita, or "learned." Nārāyaṇa's texts largely followed the structure of mathematical knowledge as expounded in the canonical *Līlāvatī* and *Bīja-gaṇita*, but also modified and expanded it in some novel ways. The rest of this section touches on some aspects of these differences.[52]

[51]See, for example, [Plo1996b], pp. 57–59, and for a somewhat similar reconstructed procedure [HayKY1990], pp. 161–169.

[52]Most of the following information on the *Gaṇita-kaumudī* is drawn from [Kus1993]; see also [DviP1936]. Verses 13.1–2, 13.25–27, 13.49–54, and 14.1–9 of the *Gaṇita-kaumudī*, with excerpts from their accompanying commentary, are reproduced from [Kus1993] in [Plo2007b], pp. 499–503. The *Bīja-gaṇitāvataṃsa* is surveyed in detail, and part of it edited and translated, in [Hay2004]; another portion of it has been published in [Shu1970].

6.3.1 The *Gaṇita-kaumudī*

Nārāyaṇa's arithmetic treatise, the "Moonlight of mathematics," states in its final verse that Nārāyaṇa ("a navigator on the ocean of mathematics") completed it on a date corresponding to 10 November 1356. It is clearly modeled (up to a point) on the content and style of the *Līlāvatī*, and it has a concise prose commentary with worked examples like the *Līlāvatī*'s (which, however, may or may not be by Nārāyaṇa himself).[53] After its opening sections on metrology and operations, it has a section on "miscellaneous methods," the standard eight procedures from mixtures up through shadows, and a section on the pulverizer. Then it departs from the *Līlāvatī* format with a section on square-nature problems (which have previously been treated exclusively in algebra texts), one on factorization of integers, one on partitions of unity into unit-fractional parts, one on combinations and permutations (called the "net of digits" as in the *Līlāvatī*), and finally one on "auspicious calculation," that is, magic squares.

Even in the earlier sections tallying more closely with Bhāskara II's arithmetic, Nārāyaṇa introduces a number of rules from other sources or of his own invention. He evidently did not share Bhāskara's objections (or incomprehension) concerning the condition that quadrilaterals treated in geometry must be cyclic, for he states a number of rules that in fact apply only to cyclic quadrilaterals, such as the following:

> [In the case] of all quadrilaterals, when the top [and an adjacent side are] exchanged, then [there is] another, third diagonal: thus there are three diagonals [for a quadrilateral with given sides and area].[54]

That is, when two adjacent sides are interchanged in a quadrilateral, a new quadrilateral is formed with (at least) one of its diagonals unchanged. This is true for cyclic quadrilaterals, but not for arbitrary ones.

Nārāyaṇa's treatments of permutations and combinations and magic squares are very comprehensive. Building on earlier rules for numbers of possible variations derived from problems in prosody and series, he lays out algorithms for determining the "extensions" or serial lists of permutations and for solving combinatorial problems, along with various types of tables to make the computations easier. The contexts of these problems range from prosody (possible combinations of syllables to form meters) to music (possible combinations of notes or beats to form melodies or rhythmic patterns, respectively) to purely numerical combinations. A typical example of the latter category is a rule for finding, for a given integer n, all the possible n-digit numbers that have a given final digit.

Nārāyaṇa seems to have been one of the first authors to introduce magic squares as a topic in Indian arithmetic, although they were certainly known

[53] See the discussion of the authorship of the commentary in [Kus1993], pp. 199–202, and [Hay2004], pp. 388–390.

[54] Quoted and discussed in [Sara1979], p. 96.

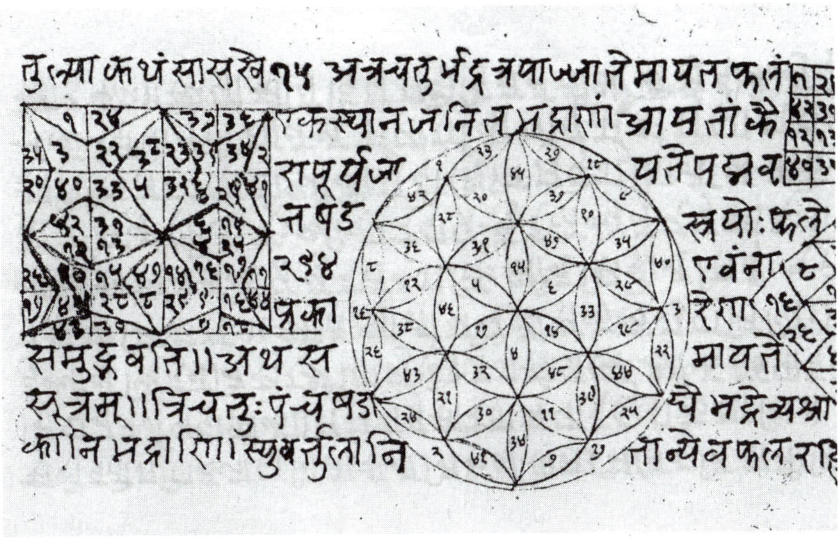

Figure 6.3 Elaborate magic "squares" in a manuscript of the *Gaṇita-kaumudī*.

in India since at least the sixth century.[55] What motivated him to do so, according to his opening verse in the section, was the desire "to surprise good mathematicians, to please those who know magic diagrams, and to dispel the pride of poor mathematicians." In a verse near the end he notes that magic squares "destroy the stupidity of mathematicians and bestow an auspicious mind."[56] It may be that exposure to Islamic magic squares, which date back to the ninth century,[57] piqued Indian mathematicians' interest in exploring the theory behind them.

Nārāyaṇa gives exhaustive rules for constructing not only ordinary magic squares but also variants made by combining multiple magic squares, as in the diagrams shown in figure 6.3.[58] The magic figure on the left, consisting of a square 4×4 grid, is the so-called "twelve-rayed" figure, where the twelve numbers in each quarter of the square add up to the same "magic sum," and so do the twelve numbers in its central block. The figure in the center is a "magic lotus," where the twelve numbers in each of the seven small circles within the large circle add up to the magic sum (which is 294 in both cases pictured). Somewhat dauntingly, Nārāyaṇa says in the above-cited closing verse that there are even more categories of such figures, "characterized by the cubes and square roots which [require] abundant calculations," that are not even mentioned in his work.

[55] See [Kus1993], pp. 161–177, for a discussion of magic squares in an astrological work of Varāhamihira and in an early fourteenth-century Prakrit mathematical work.

[56] The above quotations are from [Kus1993], p. 375 and p. 422, respectively.

[57] [Ses2004], p. 715.

[58] The image shows part of f.134v of MS Benares (Sampurnanand Sanskrit University) 104595. My thanks to Professor Takanori Kusuba for this copy of it.

6.3.2 The *Bīja-gaṇitāvataṃsa*

This "Garland of algebra," largely imitative of Bhāskara's *Bīja-gaṇita*, has fewer innovative features than the *Gaṇita-kaumudī*. It differs from the *Bīja-gaṇita*, though, in dealing with a curious subject called "series figures," treated more fully by Nārāyaṇa in the *Gaṇita-kaumudī* and apparently introduced several centuries earlier by the mathematician Śrīdhara (and possibly known in the Bakhshālī Manuscript as well).[59] These figures model computations for an arithmetic progression graphically by representing the series as an isosceles trapezium, where the altitude is the number of terms and the area is the sum of the series. The top parallel side t of the trapezium then turns out to be the first term a_1 of the series minus half its constant difference d, or $t = a_1 - d/2$, while the opposite side or base b is equal to $t+hd$, where h is the altitude or number of terms. (If the constant difference d were zero—that is, if the "progression" were just a sequence of identical terms—the figure would turn out to be a rectangle.)

A series with a negative first term, for example, is represented by "flipping" the top side so that the two nonparallel sides cross, forming two isosceles triangles like a cross section of a double cone, instead of a trapezium. One of the triangles then has positive area and the other negative area, and their difference is the net sum of the series. (Could such graphical models have been the reason that Bhāskara I, back in the seventh century [see section 5.1.2], classified series calculations as geometry instead of as numerical computation?)[60]

6.4 MATHEMATICAL WRITING AND THOUGHT

Now that we have explored a wide variety of medieval Sanskrit texts treating mathematics, it seems appropriate to pause and consider what general conclusions can be drawn about how Indian mathematicians in this period actually thought about and produced mathematics. This discussion is perforce limited to works dealing specifically with *gaṇita* in the context of the Classical Sanskrit learned sciences or *śāstras*. It is unfortunately beyond the scope of this work to explore the vast and fascinating ramifications of mathematical concepts in other areas of Indian thought, for example, in philosophical doctrines or Jain and Buddhist cosmology.

6.4.1 Subject matter and relation to other fields

First of all, what is mathematics, as defined in the Sanskrit tradition? In Mahāvīra's very comprehensive assertion at the beginning of the *Gaṇita-sāra-saṅgraha*, *gaṇita* or calculation applies to all disciplines—from religious ritual to sexology to grammar to astronomy—wherever quantification is in-

---

[59]See the discussion of series figures in [Sara1979], pp. 239–243.

[60][Hay2004], pp. 486–492; [Sara1979], pp. 238–242.

volved. In its broadest sense, then, *gaṇita* is computational procedures of any kind. Bhāskara I offered a somewhat more specific definition of "*gaṇita* as a whole," comprising particular subjects such as "figures, shadows, series, equations [and] the pulverizer," as well as the astronomical topics of planetary calculations, time reckoning, and the sphere. Other types of categories reveal different perspectives: for example, mathematics as an independent topic is generally divided into arithmetic and algebra, or known quantities versus unknown ones, while mathematical astronomy is frequently divided into *gaṇita* and *gola*, predictive algorithms versus the cosmology and geometry underlying them.

Within these broad divisions mathematical content is further subdivided, as in the operations and procedures of arithmetic and the different types of equations in algebra. Some of these classifications seem very natural and intuitive to us, because they correspond to our modern ways of categorizing mathematical topics. Others force us to think a little harder about the underlying epistemological assumptions, sometimes without arriving at a satisfactory understanding of them. For example, the construction and use of Sine tables is not discussed in nonastronomical mathematical treatises, although such treatises do explain the basic geometry of circles, triangles, and quadrilaterals that allows one to determine Sines. (We have seen how Bhāskara II in the *Līlāvatī* discusses circles, chords, arcs, and inscribed regular polygons but explicitly defers the explanation of Sines to a different treatise.) We might infer that trigonometry per se was seen as a specialized astronomical application of geometry that doesn't belong in the more abstract treatment of mathematics in general. On the other hand, rules on shadows and volumes of excavations and heaps could also be seen as particular applications of more abstract geometrical relations, but they are not excluded from arithmetic texts.

Similarly, general solution procedures such as "operations with an assumed quantity" (false position) and inversion can appear in works on arithmetic, while more complicated procedures for solving particular types of equations appear in works on algebra. But techniques for solving for unknown quantities iteratively, although they are widely employed in astronomical contexts, are never discussed anywhere in a strictly mathematical text (if we discount root-extraction algorithms). Neither are the approaches for constructing various approximation rules such as the ones we often see in *karaṇas*. What does this imply about Indian mathematicians' ideas on what belongs in mathematics proper and what constitutes a sort of applied, or esoteric, mathematical knowledge? The proverbial ocean of *gaṇita* is clearly much wider and deeper than we may have thought, and much of its extent remains uncharted.

If the internal structure of Sanskrit mathematics still contains mysteries, so does its external structure with respect to other genres of Sanskrit literature. Traditionally, *gaṇita* is not a separate *śāstra* (although Mahāvīra calls it a *śāstra* in the *Gaṇita-sāra-saṅgraha*) but rather a subdivision of the *śāstra* of *jyotiṣa*; its name is also sometimes applied to mathematical astronomy in particular. Other *śāstras* like poetic metrics or music may use some mathematical techniques, but the historical relationship between mathematics and astronomy remains uniquely close.

What was the relationship between mathematics and other areas of Sanskrit learning? Perhaps it was rather one-sided: mathematical texts were grounded in the basics of the fundamental Sanskrit disciplines such as grammar and logic, but treatises on these subjects and related ones like philosophy and Vedic exegesis generally did not invoke explicitly mathematical technical concepts or quote from mathematical treatises in their arguments. As pointed out in section 1.3, mathematical deduction seems not to have been assigned a uniquely certain kind of truth in the Sanskrit tradition as it was in the Hellenistic Greek tradition. And the study of advanced mathematics does not seem to have been common even among learned Brāhmaṇas, except for the ones whose hereditary profession was astronomy. At least, we know of no major medieval Sanskrit mathematicians or astronomers who were also renowned as, say, poets or philosophers or authorities on *dharma*, although some displayed their nonmathematical erudition now and then in their technical works.[61] We can infer, at least generally and tentatively, that mathematics remained mostly a technical specialty for members of those *jātis* professionally concerned in it, on the periphery of the core disciplines of Sanskrit learning.

6.4.2 Orality and exposition

Medieval Sanskrit mathematical treatises, starting with the earliest surviving ones around the middle of the second millennium, were routinely accompanied by long prose commentaries ill-suited for memorization. Thus it seems unlikely that many medieval mathematical authors relied only on recited or memorized texts and oral explanations of them to acquire or disseminate their technical knowledge. The bias toward writing inherent in complicated calculations too lengthy for short-term memory doubtless strengthened the need for written explanations of their procedures. Certainly from Bhāskara I in the early seventh century to Nārāyaṇa in the fourteenth, solutions to sample problems were typically presented with their "setting down" or numerical and notational layout. Mathematicians, even in the world of Sanskrit learning, must write, and written texts showed them what to write.

Nonetheless, the authors of treatises continued the tradition of composing short, easily learned texts in mnemonic verses as part of an orally transmitted textual corpus. ("Short" seems to have been a somewhat subjective term: the *Āryabhaṭīya* with its 121 verses, the *Gaṇita-sāra-saṅgraha* with its 1130-plus, and the *Tri-śatikā* with its 180 or so are all self-described as "concise" or an "epitome.") Orality and its genres evidently retained their special status as the archetype of learning, even if they could not supply all the needs of study or pedagogy in mathematics (or for that matter, in any other

[61] There are, however, a few such polymaths known later in the second millennium: see sections 7.2 and 9.1.

learned discipline).[62]

In this archetype, authors built on the work of their predecessors to form continuous although not stagnant traditions. Mathematical texts repeated many of each other's rules but also supplied new ones and presented different arrangements of mathematical subjects. Since there were undoubtedly many works that have not survived, and since Indian mathematicians do not always identify the immediate sources of their rules or claim credit for original ones, the attribution of a particular rule to a particular author usually remains to some extent hypothetical, and the same is true for attributions of particular ways to organize mathematical content.

6.4.3 The role of commentaries

The demands made on the student's comprehension by versified base texts were sometimes severe. (See section A.4 for a discussion of how the deliberate semantic ambiguity of literary Sanskrit may have been employed to test the student's competence.) However, a student hopelessly baffled by an unclear verse could generally turn to a commentary. Commentators might repeat the verse's rule in a more prolix and comprehensible form, or painstakingly step through its grammar word by word, or illustrate it by a sample problem, or explain its rationale mathematically, in more or less detail, or some or all of the above.[63] Mathematical commentaries, in their manuscript form, are where we also find the graphical figures and notational methods that orally presented verse texts cannot convey. Thus they served as a bridge between the oracular pronouncements of the base-texts and the actual manipulations that a mathematician needed to perform in order to solve problems.[64]

The prototype of Sanskrit commentaries was a canonical work from the late first millennium BCE, the *Mahābhāṣya* or "Great commentary" of Patañjali on the great grammatical codification of Pāṇini, with which all Sanskrit scholars would have been familiar. This work (or predecessors which it imitated) set the style of explaining the grammatical meanings of rules, illustrating them with examples to disambiguate their phrasing, and interpreting them when possible in multiple ways so as to draw out of them the maximum amount of information. All these features influenced the format of commentaries in other Sanskrit disciplines, including mathematics.

A mathematical commentary therefore had not only to confirm what the

---

[62]The relationship between orality and literacy in Sanskrit mathematical sciences is examined in, for example, [Yan2006] and [Fil2005].

[63]The following discussion of mathematical commentaries is largely inspired by [Kel2006], vol. 1, pp. xl–liii, and by [Bro2006]. [TuBo2006] gives a detailed technical survey of Sanskrit commentaries in general.

[64]The details of these manipulations are seldom described in the texts or attested in the manuscripts. In a worked example the given numbers are set down in the prescribed layout and the results of operations are stated, but we do not see the graphical working of such steps as, for example, taking a square root or multiplying two multidigit numbers. Various procedures for these ephemeral manipulations as reconstructed from some textual descriptions, including those in later works, are presented in [DatSi1962], vol. 1, pp. 128–197 passim.

verses literally meant but also to convince the reader that they were properly formed and appropriate for their purpose. In the standard model of Pāṇini's grammar, a *sūtra* or rule is supposed to be concise, accurate, not redundant or unnecessarily prolix, and useful in multiple contexts. A commentator's job was partly to show that the *sūtras* in his chosen base text met these criteria. Thus, writing a commentary was to some extent an act of advocacy: if you undertook to expound a treatise, you were committed to making its case, to representing the perspective of its author as well as you could. A commentator certainly might disagree with the author of his base text on some points but usually did not make an issue of them, preferring to suggest a different interpretation that harmonized better with his own views or just to pass over the topic in silence. Presumably, if you wanted to be a critic rather than an advocate, you should write your own treatise to put forth your own opinions.

Consequently it is not always possible to infer what a commentator thought about a mathematical subject just from his explanation of it in a commentary. For example, the fact that Bhāskara I in his commentary on *Āryabhaṭīya* 2.6–7 endorses Āryabhaṭa's inaccurate rules for the volumes of a triangular pyramid and a sphere may not mean that he personally believed them. And as we have seen, Bhāskara II could defend Lalla's use of Versines instead of Sines in a particular formula when writing a commentary on Lalla but attack it when presenting his own views in an independent work. The complex literary etiquette of commentaries is clearly no mere slavish subjection to textual authority as opposed to critical analysis. Rather, it represents in ways that are still not thoroughly understood a subtle interplay among the various responsibilities incurred by those who expound the mathematics of other authors.

6.4.4 The role of proof

Verse texts based on briefly stated algorithms generally did not attempt to derive or justify their procedures. The reader has doubtless been struck by the frequent repetition in the foregoing discussions of phrases like "the author does not explain how this rule was derived," "the reason for this procedure is unclear," and so forth. One of the simultaneously frustrating and fascinating aspects of Sanskrit mathematics is the mystery surrounding the origins of many of its intriguing achievements. We have seen in section 4.4.1, for example, that devising new computational rules that would be difficult for other people to figure out or imitate may have been a professional advantage for mathematical astronomers. But we have also seen that a mathematician was exhorted not just to apply predictive algorithms mechanically but also to understand their rationales in terms of the underlying spherical geometry, in order to be "free from doubt" about them and gain the respect of his peers. Can we reconcile all the available information to come up with a consistent picture of what proof meant to a medieval Indian mathematician?

It used to be commonplace for historians of mathematics to claim that traditional Indian mathematics was indifferent to proof, being more interested in numerical computations than in logical justifications. This was largely due to the fact that very few Sanskrit mathematical texts were accessible to non-Indologists at that time, and most of the ones that had been translated were verse base texts devoid of commentary and hence mostly limited to algorithmic formulas. More recent exposure to a wider range of Sanskrit mathematical texts, where many logical justifications of mathematical rules are seen, has changed this attitude considerably. Understanding the reasoning underlying Indian mathematics has required new evidence and new perspectives about the concepts of proof, truth, and validity in general.[65]

For one thing, there was apparently no single, formal set of rules for logical demonstration of mathematical statements, such as the definition-proposition-proof system of Euclid and his successors in the Arabic world and the Latin West. A Sanskrit demonstration, as in the case of Bhāskara II's spherical surface area formula, might switch from a general geometrical argument to a particular numerical instantiation of it and back again. It was an individual mathematician's ingenuity rather than a formal methodology on which he had to rely for perceiving a mathematical fact in its different guises—verbal, numerical, symbolic, or geometric.

And it was this act of perception, this seeing a mathematical relationship "like a fruit in the hand," that the mathematician strove for, instead of a mechanical sequence of approved steps leading to a logically unassailable conclusion. This may explain the somewhat disparaging tone in which explanatory details in mathematical commentaries are sometimes said to be presented "for the sake of the dull-witted." Understanding the rationales for mathematical formulas is certainly important (although doubtless some less diligent students settled, as they do in every age and clime, for simply applying the formulas without ever really comprehending them). And this understanding is also essential for evaluating mutually contradictory formulas.

But developing one's own mathematical intuition, one's ability to perceive truth, is also important: if you need to have every inference explicitly justified for you, maybe you are just not a very good mathematician. Hence, perhaps, the role of explicit proofs in mathematical texts remained flexible: a demonstration might be presented or omitted, depending on the author's decision about its necessity or appropriateness in a particular context. It is the mathematician's responsibility to know why a rule is true, but not inevitably the author's responsibility to prove it. The price of this flexibility and autonomy seems to have been the occasional loss of crucial mathematical information when rules were transmitted without accompanying rationales to fix their meaning securely.[66]

---

[65] For a more in-depth discussion of these issues, see [Srin2005] and [SarK2008], vol. 1, Epilogue, as well as [KusGeom].

[66] Such losses may be attested, for example, by Bhāskara II's failure to recognize that some of Brahmagupta's geometric formulas are meant only for quadrilaterals that are cyclic (*Līlāvatī* 169–171), and perhaps also by Bhāskara I's possible misinterpretation of an

Among at least one group of Indian mathematicians, however, composing and transmitting detailed demonstrations of mathematical results seems to have been a major focus of activity. This was the case in the fascinating and unique work of the so-called Kerala school, explored in the following chapter.

accurate Sine-difference rule in the *Āryabhaṭīya* as a different, inaccurate one (commentary on *Āryabhaṭīya* 2.12).

Chapter Seven

The School of Mādhava in Kerala

7.1 BACKGROUND

Probably the most famous school in Indian mathematics, and the one that produced many of its most remarkable discoveries, is the *guru-paramparā* or "chain of teachers" originating with Mādhava in the late fourteenth century and continuing at least into the beginning of the seventeenth. These scholars lived in the region known as Kerala on the southwestern coast of India, in its central part between modern Kochi (or Cochin) and Kozhikode (or Calicut). What survives of their work includes writings in Sanskrit and in the local Dravidian vernacular called Malayalam. In astronomy they are generally considered to be followers of the Ārya-pakṣa, but they also wrote on texts in other *pakṣas*, as well as on astronomical systems unique to Kerala.[1]

A narrow strip of land between the Western Ghat mountains and the Arabian Sea, Kerala in the mid-second millennium maintained a distinct regional culture without being entirely isolated from the neighboring parts of southern India. Moreover, its pepper production and geographical location had made it a major international hub, with trading connections stretching back for many centuries. Its political rulers were generally members of the group known as Nayars, a set of non-Brāhmaṇa castes specializing in war, defense, and governance. Less politically powerful but socially superior to Nayars were the landowning Brāhmaṇa castes, especially the preeminent Nampūtiris, who officiated in (and controlled the property of) Hindu temples. The non-Brāhmaṇa castes known as Ambalavāsis also served in temples, in subordinate capacities. As a trading center, Kerala supported significant populations of foreigners and non-Hindus. Arab merchants wielded great influence in the coastal port towns, and substantial Muslim communities of expatriates and local converts formed there. There were also numerous Nestorian Christians and a few Jews.[2]

A number of non-Brāhmaṇa Hindu groups, including the Nayars and the Ambalavāsis, followed matrilineal systems of descent and inheritance. The

[1] The terms "Keralese" and "Keralan," commonly used to identify this mathematical school and the culture of its practitioners, are widely deprecated as incorrect; the preferred adjectival derivative of "Kerala" is "Malayali," which is probably unfamiliar and potentially confusing to most readers. In this work I have relied on paraphrase to avoid the issue.

[2] [Pani1960], pp. 4–16.

Nampūtiris, on the other hand, maintained not only the more usual Indian patrilineal tradition but also a sort of extreme primogeniture custom, according to which only the eldest son could be the head of the family, enter into a formal marriage, or father legitimate children. These two types of family structure developed a sort of symbiosis whereby a Nampūtiri younger son could form a *saṃbandham* or recognized quasi-marital relationship with a woman in a matrilineal caste, without detriment to the legitimacy of her children within that caste. Professionally, too, the Nampūtiris were associated with lower caste Ambalavāsis in their joint service to the temples.[3]

These social factors allowing for collaboration and even close kinship between members of technically separate groups may help explain the rather unusual caste flexibility in the Mādhava school. Despite traditional prohibitions on low-status groups sharing the Sanskrit intellectual heritage that was considered the province of Brāhmaṇas, it appears that some Ambalavāsis, such as Śaṅkara Vāriyar and Acyuta Piṣāraṭi (see below), were accepted as not only students and commentators but even teachers of Nampūtiri Brāhmaṇas. Apparently such flexibility persisted in Kerala scholarship, judging from the complaints of an anonymous family member of the Azhvāñceri Tamprākkaḷ (the title of the leader of the Nampūtiri caste), writing probably in the mid-eighteenth century: "In the Malayalam area, those who are not eligible to learn *jyotiṣa* study it. And Brāhmaṇas ask them about [astrological topics]. Since neither of these things is proper, I, with a view to change this state of affairs through making the Brāhmaṇas learn *jyotiṣa*, summarized the most essential matters thereof."[4] As we will see, though, many Nampūtiri Brāhmaṇas were masters of *jyotiṣa* themselves. They acquired their Sanskrit learning through instruction at their family estates or *illams*, and perhaps also in temple schools.[5]

7.2 LINEAGE

Nothing is known about Mādhava's own background except that he was a Brāhmaṇa of the caste known as Emprāntiri, somewhat inferior in status to the Nampūtiris, and that his home or *illam* Ilaññipaḷḷi (Sanskrit Bakulavihāra) was in Saṅgamagrāma (modern Irinjalakuda, lat. 10°20′ N, near Kochi/Cochin). His most famous mathematical achievements are the Mādhava-Leibniz series for $\pi/4$ and the Mādhava-Newton power series for the Sine and Cosine, but this work now survives only in a few verses recorded by later members of his school. There are, however, several existing astronomical texts by Mādhava in which Śaka-era dates corresponding to 1403 and 1418 CE are mentioned, so it is generally presumed that he was born sometime around the middle of the fourteenth century. His successors frequently

[3] [Put1977], p. 26.
[4] [SarK1972], p. 71.
[5] [Put1977], pp. 25–27.

referred to him in Sanskrit as "*Gola-vid*," "one who knows the sphere."[6]

The only known direct pupil of Mādhava was a Nampūtiri named Parameśvara, whose *illam* Vaṭaśśeri (Sanskrit Vaṭaśreṇi, lat. 10°51′ N) lay on the Bharathapuzha (or Niḷā) River.[7] A very long-lived and prolific scholar, Parameśvara wrote at least twenty-five separate works on astronomy, mathematics, and astrology, several of which were commentaries on the major medieval *siddhāntas*. His work in astronomy included a long series of eclipse observations beginning around 1393, and he was apparently still active during the training of his son's student Nīlakaṇṭha (born 1443), whose writings mention certain statements that "Parameśvara told me."[8] Moreover, Nīlakaṇṭha remarks that Parameśvara's 1431 astronomical text *Dṛg-gaṇita* was composed "after studying many *śāstras*, observing [for] a period of fifty-five years, [and] observing eclipses, planetary conjunctions, and so forth by means of instruments."[9] If correct, this would push the start of Parameśvara's observational activities back to 1376, giving him a lifespan of well over ninety years from perhaps the 1360s to the 1450s.[10]

Parameśvara's chief student was his own son Dāmodara, whose works do not survive and whose dates are unknown. Dāmodara's astronomical writings are briefly quoted by one of his pupils, the abovementioned Nīlakaṇṭha. This Nīlakaṇṭha (sometimes additionally identified as Gārgyakerala, from his membership in the Garga *gotra*, and Somayājin or Somasutvan, "performer of the *soma*-sacrifice") was a Nampūtiri of the Keḷallūr *illam* (Sanskrit Keralasadgrāma) in Tṛkkaṇṭiyūr (Sanskrit Kuṇḍapura) near modern Tirur. He pursued his *jyotiṣa* studies with Dāmodara at nearby Vaṭaśśeri and was apparently a protégé of the contemporary Azhvāñceri Tamprākkaḷ, one Netranārāyaṇa, who was interested in astronomical theories. Nīlakaṇṭha's own dozen or so compositions all dealt with astronomy; in the best known of these, the *Tantra-saṅgraha*, he gives a date for his birth that corresponds to about 14 June 1444. Nīlakaṇṭha quotes the *Tantra-saṅgraha* in another and hence later work, a commentary on the *Āryabhaṭīya*, which he says he wrote when "advanced in age"—but evidently not too advanced to write at least two more books thereafter, both of which quote this commentary. In fact, if it is correct to identify Nīlakaṇṭha with one "Keḷallūr" who is mentioned as a contemporary authority in a 1543 Malayalam astrological text, he remained active in the field even as a near-centenarian![11]

---

[6] [Pin1970a], vol. 4, pp. 414–415, and [SarK1972], pp. 51–52.

[7] The location of Vaṭaśreṇi, Aśvatthagrāma, is frequently identified with the village of Ālattūr [Pin1985], [Pin1970a] 4, p. 187, [SarK1972], p. 52. [Kup1957], p. l, describes Vaṭaśreṇi as located where the Niḷā "joins the sea," from Parameśvara's own phrase, "on the north bank of the Niḷā on the ocean" (or "lake"?). Modern Ālattūr, however, is located not at the mouth of the Niḷā but inland on the Gayatripuzha, a tributary of the Niḷā.

[8] *Āryabhaṭīya-bhāṣya* Gola 48, [Kup1957], p. lii.

[9] *Āryabhaṭīya-bhāṣya* Gola 48, [Kup1957], p. lii, and [Will2005], sect. 4.14.4, p. 261.

[10] Slightly varying estimates of Parameśvara's dates are given in [Pin1970a], vol. 4, pp. 187–192, [SarK1972], pp. 52–54, and [Kup1957], pp. l–liv.

[11] [Pin1970a], vol. 3, pp. 173–177, and vol. 4, 142; [SarK1977b], pp. xxiv–xxxviii; and [SarK1972], pp. 55–57.

Dāmodara had a second student, named Jyeṣṭhadeva, also a Nampūtiri Brāhmaṇa, from the Parañ̃oṭṭu (Parakroḍa) *illam* in Ālattūr. His major (and possibly only) work, the *Yukti-bhāṣa* or "Vernacular [exposition] of rationales," explains in Malayalam prose detailed proofs of the mathematical procedures stated in the *Tantra-saṅgraha*. Although this is the first known work in the Mādhava school to preserve such rationales in detail, it seems reasonable to think that the reasoning involved in some of them must have been worked out much earlier, especially the ones that deal with rules attributed to Mādhava himself.[12]

Nīlakaṇṭha and perhaps also Jyeṣṭhadeva were the primary teachers of Śaṅkara, an Ambalavāsi of the Vāriyar caste, whose family were employed at the shrine of Śiva at Tṛkkuṭaveli (Sanskrit Śrīhutāśa) near modern Ottappalam on the Nilā River. Śaṅkara wrote an extensive verse commentary on the *Tantra-saṅgraha*, the *Yukti-dīpikā*, or "Lamp of rationales," which is largely based on the *Yukti-bhāṣa* of Jyeṣṭhadeva, as well as a concise prose one, the *Laghu-vivṛti*, or "Short commentary," said to be his last composition. Other works known or thought to be by Śaṅkara include a lengthy but unfinished mostly prose commentary *Kriyā-kramakarī* ("Performing the steps for calculation") on the *Līlāvatī*, which mentions a date in 1534; an astronomical handbook composed probably about the time of its epoch date in 1554; and an astronomical commentary dated to 1529. Śaṅkara's working career thus seems to have roughly coincided with the middle third of the sixteenth century.[13] Śaṅkara's other acknowledged teachers included Nīlakaṇṭha's patron Netranārāyaṇa and another Nampūtiri Brāhmaṇa, Citrabhānu, known as the author of an astronomical work dated to 1530 and a small treatise with solutions and proofs for algebraic equations.[14]

Other, more obscure scholars in this pedagogical lineage include Nīlakaṇṭha's younger brother, Śaṅkara, and several anonymous pupils and commentators.[15] In addition, one Nārāyaṇa of a family named Mahiṣamaṅgala completed Śaṅkara's unfinished *Kriyā-kramakarī* on the *Līlāvatī*, as well as a *Līlāvatī*-commentary of his own and a few astronomical works, one of which mentions the year 1607. This Nārāyaṇa was the protégé of local mahārājas who ruled in the mid- and late sixteenth century and is much better known for his poetic and liturgical writings than for his achievements in *gaṇita*.[16] The Mādhava school's last identifiable member is another Ambalavāsi, Acyuta Piṣāraṭi, from Nīlakaṇṭha's village of Tṛkkaṇṭiyūr. Acyuta was a pupil of Jyeṣṭhadeva and a protégé of the local ruler Ravivarman of Veṭṭatunād (Prakāśa). His ten or so astronomical works include one composed in 1593, in which he refers to "the good teacher Jyeṣṭhadeva." Acyuta was also learned

---

[12][Pin1970a], vol. 3, pp. 76–77, [Pin1985], [SarK1972], pp. 59–60, and [Srin2004b], Appendix 3. The *Yukti-bhāṣa* or *Gaṇita-Yukti-bhāṣa* has recently been edited and translated in [SarK2008], and discussed in [Div2007].

[13][Pin1985], [SarK1972], pp. 58–59, [SarK1977b], pp. xliv–lxx.

[14][Pin1970a], vol. 3, p. 47, [SarK1972], p. 57, [Hay1998].

[15][SarK1972], pp. 57–61, [GolPi1991].

[16][Pin1970a], vol. 3, pp. 150–151; [Pin1970a], vol. 4, p. 137; [SarK1975], pp. xxii–xxiv; [SarK1972], pp. 57–58.

in grammar, medicine, and poetics; he instructed the famous poet and grammarian Nārāyaṇa Bhaṭṭatiri, whose obituary verse on Acyuta gives a date for his death corresponding to 7 July 1621.[17] Many other Kerala scholars wrote on astronomy and mathematics during this period, but most of them cannot be definitely identified as part of the pedagogical family tree originating with Mādhava.[18]

7.3 INFINITE SERIES AND OTHER MATHEMATICS

The crest-jewel of the Kerala school is generally considered to be the infinite series for trigonometric quantities discovered by its founder, Mādhava. These are preserved in about a dozen verses explicitly attributed to Mādhava in various works by later members of the school, and translated in the following sections. Many of the accompanying rationales and explanations of these results are probably also originally due to Mādhava, even if not directly credited to him.

The quotations translated here are taken mostly from the commentaries *Yukti-dīpikā* and *Kriyā-kramakarī* by Śaṅkara: the *Yukti-dīpikā* on the first several verses of the second chapter of the *Tantra-saṅgraha* dealing with trigonometry, and the *Kriyā-kramakarī* on verse 199 of the *Līlāvatī*, where Bhāskara introduces the geometry of the circle. Some of the sources cited in the notes also draw on the Malayalam *Yukti-bhāṣa* of Jyeṣṭhadeva. For convenience, the topic headings in sections 7.3.1 and 7.3.2 identify their subjects in terms of modern trigonometric concepts, such as π and arctangent, and the names under which the results are most familiar in modern mathematics (e.g., "Leibniz's series," "Gregory's series").

7.3.1 Mādhava on the circumference and arcs of the circle

Mādhava's numerical value for π. Several verses ascribed to Mādhava deal with the computation of what we now call π, the ratio of the circumference C of a circle to its diameter D, and with the determination of arbitrary arcs from their Sines and Cosines. One of these verses, quoted in Śaṅkara's *Kriyā-kramakarī* on the *Līlāvatī*, contains a numerical value for the ratio C/D, encoded in the concrete number notation:

> The teacher Mādhava also mentioned a value of the circumference closer [to the true value] than that:
>
> "Gods [thirty-three], eyes [two], elephants [eight], serpents [eight], fires [three], three, qualities [three], Vedas [four], *nakṣatras* [twenty-seven], elephants [eight], arms [two] (2,827,433,388,233)—the wise

[17] [Pin1970a], vol. 1, pp. 36–38, [SarK1972], pp. 64–65.

[18] [SarK1972] contains biographical and bibliographical information on a number of these other Kerala scientists.

> said that this is the measure of the circumference when the diameter of a circle is nine *nikharva* [10^{11}]."[19]

Śaṅkara says here that Mādhava's value 2,827,433,388,233/900,000,000,000 is more accurate than "that," that is, more accurate than the traditional value for C/D of 355/113. (Mādhava's ratio corresponds to a value of π accurate to the eleventh decimal place, or 3.14159265359.)

Mādhava's derivation of the circumference by means of polygons. Śaṅkara then cites a set of four verses by Mādhava that prescribe a geometric method for computing the value of the circumference. This technique involves calculating the perimeters of successive regular circumscribed polygons, beginning with a square. The values of the polygonal sides in the first few cases are determined as follows:

> When in this way a very accurate circumference [is desired], if [one says], "how is it possible to calculate [that]?," Mādhava then said:
>
> "The square-root of an eighth part of the square of the side in a four-sided [figure] is the divisor. And [take] the quotient by the divisor from the product of the side [times] the difference of the divisor and one-fourth the side. Having extended [that] from the corner, the eight sides to be drawn are to be shown.
>
> "The square of half the side of the eight-sided [figure] is to be added to the square of the half-diameter. The square-root here is the hypotenuse. Divide by that the square of the half-diameter diminished by the square of half the side. Whatever is the result from that,
>
> "the hypotenuse diminished by that [result and] halved is called the divisor. And the multiplier is the hypotenuse minus the half-diameter. Having multiplied the half-side by this multiplier and divided [it] by the divisor, whatever is the quotient in this case too,
>
> "having extended that from the corner at the juncture of the sides, there should be in the intervening space here a sixteen-sided [figure]. And hence in this way there should be a tooth-[i.e., thirty-two-] sided [figure], and hence [the circumference of] the circle is to be determined."[20]

The beginning of Śaṅkara's commentary on these verses explains how to set up the construction for calculating the polygon sides:

---

[19] *Kriyā-kramakarī* on *Līlāvatī* 199, [SarK1975], p. 377. Translated in [IkeSIM], p. 4, and [Plo2007b], pp. 481–482. See also [Sara1979], p. 157.

[20] *Kriyā-kramakarī* on *Līlāvatī* 199, [SarK1975], pp. 377–378. See also [Srin2004b], Appendix 3; [Sara1979], pp. 156–157; and [SarK2008], vol. 1, sect. 6.2.

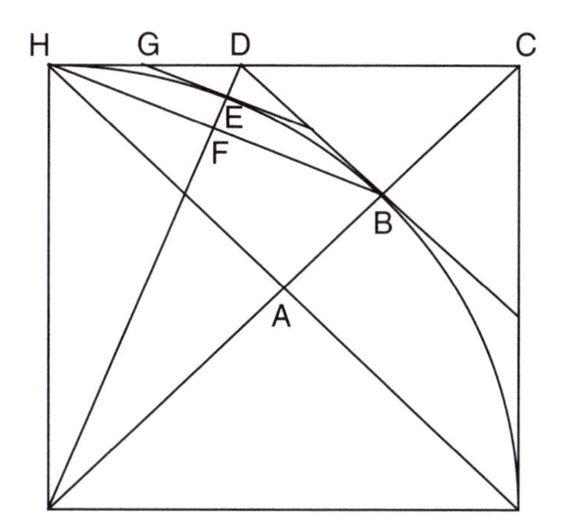

Figure 7.1 Mādhava's computation of the circumference by successive polygons.

> This is the meaning of those four verses: There at the beginning, having drawn a square figure [with side] equal to the desired diameter, [and] divided by eight the square of that side, [take] the square-root of whatever is the quotient. [Its] name is the "divisor."[21]

That is, when the circumference of a circle with radius R is sought, first circumscribe about it a square of side $s = 2R$ (one quadrant of which is shown in figure 7.1, where CH is equal to R, and B and E are points on the circle bisecting tangent line segments). Then compute the difference CD between R and the half-side DH of the circumscribed regular octagon, using Mādhava's "divisor" $\sqrt{s^2/8} = s/(2\sqrt{2})$. Mādhava says that CD is found from "the quotient by the divisor from the product of the side [times] the difference of the divisor and one-fourth the side." This is validated by the similarity of right triangles DBC and HAC, as follows:

$$CD = BC \cdot \frac{CH}{AC} = \left(\frac{s}{\sqrt{2}} - \frac{s}{2}\right) \cdot \frac{s/2}{s/(2\sqrt{2})} = \left(\frac{s}{2\sqrt{2}} - \frac{s}{4}\right) \cdot \frac{s}{s/(2\sqrt{2})}.$$

When this quantity is "extended from the corner" C, the remaining segment DH is the desired half-side of the octagon. This is then used to compute the difference DG between DH and the half-side GH of the circumscribed

[21] *Kriyā-kramakarī* on *Līlāvatī* 199, [SarK1975], p. 378.

sixteen-sided polygon. Mādhava defines the following quantities to derive it:

$$\sqrt{DH^2 + R^2} = \text{hypotenuse},$$

$$\frac{R^2 - DH^2}{\sqrt{DH^2 + R^2}} = \text{result},$$

$$\left(\sqrt{DH^2 + R^2} - \frac{R^2 - DH^2}{\sqrt{DH^2 + R^2}}\right) \cdot \left(\frac{1}{2}\right) = \frac{DH^2}{\sqrt{DH^2 + R^2}} = \text{divisor},$$

$$\sqrt{DH^2 + R^2} - R = \text{multiplier}, \quad \text{and}$$

$$DG = DH \cdot \frac{\text{multiplier}}{\text{divisor}} = DH \cdot \frac{\sqrt{DH^2 + R^2} - R}{DH^2/\sqrt{DH^2 + R^2}}.$$

Since in figure 7.1 the "multiplier" $(\sqrt{DH^2 + R^2} - R)$ is DE, and the "divisor" $DH^2/\sqrt{DH^2 + R^2}$ is DF, it is clear from the similar right triangles GED and HFD that

$$DG = DE \cdot \frac{DH}{DF} = (\sqrt{DH^2 + R^2} - R) \cdot \frac{DH}{DH^2/\sqrt{DH^2 + R^2}},$$

which is equivalent to Mādhava's result. If DG is then "extended" from the "juncture" D, the remainder GH is the desired half-side of the regular hexadecagon. The procedure can then be repeated to compute the half-side of the 32-gon, and so forth to approximate the circumference as precisely as desired.

The Mādhava-Leibniz series for π. An alternative rule for the circumference is then described by Śaṅkara as "easier" than the foregoing. It corresponds to finding a sum C of some arbitrary number of terms in a series, plus a final correction term, depending on the diameter $D = 2R$ as follows:

$$C \approx \frac{4D}{1} - \frac{4D}{3} + \frac{4D}{5} - \ldots + (-1)^{n-1}\frac{4D}{2n-1} + (-1)^n \frac{4Dn}{(2n)^2 + 1}.$$

This method, too, is due to Mādhava, as Śaṅkara states:

> An easier way to get the circumference is mentioned by him (Mādhava). That is to say:
>
> "Add or subtract alternately the diameter multiplied by four and divided in order by the odd numbers like three, five, etc., to or from the diameter multiplied by four and divided by one.
>
> "Assuming that division is completed by dividing by an odd number, whatever is the even number above [next to] that [odd number], half of that is the multiplier of the last [term].
>
> "The square of that [even number] increased by 1 is the divisor of the diameter multiplied by 4 as before. The result from these two (the multiplier and the divisor) is added when [the previous term is] negative, when positive subtracted.

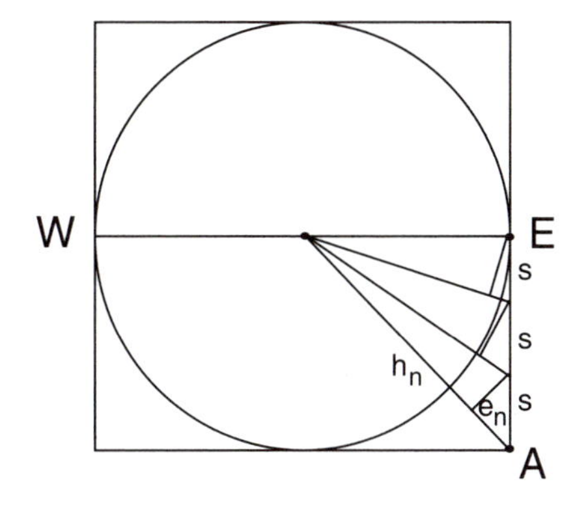

Figure 7.2 Subdivision of the quadrant for the circumference series

> "The result is an accurate circumference. If division is repeated many times, it will become very accurate."[22]

Śaṅkara lays out an elaborate geometrical rationale for the derivation of this series, starting with the procedure illustrated in figure 7.2. As before, it relies on the construction of a circle in a circumscribed square, which is here oriented to the cardinal directions (a nice use of the physical context of the problem as a sort of graphical symbolism to help explain the construction). The square's vertical half-side $EA = R$ in this case is subdivided into some number n of equal segments of length s. The endpoints of the segments are connected to the center of the circle by "hypotenuse-lines." The length of each "hypotenuse-line" h_i, and that of the perpendicular "edge" e_i dropped on h_i from the end of the previous hypotenuse-line are then to be determined. As Śaṅkara puts it,

> For demonstrating this, having placed a square whose four sides are equal to a desired diameter, draw a circle in it by means of a compass [whose opening] is equal to half of the diameter and draw the east-west and north-south line(s). Then the circumference touches the middle of each of the four sides of the square in all four directions. Having placed as many dots as you wish with equal intervals at that end of the eastern side which is south from the end of the east-west line, draw lines from the center of the circle to the dots.
>
> Then add the square of the distance between the end of that line and the end of the east-west line to the square of the half-diameter, and take the square root of each. The measure of

[22] *Kriyā-kramakarī* on *Līlāvatī* 199, [SarK1975], p. 379; the same quoted verses appear in *Yukti-dīpikā* 271–274 on *Tantra-saṅgraha* 2.1, [SarK1977b], p. 101. Translated in [IkeSIM], p. 11, and [Plo2007b], p. 482. See also, for the following derivation, [Sara1979], pp. 159–166, and [Srin2004b], Appendix 3.

> the length of the hypotenuse-line is produced.... That [distance] passing from the end of the left line to the hypotenuse [which is] the southern line, is placed with its [perpendicular] direction opposite to its (the hypotenuse's). This is the edge of the segment.... This figure is similar in shape to the figure whose hypotenuse is the southern line and edge is the half-diameter.[23]

In other words, the hypotenuse-line h_i is found from R and the vertical segment is by the Pythagorean theorem. Then the edge e_i is known from the pair of similar right triangles with hypotenuses h_i and s. That is,

$$h_i = \sqrt{R^2 + (is)^2} \quad \text{and} \quad e_i = \frac{s \cdot R}{h_i}.$$

Śaṅkara's next step is to compute the value of each small arc c_i of the circumference, which he takes to be equal to its corresponding Sine g_i, if the arcs are small enough. As illustrated in figure 7.3, g_i is derived in terms of the known edge e_i from similar right triangles contained between each pair of adjacent hypotenuses. It is then approximated in terms of the "southern" hypotenuse h_i, after substituting in the above expression for e_i:

> Having multiplied the distance derived in this way by the half-diameter, divide by [the length of] the left one of two adjacent lines. The result is the Sine of the part of the circumference between the lines. [That] is also a part of the circumference when the segment is small. Therefore, multiply each segment by the square of the half-diameter.... There, though it should be divided by the product of the two adjacent lines, even if it is divided by the square of one of those two, the error will not be large in the calculation of a part of the circumference.... Therefore [this] is said assuming the division of the parts multiplied by the square of half of the diameter by the squares of the southern lines.[24]

That is,

$$c_i \approx g_i = \frac{e_i \cdot R}{h_{i-1}} = \frac{sR^2}{h_i h_{i-1}} \approx \frac{sR^2}{{h_i}^2},$$

since ${h_i}^2$ is assumed to be close to $h_i h_{i-1}$.

The following very condensed remarks give a rough sketch of Śaṅkara's (presumably originally Mādhava's) line of argument for the remainder of his lengthy proof. The basic strategy is to find infinite series expressions for the individual c_i, and a way of adding them up term by term to produce a single series for the total arc of the half-quadrant $C/8$. Śaṅkara says:

---

[23] *Kriyā-kramakarī* on *Līlāvatī* 199, [SarK1975], p. 380; translated in [IkeSIM], pp. 12–13, and [Plo2007b], pp. 483–484.

[24] *Kriyā-kramakarī* on *Līlāvatī* 199, [SarK1975], p. 380; translated in [IkeSIM], p. 13, and [Plo2007b], p. 484.

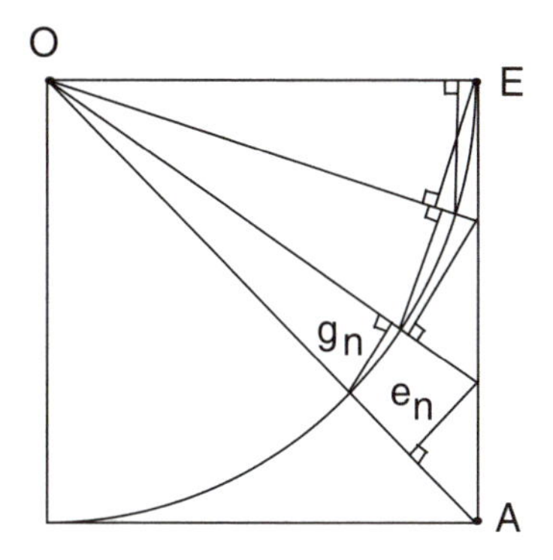

Figure 7.3 Detail of the subdivided quadrant.

If the multiplicands and the divisors were of one kind, then, after multiplying [the multiplicands] by the sum of the multipliers and dividing by the divisor once, the sum of the quotients would result. Therefore a method for deriving divisors of one kind should be sought.... There in the calculation of the first results multiply [the multiplicand] by the difference between the square of each line and the square of the half-diameter and divide by the square of the half-diameter. [Then] make the sum of the quotients....

Here in the calculation of the sum of each result the sum of the multipliers is assumed to be the multiplier.... Therefore the method of calculating the sum of the squares, the sum of the squares of squares, etc., should be investigated here....

[H]alf of [the product of] the half-diameter multiplied by the number of the sides increased by 1 is the sum of the sides.... The product of the portion and the number of the sides increased by 1 is greater than the half-diameter by a portion. Only when the portion is small will the result be accurate. In such a case half of [the product of] the half-diameter multiplied by the half-diameter is the product of the portion and the sum of the sides.

Now the calculation of the sum of the squares [of the sides] should be investigated.... Here, however, only one multiplier is equal to the half-diameter, and the others are somewhat less.... The sum of the sides multiplied by these in order should be subtracted from the sum [of the sides] multiplied by the half-diameter in order to arrive at the sum of the squares.

Here whatever are the products of the shortened parts and the successively previous sides equal to the radius diminished by its [portions] taken in reversed order beginning with the last term, the sum of these is the sum of the sums multiplied by

> the portion.... The sum multiplied by the half-diameter is the sum of the squares and the sum of half of the squares.... Therefore, it is arrived at that a third of the cube of the half-diameter is the sum of the squares multiplied by the portion....
>
> In like manner, each successive [result] from the successive powers involving the half-diameter divided by each of the successive divisors increased by [its own] third, fourth, etc., is the sum of the cubes, etc., multiplied by the portion. Therefore divisors are increased by one at a time.... Therefore just the half-diameter should be divided separately by odd numbers like three, five, etc. The results should be subtracted from or added to the half-diameter in order.[25]

First, the expression for each small arc c_i in the half-quadrant $C/8$ is rewritten so that the divisors are "of one kind," instead of each depending on a different h_i. The resulting expression involves a series of even powers of "each line" is:

$$c_i \approx \frac{sR^2}{{h_i}^2} = s - \frac{s({h_i}^2 - R^2)}{R^2} + \frac{s({h_i}^2 - R^2)^2}{R^4} - \frac{s({h_i}^2 - R^2)^3}{R^6} + \dots$$
$$= s - \frac{s((is)^2)}{R^2} + \frac{s((is)^4)}{R^4} - \frac{s((is)^6)}{R^6} + \dots$$

This is simply a special case of a more general infinite series for any expression involving some quantity a, a "multiplier" p, and a "divisor" q:

$$\frac{ap}{q} = a + \frac{a(p-q)}{p} + \frac{a(p-q)^2}{p^2} + \frac{a(p-q)^3}{p^3} + \dots.$$

The above series is derived by recursively rewriting its successive terms as follows:

$$\begin{aligned}\frac{ap}{q} &= a + \frac{a(p-q)}{q} \\ &= a + \frac{a(p-q)}{p} \cdot \frac{p}{q} \\ &= a + \frac{a(p-q)}{p} + \frac{a(p-q)}{p} \cdot \frac{(p-q)}{q} \\ &= a + \frac{a(p-q)}{p} + \frac{a(p-q)^2}{p^2} \cdot \frac{p}{q},\end{aligned}$$

and so forth.

[25] *Kriyā-kramakarī* on *Līlāvatī* 199, [SarK1975], pp. 381–385; translated in [IkeSIM], pp. 15–23, and [Plo2007b], pp. 485–490.

Now we can add up all the corresponding terms in each of our expressions for c_i, arriving at "the sum of the multipliers" in each case.

$$c_1 \approx \frac{sR^2}{{h_1}^2} = s - \frac{s(1s)^2}{R^2} + \frac{s(1s)^4}{R^4} - \frac{s(1s)^6}{R^6} + \dots,$$

$$c_2 \approx \frac{sR^2}{{h_2}^2} = s - \frac{s(2s)^2}{R^2} + \frac{s(2s)^4}{R^4} - \frac{s(2s)^6}{R^6} + \dots,$$

$$\vdots$$

$$c_n \approx \frac{sR^2}{{h_n}^2} = s - \frac{s(ns)^2}{R^2} + \frac{s(ns)^4}{R^4} - \frac{s(ns)^6}{R^6} + \dots$$

$$\frac{C}{8} = \sum_{i=1}^{n} c_i \approx R - \frac{s}{R^2}\sum_{i=1}^{n}(is)^2 + \frac{s}{R^4}\sum_{i=1}^{n}(is)^4 - \frac{s}{R^6}\sum_{i=1}^{n}(is)^6 + \dots$$

As Śaṅkara says, this requires us to find more manageable ways to express the sums of even powers of successive integers. The "product of the portion [s] and the sum of the sides [is]" is easily approximated by means of the well-known rule for the sum of an arithmetic progression:

$$s \cdot \sum_{i=1}^{n} is = s \cdot \frac{s \cdot n(n+1)}{2} = \frac{ns \cdot s(n+1)}{2} \approx \frac{R^2}{2},$$

if s is small and n is large, because $R = ns$.

Computing "the sum of the squares multiplied by the portion" is somewhat trickier. The squares of the sides $(is)^2$ are first rewritten as products of each side is with its corresponding complementary segment $R - s(n-i)$. Then the sum of these products is broken out into the sum of the sides "multiplied by the half-diameter" minus the sum of successive sums of sides "multiplied by the portion":

$$\begin{aligned}\sum_{i=1}^{n}(is)^2 &= \sum_{i=1}^{n}(is)(R - s(n-i)) \\ &= R\sum_{i=1}^{n} is - s\sum_{i=1}^{n}(is)(n-i)) \\ &= R\sum_{i=1}^{n} is - s\sum_{j=1}^{n-1}\sum_{i=1}^{j} is, \quad \text{and thus} \\ R\sum_{i=1}^{n} is &= \sum_{i=1}^{n}(is)^2 + s\sum_{j=1}^{n-1}\sum_{i=1}^{j} is.\end{aligned}$$

We can then use our previous approximation for the portion times the sum of the sides to show that "the sum multiplied by the half-diameter is the sum of the squares and the sum of half of the squares":

$$\begin{aligned}
R\sum_{i=1}^{n} is &= \sum_{i=1}^{n}(is)^2 + s\sum_{j=1}^{n-1}\sum_{i=1}^{j} is \\
&\approx \sum_{i=1}^{n}(is)^2 + \sum_{j=1}^{n-1}\frac{(js)^2}{2} \\
&\approx \sum_{i=1}^{n}(is)^2 + \frac{1}{2}\sum_{i=1}^{n}(is)^2 \\
&= \frac{3}{2}\sum_{i=1}^{n}(is)^2.
\end{aligned}$$

Hence "the sum of the squares multiplied by the portion" is "a third of the cube of the half-diameter":

$$\begin{aligned}
s\sum_{i=1}^{n}(is)^2 &\approx s\cdot\frac{2}{3}\cdot R\sum_{i=1}^{n} is \\
&\approx \frac{2}{3}\cdot R\cdot\frac{R^2}{2} \\
&= \frac{R^3}{3}.
\end{aligned}$$

We can use similar arguments to approximate the sums of other integer powers as "successive powers involving the half-diameter divided by each of the successive divisors increased by [its own] third, fourth, etc.":

$$s\cdot\sum_{i=1}^{n}(is)^k \approx \frac{R\cdot R^k}{k+k/k} = \frac{R^{k+1}}{k+1}.$$

Finally, these expressions for even powers are substituted for the corresponding terms in our series for $C/8$:

$$\begin{aligned}
\frac{C}{8} &\approx R - \frac{s}{R^2}\sum_{i=1}^{n}(is)^2 + \frac{s}{R^4}\sum_{i=1}^{n}(is)^4 - \frac{s}{R^6}\sum_{i=1}^{n}(is)^6 + \ldots \\
&\approx R - \frac{1}{R^2}\frac{R^3}{3} + \frac{1}{R^4}\frac{R^5}{5} - \frac{1}{R^6}\frac{R^7}{7} + \ldots,
\end{aligned}$$

which is equivalent to Mādhava's prescribed series for C excluding its correction term (discussed below).

The Mādhava-Gregory series for the arctangent. Śaṅkara goes on in the commentary *Kriyā-kramakarī* to quote another set of verses, also generally ascribed to Mādhava, generalizing this series result for an arbitrary arc of the circumference:

> Now, by just the same argument, the determination of the arc of a desired Sine can be [made]. That is as [follows]:
>
> "The first result is the product of the desired Sine and the radius divided by the Cosine. When one has made the square of the Sine the multiplier and the square of the Cosine the divisor,
>
> "now a group of results is to be determined from the [previous] results beginning with the first. When these are divided in order by the odd numbers 1, 3, and so forth,
>
> "and when one has subtracted the sum of the even[-numbered results] from the sum of the odd [ones], [that] should be the arc. Here, the *smaller* of the Sine and Cosine is required to be considered as the desired [Sine].
>
> "Otherwise there would be no termination of the results even if repeatedly [computed]."
>
> By means of the same stated argument, the circumference can be determined in another way too. That is as [follows]:
>
> "The first result should be the square root of the square of the diameter multiplied by twelve. From then on, the result should be divided by three, [in] each successive [case].
>
> "When these are divided in order by the odd numbers beginning with 1, and when one has subtracted the even [results] from the sum of the odd, [that] should be the circumference."[26]

The above sets of quoted verses also appear in Śaṅkara's *Yukti-dīpikā*, separated by a brief description that sets up their rationale:

> Whatever is the end of the desired arc [measured] from the direction-string in the circle, it is extended outside from the intersection with the Sine, [falling] upon the side of the quadrilateral [figure].
>
> The portion of the circumference [up to] that intersection is to be determined, [in accordance] with its being an arc, by means of results from the corresponding sum of squares and so forth, as previously stated.
>
> Now, determine such a portion of that circumference from the Sine of a [zodiacal] sign [i.e., 30° or $C/12$].[27]

The above formulas state that for an arbitrary arc θ whose Sine and Cosine are known (and assuming $\mathrm{Sin}\,\theta < \mathrm{Cos}\,\theta$),

$$\theta = \frac{R\,\mathrm{Sin}\,\theta}{1\,\mathrm{Cos}\,\theta} - \frac{R\,\mathrm{Sin}^3\,\theta}{3\,\mathrm{Cos}^3\,\theta} + \frac{R\,\mathrm{Sin}^5\,\theta}{3\,\mathrm{Cos}^5\,\theta} - \frac{R\,\mathrm{Sin}^7\,\theta}{3\,\mathrm{Cos}^7\,\theta} + \ldots$$

[26] *Kriyā-kramakarī* on *Līlāvatī* 199, [SarK1975], pp. 385–386; the same quoted verses (with minor variants) appear in *Yukti-dīpikā* 206–209 and 212–214 on *Tantra-saṅgraha* 2.1, [SarK1977b], pp. 95–96. Also translated in [IkeSIM], pp. 25–26. See also [Sara1979], pp. 182–184; [Srin2004b], Appendix 3; and [SarK2008], vol. 1, sect. 6.6–6.7.

[27] *Yukti-dīpikā* 210–212 on *Tantra-saṅgraha* 2.1, [SarK1977b], p. 96. See also [Sara1979], pp. 183–184.

In particular,

$$C = \frac{\sqrt{12D^2}}{1} - \frac{\sqrt{12D^2}}{3\cdot 3} + \frac{\sqrt{12D^2}}{5\cdot 3^2} - \frac{\sqrt{12D^2}}{7\cdot 3^3} + \dots$$
$$= \frac{12R}{1\sqrt{3}} - \frac{12R}{3\sqrt{3}^3} + \frac{12R}{5\sqrt{3}^5} - \frac{12R}{7\sqrt{3}^7} + \dots,$$

since the ratio of $\mathrm{Sin}(C/12)$ to $\mathrm{Cos}(C/12)$ is $1/\sqrt{3}$. And these results can be justified geometrically "as previously stated," just like our earlier series for the half-quadrant

$$\frac{C}{8} = R - \frac{1}{R^2}\frac{R^3}{3} + \frac{1}{R^4}\frac{R^5}{5} - \frac{1}{R^6}\frac{R^7}{7} + \dots$$

For an arbitrary arc θ, though, we would subdivide into portions not the entire half-side of the circumscribing square or R but rather a segment of it cut off by a radius "extended outside from the intersection with the Sine." This segment is the Tangent of θ, equal to $R \,\mathrm{Sin}\,\theta / \mathrm{Cos}\,\theta$. Therefore our corresponding expression for arbitrary θ will be

$$\theta = \frac{R\,\mathrm{Sin}\,\theta}{1\,\mathrm{Cos}\,\theta} - \frac{R^3\,\mathrm{Sin}^3\,\theta}{3R^2\,\mathrm{Cos}^3\,\theta} + \frac{R^5\,\mathrm{Sin}^5\,\theta}{3R^4\,\mathrm{Cos}^5\,\theta} - \frac{R^7\,\mathrm{Sin}^7\,\theta}{3R^6\,\mathrm{Cos}^7\,\theta} + \dots .$$

The correction term for Mādhava's π series. The final term $\frac{4Dn}{(2n)^2+1}$ in the series for C compensates for the inaccuracy caused by the slow convergence of its terms. Historians have suggested that Mādhava found this term by experimenting with "pulverizer-like" continued fraction approximations to the error produced by successive partial sums of the series, when compared to a recognized value of π such as 355/113.[28] That is, if $C(k)$ represents the kth partial sum of the Mādhava-Leibniz infinite series for the circumference, then the errors produced by $C(k)$ as compared to $\pi/4$ for the first few values of k are the following:

---

[28]For a detailed explanation of the conjectured derivation of the various correction terms and also of Śaṅkara's rationale for them, see [HayKY1990]. [Gup1992c] argues for essentially the same reconstruction, but using a different standard value of π. See also [SarK2008], vol. 1, sect. 6.8–6.10.

$$\frac{C(1)}{4D} - \frac{1}{4} \cdot \frac{355}{113} = 1 - \frac{355}{452} = \frac{97}{452} = \cfrac{1}{4 + \cfrac{1}{1 + \cfrac{1}{1 + \cfrac{31}{33}}}} \approx \cfrac{1}{4 + \cfrac{1}{1}},$$

$$\frac{C(2)}{4D} - \frac{1}{4} \cdot \frac{355}{113} = \left(1 - \frac{1}{3}\right) - \frac{355}{452} = -\frac{161}{1356} \approx -\cfrac{1}{8 + \cfrac{1}{2}},$$

$$\frac{C(3)}{4D} - \frac{1}{4} \cdot \frac{355}{113} = \left(1 - \frac{1}{3} + \frac{1}{5}\right) - \frac{355}{452} = \frac{551}{6780} \approx \cfrac{1}{12 + \cfrac{1}{3}}.$$

From such examples, apparently, Mādhava simply extrapolated the general error term $\cfrac{1}{4k + \cfrac{1}{k}} = \frac{k}{(2k)^2 + 1}$. When multiplied by $4D$ and made positive or negative depending on the parity of k, this becomes the correction term tacked on to the end of $C(k)$. If this reconstruction of Mādhava's reasoning is accurate, it indicates that he was as skilled in working with bold numerical intuitions as with painstaking geometric rationales.

Mādhava's arc-difference rule. Another procedure for finding an unknown arc from known Sines and Cosines is explained in the verses of Nīlakaṇṭha's second chapter of the *Tantra-saṅgraha* itself, along with a rule for computing the Sine and Cosine (discussed below). Both rules are directly attributed to Mādhava. The formula for the arc is as follows:

> "The divisor [derived] from the sum of the Cosines is divided by the difference of the two given Sines. The radius multiplied by two is divided by that [result]. That is the difference of the arcs."
>
> Thus is made the determination of the Sine and arc as stated by Mādhava.[29]

In other words, if the Sines and Cosines of the known arc θ and the unknown one $\theta + \Delta\theta$ are given, then their difference $\Delta\theta$ is found from

$$\Delta\theta \approx \frac{2R}{\dfrac{\operatorname{Cos}\theta + \operatorname{Cos}(\theta + \Delta\theta)}{\operatorname{Sin}(\theta + \Delta\theta) - \operatorname{Sin}\theta}}.$$

Śaṅkara remarks on this in his other *Tantra-saṅgraha* commentary, the *Laghu-vivṛti*, explaining why a similar-triangle relation justifies this approximation:

---

29 *Tantra-saṅgraha* 2.14–15, [SarK1977b], pp. 112, 120, and [Pil1958], p. 21.

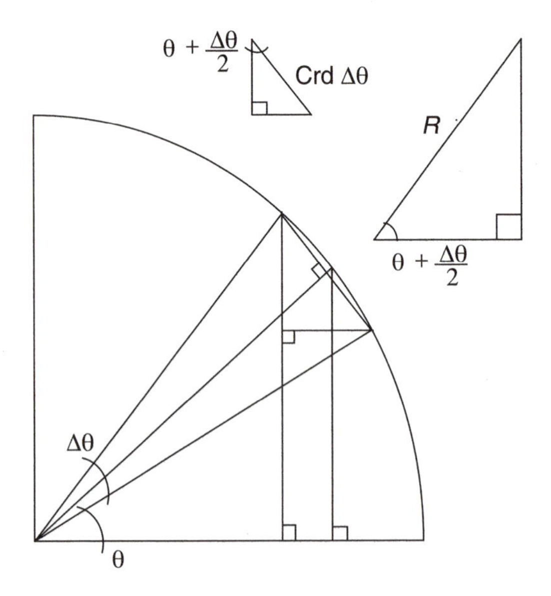

Figure 7.4 The rationale for the arc-difference rule.

> Here, where the divisor should be made from the Cosine of the medial arc [i.e., $(\theta+\Delta\theta/2)$], it is said [to be made] with the sum of the Cosines of both full [arcs], by assuming that that [sum] equals twice the medial Cosine. But in reality, the sum of the Cosines of the two full [arcs] is somewhat less than twice the medial Cosine. Because of the deficiency of that divisor, the result of that [is] somewhat too big. But actually, that is what is desired: for that result is really the chord [of $\Delta\theta$], which is a little less than its arc, so the arc is what is desired. So an excess of the result is [in fact] right. And the radius is said to be doubled because of the doubling of the Cosine forming the divisor. Again, as a rule there is equality of the Chord and its arc. Thus the determination of the Sine [and arc] is called accurate.[30]

It is evident from the similar triangles in figure 7.4 that

$$\frac{\mathrm{Crd}\,\Delta\theta}{R} = \frac{\mathrm{Sin}(\theta+\Delta\theta) - \mathrm{Sin}\,\theta}{\mathrm{Cos}(\theta+\Delta\theta/2)}, \quad \text{so} \quad \frac{\mathrm{Crd}\,\Delta\theta}{2R} = \frac{\mathrm{Sin}(\theta+\Delta\theta) - \mathrm{Sin}\,\theta}{2\,\mathrm{Cos}(\theta+\Delta\theta/2)}.$$

As Śaṅkara explains, the rationale for Mādhava's rule for $\Delta\theta$ depends on the fact that $\Delta\theta$ is a little larger than $\mathrm{Crd}\,\Delta\theta$. This justifies replacing "twice the medial Cosine," or $2\,\mathrm{Cos}(\theta+\Delta\theta/2)$, with the slightly smaller "sum of the Cosines of both full [arcs]," $\mathrm{Cos}\,\theta + \mathrm{Cos}(\theta+\Delta\theta)$, to more or less balance the substitution of $\Delta\theta$ for $\mathrm{Crd}\,\Delta\theta$ in the other ratio.

[30] *Laghu-vivṛti* on *Tantra-saṅgraha* 2.10–14, [Pil1958], pp. 21–21; translated in [Plo2001]. See also [Plo2005b].

7.3.2 Mādhava on the Sine and Cosine

Mādhava's "Taylor series approximations" for Sine and Cosine. In addition to working out rules for circumferences and arcs, Mādhava also apparently left a substantial body of work on determining Sines and Cosines, traces of which survive in later quotations. For example, just prior to quoting the abovementioned rule for $\Delta\theta$, Nīlakaṇṭha in the *Tantra-saṅgraha* states a formula for determining the Sine and Cosine of a given arc θ. This, the "determination of the Sine as stated by Mādhava," employs the Sine and Cosine of an arc α tabulated in a Sine table, as follows:

> "Having set down the two composite Sines [i.e., the Sine and Cosine of the tabulated arc α] closest to the arc [θ] whose Sine and Cosine are sought, one should compute the arc [of] deficiency or excess.
>
> And set down as a divisor 13751 divided by twice the arcminutes of that [difference arc], for the purpose of the mutual correction of those [quantities]. Having first divided [by that divisor] one [of the tabulated Sine or Cosine], add or subtract [the result] with respect to the other, according as the [difference] arc is excessive or deficient.
>
> Now in the same way, apply that [corrected quantity] times two to the other: this is the correction."[31]

The Sine and Cosine of α are "composite" in the sense that they are accumulated from tabulated Sine-differences. The "arc of deficiency or excess," $(\theta - \alpha)$ in arcminutes, is used to compute a "divisor" D:

$$D = \frac{13751}{2(\theta-\alpha)} = \frac{4 \cdot 3437;45}{2(\theta-\alpha)} \approx \frac{2R}{\theta-\alpha}$$

when the trigonometric radius R is taken to be $3437;45 \approx 360 \cdot 60/2\pi$, a more precise form of the traditional Indian value 3438. Then the known Sine and Cosine are "corrected" by each other:

$$\operatorname{Sin}\theta \approx \operatorname{Sin}\alpha + \frac{\operatorname{Cos}\alpha}{D}, \qquad \operatorname{Cos}\theta \approx \operatorname{Cos}\alpha - \frac{\operatorname{Sin}\alpha}{D},$$

and subsequently these "corrected" terms, multiplied by two, are applied "in the same way"—that is, divided by the divisor D and then added or

[31] *Tantra-saṅgraha* 2.10–13, [SarK1977b], p. 112; translated in [Plo2005b] and in [Gup1969], pp. 92–94. The same rule is cited in a different form by Parameśvara, including a variant with a third-order form ([Gup1969], pp. 94–96, [Gup1974], [Plo2001]). See also [SarK2008], vol. 1, sect. 7.4.3.

subtracted:

$$\begin{aligned}\operatorname{Sin}\theta &\approx \operatorname{Sin}\alpha + \left(\operatorname{Cos}\alpha - \frac{\operatorname{Sin}\alpha}{D}\right)\frac{2}{D}\\ &= \operatorname{Sin}\alpha + \frac{(\theta-\alpha)\operatorname{Cos}\alpha}{R} - \frac{(\theta-\alpha)^2\operatorname{Sin}\alpha}{2R^2},\\ \operatorname{Cos}\theta &\approx \operatorname{Cos}\alpha - \left(\operatorname{Sin}\alpha + \frac{\operatorname{Cos}\alpha}{D}\right)\frac{2}{D}\\ &= \operatorname{Cos}\alpha - \frac{(\theta-\alpha)\operatorname{Sin}\alpha}{R} + \frac{(\theta-\alpha)^2\operatorname{Cos}\alpha}{2R^2},\end{aligned}$$

which is equivalent to a modern second-order Taylor series expansion.

The Mādhava-Newton series for the Sine and Cosine. Other often-cited verses by Mādhava encode in the *kaṭapayādi* alphanumeric notation coefficients for the first few terms of other power series for the Sine and Cosine. Śaṅkara includes them in his commentary on *Tantra-saṅgraha* 2:

> "*vidvāṃs tunnabalaḥ kavīśanicayaḥ sarvārthaśīlasthiro nirviddhāṅganarendraruṅ* ...
>
> "(The wise ruler whose army has been struck down gathers together the best of advisors and remains firm in his conduct in all matters; then he shatters the king whose army has not been destroyed). When these five [numbers] have been spoken in order, [starting] from the one which is at the bottom multiplied by the square of the given arc, when one has divided [by the square of the arc of 90 degrees], the quotient [of each] is to be subtracted, proceeding upwards, [but the quotient] of the last [is multiplied and divided] by the cube; [the quotient is to be subtracted] from the end on the arc.
>
> "*stenaḥ strīpiśunaḥ sugandhinaganud bhadrāṅgabhavyāsano mīnāṅgo narasiṃha ūnadhanakṛdbhūr eva* ...
>
> "([Viṣṇu is] the thief, the betrayer of women, the mover of the fragrant mountain, with shining limbs and auspicious pose, the fish-limbed, Narasiṃha, and the maker [?] of decrease and increase). When these six [numbers are known], [starting] from the one which is at the bottom multiplied by the square of the given arc, when one has divided [by the square of the arc of 90 degrees], the quotient [of each] is to be subtracted, proceeding upwards. Then the final result should be [the value] of the Versine."[32]

[32] *Yukti-dīpikā* 437–438 on *Tantra-saṅgraha* 2, [SarK1977b], pp. 117–118, [Pin1985]. [GolPi1991], p. 52, notes that the same verses are explicitly ascribed to Mādhava by Nīlakaṇṭha in *Āryabhaṭīya-bhāṣya* on *Āryabhaṭīya* 2.17.

In the words of the first Sanskrit sentence, the significant *kaṭapayādi* consonants (those immediately preceding vowels) correspond to the following numerical equivalents:

v	*v*	*t*	*n*	*b*	*l*	*k*	*v*	*ś*	*n*	*c*	*y*	*s*	*v*
4	4	6	0	3	3	1	4	5	0	6	1	7	4
th	*ś*	*l*	*th*	*r*	*n*	*v*	*dh*	*g*	*n*	*r*	*r*	*r*	
7	5	3	7	2	0	4	9	3	0	2	2	2	

This represents the following set of five coefficients for the Sine series: [0; 0,]44. [0;]33, 06. 16; 05, 41. 273; 57, 47. 2220; 39, 40.

In the second *kaṭapayādi* sentence, the consonants are numerically interpreted as follows:

t	*n*	*r*	*p*	*ś*	*n*	*s*	*g*	*dh*	*n*	*g*	*n*	*bh*	*r*	*g*	*bh*	*y*
6	0	2	1	5	0	7	3	9	0	3	0	4	2	3	4	1
s	*n*	*m*	*n*	*g*	*n*	*r*	*s*	*h*	*n*	*dh*	*n*	*k*	*bh*	*r*	*v*	
7	0	5	0	3	0	2	7	8	0	9	0	1	4	2	4	

These constitute a set of six Versine series coefficients, namely [0; 0,]06. [0;]05, 12. 03; 09, 37. 071; 43, 24. 872; 03, 05. 4241; 09, 0.

The application of these coefficients is explained by Śaṅkara in his subsequent verses describing how the terms of the desired series are to be constructed:

> Having multiplied the arc and the results of each [multiplication] by the square of the arc, divide by the squares of the even [numbers] together with [their] roots, multiplied by the square of the Radius, in order. Having put down the arc and the results one below another, subtract going upwards. At the end is the Sine; the epitome of that is made by means of [the verses] beginning "*vidvān*," etc.
>
> Having multiplied unity and the results of each [multiplication] by the square of the arc, divide by the squares of the even [numbers] minus their roots, multiplied by the square of the Radius, in order. But divide the first [instead] by twice the Radius. Having put down the results one below another, subtract going upwards. At the end is the Versine; the epitome of that is made by means of [the verses] beginning "*stena strī*," etc.[33]

Here Śaṅkara begins the computation for the Sine of an arc θ by computing a sequence of "results" of multiplication: the first is $\theta \cdot \theta^2$, the second $\theta \cdot \theta^2 \cdot \theta^2$, the third $\theta \cdot \theta^2 \cdot \theta^2 \cdot \theta^2$, and so forth. For $i = 1, 2, 3, \ldots$, the ith result is divided by R^2 times a factor equal to "the squares of the even numbers [$2i$]

[33] *Yukti-dīpikā* 440–444ab on *Tantra-saṅgraha* 2, [SarK1977b], p. 118, [GolPi1991]. See also, for these verses and the discussion below of their rationale, [Plo2005b], and [SarK2008], vol. 1, sect. 7.5.

minus their roots," or $(2i)^2 + 2i$. The resulting sequence of results will be

$$\frac{\theta^3}{R^2(2^2+2)} = \frac{\theta^3}{R^2 \cdot 3!},$$
$$\frac{\theta^5}{R^4(2^2+2)(4^2+4)} = \frac{\theta^5}{R^4 \cdot 5!},$$
$$\frac{\theta^7}{R^6(2^2+2)(4^2+4)(6^2+6)} = \frac{\theta^7}{R^6 \cdot 7!},$$

and so forth. Taking "the arc" as the first term, these results are then to be "put down one below another" and successively subtracted "going upwards": that is, each term is diminished by the difference of the two following terms to ultimately produce the Sine, as follows:

$$\operatorname{Sin}\theta = \theta - \left(\frac{\theta^3}{R^2 \cdot 3!} - \left(\frac{\theta^5}{R^4 \cdot 5!} - \left(\frac{\theta^7}{R^6 \cdot 7!} - \cdots\right)\right)\right).$$

In the same way, the series for the Versine is also determined by a recursive subtraction of the prescribed "results":

$$\begin{aligned}\operatorname{Vers}\theta &= \frac{\theta^2}{R \cdot 2} - \left(\frac{\theta^4}{R^3 \cdot 2(4^2-4)} - \left(\frac{\theta^6}{R^5 \cdot 2(4^2-4)(6^2-6)} - \cdots\right)\right)\\ &= \frac{\theta^2}{R \cdot 2!} - \left(\frac{\theta^4}{R^3 \cdot 4!} - \left(\frac{\theta^6}{R^5 \cdot 6!} - \cdots\right)\right).\end{aligned}$$

Each of Mādhava's *kaṭapayādi* coefficients, when multiplied by the appropriate power of the given arc expressed as $\theta^{(')}/(90 \cdot 60)$, produces one of the terms in the above Sine or Versine series. For example,

$$2220;39,40 = \frac{5400^3}{3!R^2}, \quad 273;57,47 = \frac{5400^5}{5!R^2},$$

and so forth. These values imply that R is again taken approximately equal to $360 \cdot 60/2\pi$, or more precisely 3437; 44, 48 arcminutes. The powers of 5400 in their numerators (requiring θ to be divided by 5400 to cancel them out) are presumably just scale factors to keep the coefficients at a convenient order of magnitude.

Śaṅkara's geometric rationale for these series depends ultimately on the same similar-triangle relationship, linking the difference of two consecutive Sines with the Cosine of the "medial" arc halfway between them, that we saw earlier in the explanation of Mādhava's arc-difference rule. By this relationship, Sine-differences depend on Cosines, and Cosine-differences (which are equivalent to Sine "second-differences," or the differences between consecutive Sine-differences) on Sines. (A modern mathematical parallel is the

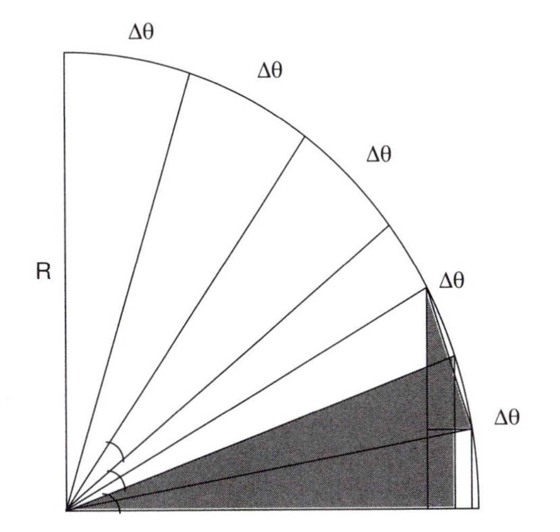

Figure 7.5 Similar triangles relating the Sine and Cosine and their differences.

concept of the cosine as the first derivative of the sine and vice versa, meaning that the sine is its own second derivative.)

More specifically, consider the quadrant in figure 7.5, divided into n unit arcs $\Delta\theta$, the Chord of which is denoted by Crd. Let the Sine and Cosine of the ith consecutive arc $i\Delta\theta$ be denoted Sin_i and Cos_i, and let the consecutive differences $(\mathrm{Sin}_i - \mathrm{Sin}_{i-1})$ and $(\mathrm{Cos}_{i-1} - \mathrm{Cos}_i)$ be denoted $\Delta\,\mathrm{Sin}_i$ and $\Delta\,\mathrm{Cos}_i$, respectively. Likewise, let the Sine and Cosine of each "medial" arc $(i\Delta\theta + \Delta\theta/2)$ be represented by $\mathrm{Sin}_{i.5}$ and $\mathrm{Cos}_{i.5}$, while $\Delta\,\mathrm{Sin}_{i.5} = \mathrm{Sin}_{i.5} - \mathrm{Sin}_{i.5-1}$ and $\Delta\,\mathrm{Cos}_{i.5} = \mathrm{Cos}_{i.5-1} - \mathrm{Cos}_{i.5}$ (always positive). From the pairs of similar right triangles such as the shaded triangles in figure 7.5, we have the following relationships between Sines, Cosines, and their differences:

$$\Delta\,\mathrm{Sin}_{i.5} = \mathrm{Cos}_i \cdot \frac{\mathrm{Crd}}{R}, \quad \Delta\,\mathrm{Cos}_{i.5} = \mathrm{Sin}_i \cdot \frac{\mathrm{Crd}}{R};$$

$$\Delta\,\mathrm{Sin}_{i+1} = \mathrm{Cos}_{i.5} \cdot \frac{\mathrm{Crd}}{R}, \quad \Delta\,\mathrm{Cos}_{i+1} = \mathrm{Sin}_{i.5} \cdot \frac{\mathrm{Crd}}{R}.$$

These proportions and their relationship to the second-differences are explained by Śaṅkara, roughly as follows:

> Divide the Sine produced to the junction of the first arc, multiplied by the Chord, by the Radius: [it is] the difference of the Cosines in the centers of those arcs. Having multiplied that [Cosine-difference, $\Delta\,\mathrm{Cos}_{1.5}$] by the Chord, divide again by the Radius. [That] should be the difference of the differences of the Sines produced to the junctions of the first and second [arcs]. Therefore, divide the first Sine, multiplied by the square of the

> Chord, by the square of the Radius. [The] quotient is the difference of the differences of the first and second [Sines]....
>
> In the same way, the division by the square of the Radius of every Sine multiplied by the square of the Chord should be similarly the difference of [its own difference and] the difference after that....
>
> However many differences of [Sine-]differences there are, when the sum of that many, beginning with the first, is subtracted from the first difference, the desired difference is obtained. When it is desired to obtain the total of the differences of [Sine-]differences, [the quotient] from the sum of Sines multiplied by the square of the Chord, divided by the square of the Radius, should be the sum of the differences of the differences. Therefore, the sum of the differences of the differences is [derived] from the sum of the differences of the Cosines produced to the centers of their own arc-portions.[34]

That is, if the second-difference $\Delta\Delta\,\mathrm{Sin}_i$ is defined as $\Delta\,\mathrm{Sin}_i - \Delta\,\mathrm{Sin}_{i+1}$, then from the above proportions,

$$\Delta\Delta\,\mathrm{Sin}_i = \Delta\,\mathrm{Sin}_i - \Delta\,\mathrm{Sin}_{i+1} = \mathrm{Cos}_{i.5-1}\cdot\frac{\mathrm{Crd}}{R} - \mathrm{Cos}_{i.5}\cdot\frac{\mathrm{Crd}}{R} = \Delta\,\mathrm{Cos}_{i.5}\cdot\frac{\mathrm{Crd}}{R}$$
$$= \mathrm{Sin}_i\cdot\frac{\mathrm{Crd}^2}{R^2}.$$

And since each Sine-difference is the difference of the preceding Sine-difference and their second-difference,

$$\begin{aligned}
\Delta\,\mathrm{Sin}_{i+1} &= \Delta\,\mathrm{Sin}_i - \Delta\Delta\,\mathrm{Sin}_i \\
&= \Delta\,\mathrm{Sin}_{i-1} - \Delta\Delta\,\mathrm{Sin}_{i-1} - \Delta\Delta\,\mathrm{Sin}_i \\
&\vdots \\
&= \Delta\,\mathrm{Sin}_1 - \Delta\Delta\,\mathrm{Sin}_1 - \ldots - \Delta\Delta\,\mathrm{Sin}_i \\
&= \Delta\,\mathrm{Sin}_1 - \sum_{k=1}^{i}\Delta\Delta\,\mathrm{Sin}_k \\
&= \Delta\,\mathrm{Sin}_1 - \sum_{k=1}^{i}\mathrm{Sin}_k\cdot\frac{\mathrm{Crd}^2}{R^2} \\
&= \Delta\,\mathrm{Sin}_1 - \sum_{k=1}^{i}\Delta\,\mathrm{Cos}_{k.5}\cdot\frac{\mathrm{Crd}}{R}.
\end{aligned}$$

The next step is to note that any desired Sine Sin_i is just a sum of cumulative Sine-differences that can be expressed as above. And since the Versine is just

[34] *Yukti-dīpikā* 352–364 on *Tantra-saṅgraha* 2, [SarK1977b], pp. 109–110.

R minus the Cosine, we can replace Cosine-differences in such expressions with the corresponding Versine-differences, since

$$\Delta \operatorname{Vers}_i = \operatorname{Vers}_i - \operatorname{Vers}_{i-1} = R - \operatorname{Cos}_i - (R - \operatorname{Cos}_{i-1}) = \operatorname{Cos}_{i-1} - \operatorname{Cos}_i = \Delta \operatorname{Cos}_i .$$

Consequently,

$$\begin{aligned} \operatorname{Sin}_i &= \Delta \operatorname{Sin}_1 + \Delta \operatorname{Sin}_2 + \ldots + \Delta \operatorname{Sin}_i \\ &= i\Delta \operatorname{Sin}_1 - \sum_{k=1}^{i-1} \sum_{j=1}^{k} \operatorname{Sin}_j \cdot \frac{\operatorname{Crd}^2}{R^2} \\ &= i\Delta \operatorname{Sin}_1 - \sum_{k=1}^{i-1} \sum_{j=1}^{k} \Delta \operatorname{Cos}_{j.5} \cdot \frac{\operatorname{Crd}}{R} \\ &= i\Delta \operatorname{Sin}_1 - \sum_{k=1}^{i-1} \sum_{j=1}^{k} \Delta \operatorname{Vers}_{j.5} \cdot \frac{\operatorname{Crd}}{R} . \end{aligned}$$

As Śaṅkara puts it,

> Therefore, many sums of Sines one below another, ending with the sum of the first and second, are to be made here successively. [The quotient] from the sum of all those [sums], multiplied by the square of the Chord, divided by the square of the Radius—which is the result produced from the sums of the Sines—is equal to the quotient [from division] by the Radius of the sum of [the successive sums of] the Versine-differences, times the Chord. However many differences are considered in the case of [any] desired arc, subtract that [sum of sums] from the first difference times that number of differences.[35]

The remaining steps in Śaṅkara's demonstration involve juggling various simplifying approximations and recursive corrections to them, as follows:

1. Assume that small quantities are equal to unity, in particular, that the unit arc $\Delta\theta$ equals its Chord and Sine, and approximate successive Sines or Versines by successive arcs or integers.

2. Find approximate expressions for sums of successive integers and their sums.

3. Recursively correct the approximations in step (1) to produce the successive terms of infinite series for the Sine and Versine.

The following brief excerpts from Śaṅkara's next fifty-odd verses indicate how these steps are followed. For step 1:

[35] *Yukti-dīpikā* 367–370 on *Tantra-saṅgraha* 2, [SarK1977b], p. 111.

> Because of the uniform smallness of the arc-portions, the first Sine-difference is assumed [instead].... Therefore, subtract from the desired arc the result produced as stated.... Whatever is the sum-of-sums of the arcs—owing to ignorance of [the values of] the Sines—is to be considered the sum [of the sums] of the Sines. But in that case, the last Sine should be the desired arc.... Because of the smallness [of $\Delta\theta$], those arc-portions are considered [to be composed] with unity [i.e., as integers]. So the integers in the desired arc are equal to that [arc].

> Therefore the sum of the Sines is assumed from the sum of the numbers having one as their first term and common difference. That, multiplied by the Chord [between] the arc-junctures, is divided by the Radius. The quotient should be the sum of the differences of the Cosines drawn to the centers of those arcs.... The other sum of Cosine-differences, [those] produced to the arc-junctures, [is] the Versine.[But] the two are approximately equal, considering the minuteness of the arc-division. On account of considering the arc-units [equal] to unity because of [their] minuteness, the Chord too is equal to that [unity], which makes no difference in the multiplier.[36]

In other words, if $\Delta \operatorname{Sin}_1 \approx \Delta\theta$, then $i\Delta \operatorname{Sin}_1$ can be replaced by the arc $i\Delta\theta$. And if $\Delta\theta$ is the unit arc, then $i\Delta\theta$ is just i. More radically, Śaṅkara suggests that a sum of i sums of arcs can be substituted for a sum of $(i-1)$ sums of the corresponding Sines. (He notes that those Sines can also be expressed in terms of medial Cosine- or Versine-differences, and owing to the "minuteness of the arc-division," medial differences can be approximated by the adjacent nonmedial ones.) Finally, Crd $\Delta\theta$ too is assumed equal to unity.

[36] *Yukti-dīpikā* 375–385 on *Tantra-saṅgraha* 2, [SarK1977b], pp. 111–113.

So, starting from our immediately preceding expression for Sin_i in terms of a double sum of Sines, we modify it as follows:

$$
\begin{aligned}
\mathrm{Sin}_i &= \Delta\,\mathrm{Sin}_1 + \Delta\,\mathrm{Sin}_2 + \ldots + \Delta\,\mathrm{Sin}_i \\
&= i\Delta\,\mathrm{Sin}_1 - \sum_{k=1}^{i-1}\sum_{j=1}^{k} \mathrm{Sin}_j \cdot \frac{\mathrm{Crd}^2}{R^2} \\
&\approx i\Delta\theta - \sum_{k=1}^{i}\sum_{j=1}^{k} j\Delta\theta \cdot \frac{\mathrm{Crd}^2}{R^2} \\
&\approx i - \sum_{k=1}^{i}\sum_{j=1}^{k} j \cdot \frac{\mathrm{Crd}^2}{R^2} \\
&\approx i - \sum_{k=1}^{i}\sum_{j=1}^{k} \Delta\,\mathrm{Vers}_j \cdot \frac{\mathrm{Crd}}{R} \\
&= i - \sum_{j=1}^{i} \mathrm{Vers}_j \cdot \frac{\mathrm{Crd}}{R} \\
&\approx i - \sum_{j=1}^{i} \mathrm{Vers}_j \cdot \frac{1}{R}.
\end{aligned}
$$

Hence a double sum of Sines is essentially the same as a double sum of integers, which (times a factor of R) is the same as a sum of Versines. Now we need a more convenient way to express this double sum of integers, which brings us to step 2. First, nested sums are recast as products, which are then approximated as sums of integer powers:

> Whatever is the product of however many numbers beginning with the first-term and increasing [successively] by one, that [product] is divided by the product of that many numbers beginning with one and increasing by one. The results one after another are the sums-of-sums of those [numbers]....
>
> Half that square of the arc should be the sum of the [consecutive] arcs. Because half the product of the first-term and the first-term plus one is a sum: therefore, from the cube and squared-square [etc.] of the desired arc, divided by the product of numbers beginning with one and increasing by one, there are many resulting sums one after another.[37]

So if we have q recursive sums of integers beginning with 1, and the last term of the qth sum is p, then they can be expressed in terms of products of

[37] *Yukti-dīpikā* 386–391 on *Tantra-saṅgraha* 2, [SarK1977b], p. 113.

$(q+1)$ successive integers:

$$\sum_{j_q=1}^{p} \cdots \sum_{j_2=1}^{j_3} \sum_{j_1=1}^{j_2} j_1 = \frac{p \cdot (p+1) \cdot \ldots \cdot (p+q)}{1 \cdot 2 \cdot \ldots \cdot (q+1)} = \frac{(p+q)!}{(p-1)!(q+1)!}.$$

Now we assume that a single sum or arithmetic progression of i integers can be approximated by half the square of i. And the same approximation gives us corresponding expressions for the double sum, triple sum, and so on, which can also be considered single sums of successive powers of integers, as in our previous derivation of the Mādhava-Leibniz series:

$$\sum_{j_1=1}^{i} j_1 = \frac{i(i+1)}{1 \cdot 2} \approx \frac{i^2}{2},$$

$$\sum_{j_2=1}^{i} \sum_{j_1=1}^{j_2} j_1 = \frac{i(i+1)(i+2)}{1 \cdot 2 \cdot 3} \approx \frac{i^3}{6} \approx \sum_{j=1}^{i} \frac{j^2}{2},$$

$$\sum_{j_3=1}^{i} \sum_{j_2=1}^{j_3} \sum_{j_1=1}^{j_2} j_1 = \frac{i(i+1)(i+2)(i+3)}{1 \cdot 2 \cdot 3 \cdot 4} \approx \frac{i^4}{24} \approx \sum_{j=1}^{i} \frac{j^3}{6},$$

and so on.

The final step of the rationale uses these approximations to modify our expressions for the Sine and Versine derived in step 1. These involve deriving successive terms of their series from mutual corrections to their previous terms. As Śaṅkara explains,

> The difference of the desired Sine and [its] arc is from the sum of the sum of the Sines. [Half] the square of the desired arc should be divided by the Radius; thence is [found] the Versine. But thence whatever [results] from the cube of the same desired arc [divided] by the square of the Radius, from that the quotient with six is approximately inferred [to be] the difference of the arc and Sine.…
>
> To remove the inaccuracy [resulting] from producing [the Sine and Versine expressions] from a sum of arcs [instead of Sines], in just this way one should determine the difference of the [other] Sines and [their] arcs, beginning with the next-to-last. And subtract that [difference each] from its arc: [those] are the Sines of each [arc]. Or else therefore, one should subtract the sum of the differences of the Sines and arcs from the sum of the arcs. Thence should be the sum of the Sines. From that, as before, determine the sum of the Versine-differences.…
>
> [W]hatever sum-of-sums is inferred from the determination of the Versine, the difference of the Sine and [its] arc is deduced from the sum-of-sums after that [one]. Thence in this case, the quotient from the product of the cube and square of the desired arc,

> [divided] by whatever is the product of five numbers beginning with one and increasing by one, [is] also divided by the squared-square of the Radius. The difference of the Sine and [its] arc determined in this way should be more accurate....
>
> The sum of Sines is the Versine. So the [new] difference of Sine and arc [is produced] from the [sum-of-sums] after that. And thus in this way it is made from the sum-of-sums after that, again and again.[38]

From our earlier expression of Sin_i as the difference between i and a double sum of integers or a sum of Versines, we may write equivalent expressions involving sums of integer powers:

$$i - \text{Sin}_i \approx \sum_{k=1}^{i} \sum_{j=1}^{k} j \cdot \frac{1}{R^2} \approx \sum_{j=1}^{i} \text{Vers}_j \cdot \frac{1}{R} \approx \sum_{j=1}^{i} \frac{j^2}{2} \cdot \frac{1}{R} \approx \frac{i^3}{6} \cdot \frac{1}{R^2}.$$

But since each Vers/R term initially represented a sum of Sin/R^2 terms which we approximated by a sum of arcs, each Vers/R term is off by an amount corresponding to the difference between the arc and Sine. So we can "remove the inaccuracy" by subtracting out just this difference:

$$\text{Vers}_i \approx \sum_{j=1}^{i} \left(j - \frac{j^3}{6R^2} \right) \cdot \frac{1}{R} = \sum_{j=1}^{i} \frac{j}{R} - \sum_{j=1}^{i} \frac{j^3}{6R^3} \approx \frac{i^2}{2R} - \frac{i^4}{24 \cdot R^3}.$$

Then this modified expression for the Versine can be substituted back into the approximation for the difference between the arc and Sine:

$$\begin{aligned} \text{Sin}_i &\approx i - \sum_{j=1}^{i} \text{Vers}_j \cdot \frac{1}{R} \approx i - \sum_{j=1}^{i} \left(\frac{j^2}{2R} - \frac{j^4}{4!R^3} \right) \cdot \frac{1}{R} \\ &= i - \sum_{j=1}^{i} \frac{j^2}{2R^2} + \sum_{j=1}^{i} \frac{j^4}{4!R^4} \\ &\approx i - \frac{i^3}{3!R^2} + \frac{i^5}{5!R^4}. \end{aligned}$$

If we had used this more accurate expression for the difference between arc and Sine, instead of the previous one, to correct the Versine, it would produce a more accurate sixth-order expression for the Versine, which in turn could be used to produce a still more accurate difference of the arc and Sine, and so back and forth indefinitely to give the terms of the infinite series for Sine and Versine.

These series were evidently used by Mādhava to produce more accurate values of Sines, encoded in *kaṭapayādi* notation in the following collection of mnemonic phrases that total six verses:

[38] *Yukti-dīpikā* 401–403, 406–408, 417–419, 424 on *Tantra-saṅgraha* 2, [SarK1977b], pp. 114–116.

śreṣṭhaṃ nāma variṣṭhānāṃ himādrir vedabhāvanaḥ |
tapano bhānuḥ sūktajño madhyamaṃ viddhidohanam ||

dhigājyo nāśanaṃ kaṣṭaṃ channabhogāśayāmbikā |
mṛgāhāro nareśo 'yaṃ vīro raṇajayotsukaḥ ||

mūlaṃ viśuddhaṃ nālasya gāneṣu viraĺā narāḥ |
aśuddhiguptā coraśrīḥ śaṅkukarṇo nageśvaraḥ ||

tanujo garbhajo mitraṃ śrīmān atra sukhī sakhe |
śaśī rātrau himāhāro vegajñaḥ pathisindhuraḥ ||

chāyālayo gajo nīlo nirmalo nāsti satkule |
rātrau darpaṇam abhrāṅgaṃ nagas tuṅganakho balī ||

dhīro yuvā kathā lolaḥ pūjyo nārījanair bhagaḥ |
kanyagāre nāgavallī deyo viśvasthalī bhṛguḥ ||[39]

The meaning of these phrases as ordinary Sanskrit verse is secondary, and not totally coherent. The first one and a half lines might be translated as the following rather epigrammatic remarks:

> The name of the most excellent ones is best.
> The creator of knowledge is the Himalaya mountain.
> The knower of the Vedic hymns is the burning sun.

A truly learned commentator could doubtless provide a meaningful literary interpretation for the entire verse collection, but a reader concerned primarily with its mathematical meaning would probably not care.

Each line or half-verse consists of two quarter-verses with eight syllables (and eight *kaṭapayādi* consonants) apiece. Each quarter-verse represents one of the twenty-four Sines, with the special semivocalic consonant represented here by *ĺ* taken to signify 9. Thus the six verses state the following Sine values accurate to the second sexagesimal place:

1	0224;50,22	2	0448;42,85	3	0670;40,16	4	0889;45,15
5	1105;01,39	6	1315;34,07	7	1520;28,35	8	1718;52,24
9	1909;54,35	10	2092;46,03	11	2266;39,50	12	2430;51,15
13	2584;38,06	14	2727;20,52	15	2858;22,55	16	2977;10,34
17	3083;13,17	18	3176;03,50	19	3255;18,22	20	3320;36,30
21	3371;41,29	22	3408;20,11	23	3430;23,11	24	3437;44,48

These values are not a very remarkable improvement over those derived from Govindasvāmin's modified Sine-differences that we examined in section 4.3.3; they differ by at most half an arcminute or so. Their chief significance appears to be not so much in their superior precision as in the insightful rationales by which they were derived.

[39] Quoted in Nīlakaṇṭha's commentary on *Āryabhaṭīya* 2.12, [SasK1930], p. 55.

7.3.3 Other mathematical investigations

The above highlights from Mādhava's own work and its exegetical tradition, profound and complex as they are, represent only a fraction of the mathematical studies of Mādhava's school. The same interest in analysis and demonstration that produced exceptional discoveries in trigonometry was also applied to other mathematical topics. The remainder of this section touches on a couple of the salient features of these achievements.

Commentaries and rationales. The members of the Mādhava school carefully studied earlier mathematical works, as is clear from the various commentaries and supercommentaries that they wrote on them. Earlier works doubtless provided some of the inspiration for their own detailed analyses: consider, for example, the arguments involving small subdivisions of a sphere's surface that Bhāskara II used to demonstrate his formulas for its surface area and volume, as we saw in section 6.2.3. It is not such a long step from Bhāskara's arguments about adding up small pieces of a sphere's surface to Śaṅkara's (and presumably Mādhava's) about adding up small pieces of a circle's circumference.

But some of the Kerala school texts are particularly remarkable not just for the profundity but also for the breadth with which they apply these approaches to demonstration. The commentary of Nīlakaṇṭha on the *gaṇita* chapter of the *Āryabhaṭīya*, for example, provides detailed rationales for all its rules, many of them geometrically very ingenious. The work of Śaṅkara's teacher Citrabhānu, mentioned in section 7.2, does the same for algebraic rules for solution of equations. (Interestingly, although geometric reasoning is used extensively to demonstrate algebraic rules, the reverse does not seem to be true: the Kerala mathematicians apparently did not express elaborate geometric problems by equations and manipulate them algebraically to derive their solutions.) The most comprehensive of these works on rationales is the Malayalam *Yukti-bhāṣa* of Jyeṣṭhadeva, which explains and demonstrates the mathematics in Nīlakaṇṭha's *siddhānta*, the *Tantra-saṅgraha*, from the eight fundamental operations of arithmetic up through the spherical geometry of the moon's phases.[40]

It is evident that in Mādhava's school, explanations and rationales were considered valid subjects for mathematical creativity in their own right, not just concessions to the "slow-witted" or verifications to dispel doubt about a rule. As we have seen in our quotations from Śaṅkara, some rationales were evidently even deemed important enough to embody in verse, like the formulas of a base text.

Numerical methods. A subject in which demonstrations were not employed

[40]A couple of Nīlakaṇṭha's demonstrations for Āryabhaṭa's rules on summation of series are translated and explained in [Plo2007b], pp. 493–498. His interpretation and demonstration of Āryabhaṭa's Sine-difference rule as a trigonometrically accurate formula rather than a numerical approximation (see section 5.1.1, *Āryabhaṭīya* 2.12) is discussed in [Hay1997a]. The rationales of Citrabhānu are presented in [Hay1998], and the *Yukti-bhāṣa* with English translation and commentary in [SarK2008].

but that seems nonetheless to have inspired considerable effort and ingenuity was that of "not-just-once" computations, or iterative approximations. Parameśvara was the most active member of the Kerala school in this field, devising more general versions of several of the iterative rules of earlier authors, as well as coming up with new ones. He seems to have been interested in what are today called convergence conditions for iterative algorithms, namely, the situations when they will fail to produce successively better approximations to the desired exact value, or when the successive approximations will converge to it only slowly. His work in this area includes various ways to improve convergence for unreliable iterative rules. Like all other Indian mathematicians, Parameśvara appears to have worked with iterative methods purely empirically; no rationales for their operation are offered.[41]

7.4 ASTRONOMY AND SCIENTIFIC METHODOLOGY

Mādhava's school introduced new developments of existing traditions in astronomy as well as in mathematics. In addition to the standard *pakṣas* of Sanskrit astronomy, Kerala astronomers were familiar with some variant systems. One of these is the so-called *vākya* or "sentence" system mentioned in section 4.3.1, which consists essentially of versified tables of periodically recurring true positions of the celestial bodies, within which the user can interpolate to find true positions for any desired time. The tables are called "sentences" because they are encoded in *kaṭapayādi* notation and thus, like Mādhava's Sine values and coefficients that we saw in section 7.3.2, can have a literary meaning as well as a numerical one. Some of the surviving astronomical texts of Mādhava himself belong to this genre.[42]

Another South Indian astronomical system was adopted and modified by Parameśvara in a work called the *Dṛg-gaṇita*, or "Observation and computation." Although Indian astronomers routinely invoked the desirability of having calculations agree with observations, Parameśvara's work is one of the few places where we see direct evidence of an observational record being constructed.[43]

Parameśvara recorded a number of his own eclipse observations at the end of the *Siddhānta-dīpikā*, a supercommentary he wrote on Govindasvāmin's

---

[41] Some of Parameśvara's work on iterative approximations is discussed in [Plo1996a] and [Plo2004].

[42] [Pin1970a], vol. 4, pp. 414–415. A different Parameśvara, also writing on astronomy in Kerala in the fifteenth century, composed a *Haricarita*, or "Deeds of Hari (Viṣṇu)," that is simultaneously an account of the life of the divine hero Kṛṣṇa (an incarnation of Viṣṇu) and a set of *vākyas* for lunar motion: [Pin1970a], vol. 4, p. 192.

[43] This raises further questions about the applicability of the statistical analysis evaluations of Billard discussed in section 4.6.2, since Billard concludes that Parameśvara's parameters are "not only speculative but of very mediocre convergence," [Bil1971], p. 156. If systematic observations are the foundation of numerous sets of quite accurate parameters in Indian astronomy, then why would Parameśvara, the most systematic observer attested in the Indian tradition, have produced parameters that are only mediocre?

commentary on Bhāskara I's *Mahā-bhāskarīya*. In these observations (at least some of which were apparently made by assistants or with their help), Parameśvara notes the location, the date, and the altitude of the eclipsed body, as the following example shows:

> When the *ahar-gaṇa* [accumulated days from the start of the Kaliyuga] measured 1,655,662, the sun was seen eclipsed slightly on the banks of the Nilā river by keen observers.
>
> At the time of first contact, the "foot-shadow" [*padabhā*] measured 15. At the time of release, 10 less a half.
>
> On that day, the two eyes of the men observing the disc of the sun were not especially injured; therefore at this time a dullness of the rays is to be assumed.[44]

This corresponds to the partial solar eclipse (magnitude 0.07) visible at Ālattūr (presumably observed at Parameśvara's own *illam*) on 1 February 1432. Identifying the locations and dates of Parameśvara's recorded observations makes it possible to compute the corresponding altitudes predicted by modern astronomy for the bodies at these times. (They indicate that Parameśvara's unusual altitude measure, the so-called "foot-shadow," is equivalent to the shadow of a gnomon six feet high. This peculiar quantity is further discussed in section 7.5.) But as far as we can tell at present, the observations and astronomical parameters recorded by Mādhava, Parameśvara, and their successors did not require any extraordinary precision for values of π or Sines and Cosines, whose brilliant refinements in the Kerala school seem to have been motivated instead by pure mathematical curiosity.

Other astronomical innovations of a different sort followed in the work of Nīlakaṇṭha. In his treatises *Jyotir-mīmāṃsā* (roughly, "Analysis of astronomy") and *Siddhānta-darpaṇa* ("Mirror of *siddhāntas*"), Nīlakaṇṭha makes an extremely intriguing attempt to explain astronomical investigations metaphysically, and to synthesize *siddhānta* practices into a coherent mathematical model. As we have seen in sections 4.3.6 and 6.4, Indian astronomers frequently strove to reconcile or balance arguments from textual authority, often based on a text's status as divine revelation, with the requirements of mathematical consistency or observational accuracy. Nīlakaṇṭha, on the other hand, explicitly advocates collecting observational data to refine and modify the models and parameters of an astronomical system, irrespective of whether they agree with established tradition. Moreover, he appeals to the traditional Sanskrit discipline of *mīmāṃsā*, or the philosophy of scriptural exegesis, to analyze the concept of textual authority. In the following passage, he compares a naive view of scriptural authority with the actual, gradual, process of creating a scientific tradition:[45]

---

[44] *Siddhānta-dīpikā* 74–76 on *Mahā-bhāskarīya* 5.77, [Will2005], sect. 4.14.4, p. 263; [Kup1957], p. 330.

[45] *Jyotir-mīmāṃsā*: [SarK1977a], pp. 2–3. On this subject see also [Srin2004a] and [NarR2007].

"Pleased by [feats of] asceticism, Brahman taught to Āryabhaṭa the various numbers [of] the revolutions and circumferences, etc., existing for the computational determination of the planets. Āryabhaṭa in turn combined all that [knowledge], just as [it was] taught, with ten verses [in the chapter on parameters of the *Āryabhaṭīya*]"—this [is what] some think. "Because of Brahman's omniscience, freedom from passion and hatred etc., and certainty of truth, how [can there be] criticism [or testing] of that?" Stupid! [It is] not thus. The favor of the deity is just the cause of mental clarity. Neither Brahman nor the Sun-god [but Āryabhaṭa] himself taught it....

> "[There is] agreement between [the positions of] the moon etc. determined by computation and observation [for one] at a particular place; then [for] another with a fixed opinion, [it is] a teaching confirmed by computation; then [for] one understanding the received teaching, [it is] an authoritative tradition, [which was] to another agreement [with observation], a teaching, [etc.]. Thus authority [is established] from continuity of tradition." (from a commentary on *Mīmāṃsā-sūtra*)

This bold approach to the construction and revision of mathematical models allowed Nīlakaṇṭha to make some ingenious modifications to the Āryapakṣa system. Most strikingly, he changed the arrangement of planetary orbital circles to something that was computationally equivalent to a quasi-heliocentric system similar to the later one of Tycho Brahe, where most of the planets orbit the sun, which in turn orbits the earth. He reversed the usual order of deferents and epicycles so that the center of a planet's own orbit traveled on its *manda*-epicycle (see section 4.3.3), whose center in turn was located at the sun's longitude on the *śīghra*-epicycle, which was centered on the earth. Nīlakaṇṭha fully recognized certain physical implications of this approach, noting, for example:[46]

> Mercury completes its own revolution-circle in eighty-eight days.... And this is not suitable, because its [sidereal] revolution is perceived in one year, not in eighty-eight days.... This statement is made: the earth is not encompassed by the revolution-circle of [either of] those two [Mercury and Venus]. The earth is always outside it.

But although Nīlakaṇṭha's innovations made the star-planets' actual revolutions no longer geocentric, it is not clear that he intended them to be literally heliocentric. He appears to have made them circle about points having the

[46]Commentary on *Āryabhaṭīya* 4.3, [Pil1957], pp. 8–9. Based on the translation in [SarK2008], vol. 2, Epilogue.

same longitude as the sun, but at different distances from the earth.[47] In any case, Nīlakaṇṭha's system was an ingeniously radical rethinking that brilliantly resolved computational inconsistencies in the earlier models; it demonstrates once more, as we saw in section 4.6.3, the advantages that Indian astronomers could reap from not being locked into rigid physical preconceptions about the geometry of the cosmos.

7.5 QUESTIONS OF TRANSMISSION

We know that Greco-Islamic mathematics had a well-established tradition of detailed geometric proof. And we know that the school of Mādhava used detailed geometric rationales in brilliant ways to deduce results about trigonometry, based on manipulations of indefinitely small quantities very reminiscent of the infinitesimals and power series of later European mathematics. Bearing in mind these facts and the multicultural, cosmopolitan nature of the population that flourished on the Malabar coast, it is natural to wonder about possible connections between the Kerala, Islamic, and European mathematical traditions.

For example, it is tempting to read into some of the innovations in the work of the Mādhava school, such as their elaborate geometric demonstrations, their complicated trigonometric formulas, the observational records of Parameśvara, and the boldly rationalist methodology of Nīlakaṇṭha, traces of the Greco-Islamic mathematical science that was presumably familiar to some of their Muslim neighbors on the Kerala coast. In particular, one strong suggestion of (possibly indirect) technological influence from the Islamic tradition appears in the eclipse observations recorded by Parameśvara. Recollect that the quantity that Parameśvara uses to indicate solar altitude, the *padabhā*, is equivalent to the shadow cast by a gnomon measured in feet. Such measures are typical for the altitude scales or "shadow squares" inscribed on the back of Islamic astrolabes, suggesting that Parameśvara was reading his altitude measurements off an instrument of this type.[48] On the other hand, the use of an Islamic astrolabe, which was quite common among Indian astronomers in the second millennium (see section 8.3.1), need not indicate any acquaintance with or interest in Islamic astronomy or mathematics per se.

As for the more abstract resemblances between mathematical interests and outlooks in the two traditions, it must be borne in mind that they are unsupported by any definite evidence for Islamic influence on the content

---

[47]The similarities between Nīlakaṇṭha's system and a Tychonic model are noted and explained in [Ram1994] and [SarK2008], vol. 2, Epilogue; some of his other innovations, as well as arguments against the claim that he actually espoused physical heliocentrism, are discussed in [Pin2001].

[48]See [SarS2008], p. 246. A comparison of some features of Kerala and Islamic trigonometry, without inferring any influence from the latter on the former, is given in [Plo2002a].

or methodology of Kerala mathematics. On the contrary, the work of the Mādhava school is very deeply rooted in earlier Sanskrit learning: consider, for example, the classic similar-right-triangle proportions and sums of series in the *gaṇita* tradition that contributed to the development of the rationales for trigonometric infinite series, or the *mīmāṃsā* texts that Nīlakaṇṭha used to buttress his defense of revising astronomical models. Moreover, it is far from clear how such a hypothetical Islamic-Kerala transmission would have taken place across ethnic and cultural boundaries, with no obvious social channel of communication between Arab merchant society in the cosmopolitan port cities and the rural *illams* of the scholar-priests.

A similar struggle between speculation and prudence is provoked by the fascinating conceptual similarities between the Mādhava school's methods in infinite series and early modern European infinitesimal calculus techniques. Some scholars have proposed that the former could have been the ultimate source of the latter.[49] In broad outline, this hypothesis suggests that Jesuit missionaries in the vicinity of Cochin in the second half of the sixteenth century sought improved trigonometric and calendric methods in order to solve problems of navigation. Finding the Sine methods described in works of the Kerala school useful for that purpose, they transmitted this knowledge to their correspondents in Europe, whence it was disseminated via informal scholarly networks to the early developers of European calculus methods.

Although this idea is intriguing, it does not seem at present to have moved beyond speculation. There are no known records of sixteenth- or seventeenth-century Latin translations or summaries of these mathematical texts from Kerala. Nor do the innovators of infinitesimal concepts in European mathematics mention deriving any of them from Indian sources. It is true that historical theories about mathematical transmission in antiquity have sometimes been accepted primarily on the basis of such conceptual similarities rather than of (unavailable) documentary evidence. And the historiographic question thus raised is an interesting one: what are or should be the criteria for accepting a hypothesis of cross-cultural transmission as plausible, and are those criteria culturally dependent? However, it seems reasonable to argue that a mathematical culture as heavily documented as early modern Europe's is better mapped historically than most ones in the ancient world, and thus requires stronger evidence if it is to be significantly revised.

Other features of this hypothesis also are less than convincing. First and foremost, how would Sanskrit texts on infinitesimal methods for trigonometry and the circle, transmitted no earlier than the late sixteenth century, have inspired the quite different European infinitesimal techniques used at the start of the seventeenth century? Mathematicians like Kepler and Cavalieri focused instead on mechanical questions such as centers of gravity and areas and volumes of revolution, as well as general problems of quadrature

[49] A strong form of this view is argued in, for example, [Raju2007], pp. 267–374. Some of the similarities between Kerala school techniques and early modern calculus are examined in [Bre2002] and [Kat1995].

explicitly associated with the work of Hellenistic forerunners and of Renaissance geometers such as Nicholas of Cusa. The infinite series that Mādhava discovered did not appear in Latin mathematics until late in the seventeenth century, as suggested by the names—Newton, Leibniz, Gregory—attached to them by earlier historians.

In short, as brilliant and fascinating as the accomplishments of the Kerala school were, there is as yet no persuasive reason to think that our understanding of them will end up substantially changing the history of later European mathematics. So far, all the known similarities between the Kerala school and European works, as well as those between Islamic science and the Kerala school, can be more economically explained as cases of parallel evolution than as evidence of direct transmission.

However, that conclusion should not be taken to mean that medieval Indian mathematics as a whole was isolated or uninfluential with regard to other mathematical cultures. We turn now to the verified transmissions of mathematical ideas to and from India, particularly in its connections with the medieval Islamic world.

Chapter Eight

Exchanges with the Islamic World

Prior to about the nineteenth century, most of the Indian mathematics that came to western Asia and Europe, as well as most of the Western mathematics that found its way to India, was transmitted by Muslim intermediaries in what became the vast empire of the caliphate. Consequently, the following brief sketch of the story of these interactions is called here "exchanges with Islam," although it also covers some contacts with pre-Islamic cultures and non-Muslims in Islamic regions.

8.1 INDIAN MATHEMATICS IN THE WEST

After the heyday of the Indo-Greek kingdoms in northern and western India and the waning of Roman dominance, the Sasanian empire of pre-Islamic Iran was India's chief contact in the West. Unfortunately, very little evidence survives about the scholarly and scientific knowledge the Iranians received from India beyond some remnants of astrological texts. It seems clear, though, that Indian astronomical works in the first centuries of the Common Era influenced the Sasanian astronomical texts known in Middle Persian as *zīk* or *zīg* ("canon," literally "line" or "string").[1]

8.1.1 Decimal numerals and arithmetic

It was probably through the Sasanian empire that awareness of the Indian decimal place value numerals first spread to western Asia. There the Syrian Monophysite bishop Severus Sebokht wrote in 662 of "the science of the Indians," including "their subtle discoveries in astronomy, discoveries that are more ingenious than those of the Greeks and the Babylonians, and of their valuable methods of calculation which surpass description," which he described as "done by means of nine signs" (presumably ignoring the zero dot).[2]

After the rise of Islam, Muslim scientists may have first received the decimal place value numerals from other West Asians in Iran or elsewhere, or directly through translations of Indian astronomical texts. A few Central Asian Arabic astronomy treatises from the eighth century, known as *zīj*, in imitation of the Sasanian *zīg*, were apparently based on Sanskrit sources.

[1] The evidence for this influence is reviewed in [Pin1973a] and [Pin1996b].

[2] See the citation in [Kun2003], p. 3.

Among these were the *Zīj al-Arkand* (from *Khaṇḍa-khādyaka*?) and *Zīj al-Arjabhar* (from *Āryabhaṭīya*?).[3] In the early 770s, during the reign of the caliph al-Manṣūr, an Indian scholar is recorded to have traveled to the new capital of Baghdad with an embassy from Sind in northwest India. Surviving descriptions of this embassy recount that the visiting scholar brought along a Sanskrit work on astronomy that was translated and adapted into an Arabic astronomical treatise called the *Sindhind al-kabīr*, or "Great *Sindhind*." The Arabic name apparently conflates the words Sind and *siddhānta*, and may indicate that the lost Sanskrit original was called *Mahā-siddhānta* (not to be confused with the later work of that name composed by Āryabhaṭa II). The parameters known from the surviving fragments of the *Sindhind* indicate that its main source text must have belonged to the Brāhma-pakṣa, although it apparently was not the *Brāhma-sphuṭa-siddhānta* itself.[4]

The decimal place value numerals and decimal arithmetic contained in these texts became known in Arabic as "Indian calculation" (*ḥisāb al-hind*), which was soon a mandatory part of scientific education in the Islamic world. (The standard system for representing numerals in scientific Arabic was a non-place-value alphanumeric notation resembling that of Greek mathematics; see figure 8.3 below for an illustration.) Known Arabic texts expounding the methods of Indian calculation include that of al-Khwārizmī in the early ninth century, followed by about fifteen more by the middle of the eleventh century. Modified forms of these decimal numerals were used in the western parts of the Islamic world, where they were known as *ghubār* or "dust" numerals, presumably in reference to computations on the dust board. They were picked up by scholars writing in Latin at least by the late tenth century, after which they evolved into the "Arabic" or more correctly "Indo-Arabic" numerals universally used today.[5] The Arabic word *ṣifr* or "void, emptiness, zero," evidently a translation of Sanskrit *śūnya*, likewise became "cipher," "zero," and their various cognates in European languages.

8.1.2 Trigonometry of Sines and the *zīj*

Islamic trigonometry also is clearly derived in part from its Indian counterpart. The Hellenistic Chord function, along with other features of Ptolemaic astronomy, was known to Arab mathematicians, thanks to eighth- and ninth-century translations of Greek texts, and possibly also from Sasanian

[3]See [Pin1970b], p. 103; [Pin1996b], p. 40; and [Pin1997], p. 54. A different treatise in sixth-century Sasanian Iran mentioned in later sources as *Zīj al-Arkand* probably derived its name instead from Sanskrit *ahar-gaṇa* (see section 4.3.2, and [Pin1973a], pp. 36–37).

[4][Pin1968a], [Pin1970b]. Later Arabic texts sometimes identify this visitor from Sind with a legendary Indian scholar named Kanaka at the court of Hārūn al-Rashīd a few decades later, who is variously credited with royal descent, great erudition in astronomy and medicine, magical inventions, and the discovery of amicable numbers, to name just a few. Whoever and whatever this Kanaka may have really been, though, there seems to be no convincing reason to link him with the *Sindhind*. See [Pin1997], pp. 51–62, and [Pin1970a], vol. 2, p. 19.

[5]See [Kun2003], and for details of the development of Indo-Arabic numerals in the west, [BurC2002].

and Syriac works.[6] The Arabic Sine, Cosine, and Versine, however, were adapted from Sanskrit sources. We know this from references to the tables of Sines in Indian-influenced Arabic astronomical treatises such as the *Sindhind*, and also from their Arabic technical terminology. The standard Arabic word for Sine, *jayb* (literally "cavity," "pocket"), is apparently a misinterpretation of an earlier word *jība* using the same consonants *j-y-b*. This term *jība*, being meaningless in Arabic, was read as the more familiar word *jayb*. (It is the literal sense of *jayb* as "pocket" or "fold" that was later translated into Latin as "sinus," whence our "sine.")

But where did the mysterious word *jība* itself come from? Evidently from transliterating the Sanskrit word *jīvā*, or "bowstring," a synonym of the standard Sanskrit term *jyā* for the Sine. As the eleventh-century Muslim scientist al-Bīrūnī (whose account of Indian mathematics is discussed in more detail in section 8.2 below) explained:

> People [who know] this [trigonometry] call its scientific books *zījes*, from *al-zīq* which in Persian is *zih*, i.e., "chord." And they call the half-Chords *juyūb* [plural of *jayb*], for the name of the Chord in the Indian [language] was *jībā*, and [the name] of its half *jībārd*. But since the Indians use only the half-Chords, they applied the name of the full [Chord] to the half, for ease of expression.[7]

Sanskrit terms for Cosine and Versine also inspired Arabic equivalents meaning "Sine of the complement" and "arrow," respectively. The tangent and cotangent as separately tabulated functions first appeared in later Arabic texts under the name "shadow" (since, e.g., a horizontal gnomon's shadow is to the gnomon as the Sine of the sun's altitude is to its Cosine).[8]

Other features of some early Arabic *zīj* texts such as the *Sindhind*, as described in later sources, also show their dependence on Sanskrit mathematical astronomy. Parameters and computational algorithms from different *pakṣas* were combined with some Greek and Sasanian techniques in an eclectic mix. Even the verse format of Sanskrit mathematical texts was apparently sometimes imitated, judging from surviving references to an Arabic poem on astronomy or astrology composed by one of the compilers of the

---

[6][Pin1973a], pp. 34–35.

[7]*Al-Qānūn al-Mas'ūdī*, book 3, introduction: [Bir1954], vol. 1, p. 271. (My thanks to Jan Hogendijk for supplying the reference and for help with the translation.) See section A.4 for more on the Sanskrit terms for Chords and Sines. This well-known story of how the sine function got its name is frequently told in a slightly modified form, asserting that the Arabic transliteration *j-y-b* was derived from *jyā* itself. But the alternative term *jīvā* was established in Sanskrit mathematics alongside its more common synonym *jyā* at least by the time of Āryabhaṭa, who uses both terms in the *Āryabhaṭīya* (e.g., in verses 4.23, 4.25: [ShuSa1976], pp. 130, 133). And on phonetic grounds it seems like a much better candidate than *jyā* for the source of the Arabic word, so al-Bīrūnī's explanation is doubtless the true one.

[8]The trigonometric relationship of shadow lengths to Sines and Cosines was of course recognized in Indian applications such as those discussed in section 4.3.4, but there the shadow was not considered an independent trigonometric quantity like the Sine or Versine.

Sindhind. Over time, however, for reasons that are not entirely understood, the *zīj* tradition became much more Hellenized, adopting as its models the renowned *Almagest* and *Handy Tables* of Ptolemy. Perhaps the Greek astronomy works were preferred as a more rigorous and "rational" counterpart to the formal Hellenistic mathematics that Muslim scientists adopted. Whatever the reasons, after about the early tenth century, Islamic astronomy had become essentially Greco-Islamic except for its Indian trigonometric functions.[9]

8.1.3 Algebra and other techniques

The fragmentary evidence of these transmissions gives us an idea of what Indian mathematical techniques were accessible to Arab and Persian mathematicians of the early Islamic world, and how some of them helped shape the basic framework of early Islamic mathematical astronomy. However, it is not always easy to reconstruct from the existing record which parts of the mature Islamic mathematical tradition (besides decimal arithmetic and trigonometry) are originally Indian and which were derived from Greek works or developed independently. Of course, Arabic and Persian texts on Euclidean geometry or Ptolemaic astronomy are unmistakably Greek in origin or inspiration, just as the texts on "Indian calculation" are unmistakably Indian. Many other mathematical developments, however, are harder to trace.

In particular, we should be cautious about the claims in some popular histories of mathematics that medieval Arabic or Islamic algebra (so called from *al-jabr wa 'l-muqābala*, or "restoration and confrontation," the standard Arabic name for this topic) was taken from Indian sources. Although the methods for manipulating and finding unknown quantities that we have seen in the early seventh-century *Brāhma-sphuṭa-siddhānta* might well have been encountered by West Asian students of Sanskrit astronomical texts in the eighth century, the styles of the Arabic and Sanskrit algebra traditions are not much alike.

For one thing, Arabic algebra avoids the use of absolute negative quantities, which are routinely accepted in Indian works.[10] The six types of quadratic equations that al-Khwārizmī in his algebra treatise classified separately, so as not to include negative quantities in their solution, do not seem to have an Indian counterpart. Geometric interpretations and demonstrations of algebra techniques are commonplace in Arabic texts but not in Sanskrit ones. Moreover, medieval Indian notational features such as the

[9]See [Pin1973a] and [Pin1996b], as well as [Pin1970b], p. 104, for the extant fragment of the abovementioned Arabic poem.

[10]A late tenth-century arithmetic text by the mathematician Abū'l Wafā' al-Būzjānī, however, which interprets the result of subtracting a larger number from a smaller one as a "debt," may have been influenced by Indian arithmetic. Abū'l Wafā's text also describes what he calls an "Indian rule" relating the diameter of a circle to the side of an arbitrary regular polygon inscribed in it, which may be related to Bhāskara I's algebraic Sine rule discussed in section 4.3.3; see [Gup1992a].

tabular proto-equation format and syllabic abbreviations of the names of unknown quantities were not adopted in the purely rhetorical early expositions of Arabic algebra. In short, although early Islamic mathematicians may have been exposed to some sophisticated Sanskrit algebra methods, these methods did not become the model for Islamic algebra in the way that Sanskrit decimal arithmetic served as the model for "Indian calculation" in Arabic. The current best guess is that Islamic algebra emerged as a synthesis mainly of traditional problem-solving practices (some possibly Babylonian in origin) and Hellenistic geometry and arithmetic.[11]

Several other topics in Islamic mathematics also suggest links with Indian sources, but here too the connections are often tenuous. The technique called "double false position," based on a form of linear interpolation, is well attested in Islamic texts from the ninth century on, eventually becoming Leonardo of Pisa's "regula elchatayn" (from Arabic *al-khaṭa'ayn*, "two falsehoods"). Double false position is also the main subject of a Latin work called the *Liber Augmenti et Diminutionis*, or "Book of Increase and Decrease," probably translated from Arabic or Hebrew in the twelfth century; the book's full title ascribes this knowledge to "sapientes Indi," that is, "Indian sages." Indeed, the rule of double false position does have a close mathematical kinship with some techniques found in Sanskrit works, such as the Regula Falsi iterated interpolations mentioned in section 4.3.6 and the "operations with assumed quantities," or single false position, discussed in section 5.2.2. But the noniterated double false method as such does not appear in Sanskrit sources, and why a Western text associated it with "Indian sages" has not been satisfactorily explained.[12]

Methods of successive or iterative approximations are another subject that Islamic mathematicians may have derived at least in part from Sanskrit sources. In this case no direct attribution to India is known, but there are resemblances between various approximation rules in the mathematical astronomy of both traditions that strongly suggest transmission.[13] A similar situation exists for noniterated nonlinear interpolation techniques for improving the accuracy of computation with tables. A few Indian versions of such nonlinear algorithms have been mentioned in sections 4.3.3 and 4.4.1, and possibly related versions frequently occur in Islamic texts. However, the lines of development for these methods within each textual tradition, and the potential links between them, are still very poorly understood. Much more work needs to be done in order to draw any useful conclusions about the relationships between numerical methods in Indian and Islamic mathematics.[14]

---

[11]Islamic algebra and its sources are discussed in, for example, [BerJ1986], pp. 99–126, [BerJ2007], pp. 542–563, and [Hoy2002], pp. 410–417. Some aspects of medieval Arabic algebraic notation are surveyed in [Oak2007].

[12]The history of double false position is outlined in [Liu2002] and in [Sch2004], while its Indian aspects are discussed in [Plo2002b], pp. 182–183.

[13]Such resemblances are noted in, for example, [Ken1983c], [Ken1983a], and [Plo2002b].

[14]The two traditions of interpolation methods are surveyed separately, although in each case only briefly and incompletely, in [Gup1969] and [Ham1987], respectively.

8.2 MATHEMATICAL ENCOUNTERS IN INDIA

The establishment of Islamic military and political control over large parts of the subcontinent began around the turn of the second millennium, with the looting expeditions conducted by the Ghaznavid sultans, particularly Maḥmūd in the early eleventh century, into the Panjab from their territory in what is now Afghanistan. From the early thirteenth century to the early sixteenth, a succession of Muslim invaders from the northwest controlled the city of Delhi and parts of the surrounding regions; this sequence of dynasties is known as the Delhi Sultanate.

In 1526 the Central Asian ruler known as Babur, a descendant of Timur, seized Delhi and initiated the Mughal empire. Except for a brief period in the mid-sixteenth century when imperial power passed to the Afghan Sur dynasty, the Mughals continued at least nominally to rule much of India until the British formally annexed it in the middle of the nineteenth century. The period of growing Muslim domination in India in the second millennium coincided with the growth of Islam in the "Indianized" Southeast Asian polities as well.

The Indo-Muslim empires brought major changes to all aspects of Indian life, particularly in the northern part of the subcontinent. The centuries of tempestuous conflicts between established dynasties and new invaders, as well as between the central rulers and peripheral kingdoms or rebellious sultanates, took a toll on the stability and prosperity of many inhabitants. Military conquest of a local kingdom often involved the looting or desecration of important temples that symbolized the sacredness of the conquered king's authority. Numerous temples, schools, and other institutions of Hindus, Buddhists, and Jains were destroyed, sometimes to the accompaniment of militant monotheist rhetoric attacking polytheistic beliefs and practices. The immense waste of lives and resources must have inflicted great damage on the state of learning, including scientific and mathematical knowledge, in many places.

However, it would be a severe distortion to assume that the Indo-Muslim era consisted chiefly of antagonism and violence. Some Muslim leaders, most notably the Mughal emperor Akbar in the sixteenth century, promoted religious tolerance and took a great interest in non-Muslim learning. (In fact, some of the most furious polemics of Muslim chauvinists in India were directed not at the "unbelievers" but at their own Muslim rulers for refusing to persecute the unbelievers.) Hindu rulers often held important positions under Muslim sultans, and some converted to the religion of the new elites. Religious syncretism was also present at the popular level. Many Indian converts, attracted by aspects of Islam such as its social egalitarianism and rejection of caste, brought to their new faith customs and kinship ties of the Hindu world. Some heterodox Muslim sects, including some Sufi orders, blended Hindu devotional traditions into their worship. Many saints, shrines, and temples were objects of pilgrimage for Hindus and Muslims alike, and some temples received protection and even funding from Muslim rulers. Possibly the ultimate in Indian Hindu-Muslim ecumenism was the founding of the Sikh religion, which combined elements of both belief systems, in the Panjab at the start of the sixteenth century.[15]

At royal and imperial courts, many scholars writing in Persian or Arabic and in Sanskrit shared their learning with one another, sometimes at the command of a Muslim or Hindu monarch trying to gain insight into the ways of his subjects or his liege lord, respectively. The transmission of both Sanskrit and Persian works was facilitated by the custom, followed by many Mughal rulers, of maintaining a non-Muslim court astronomer-astrologer with the official title "King of *jyotiṣa*," as well as Muslim ones to handle the Islamic astronomy side. (The professional collaboration of the two groups is illustrated, literally, in a Mughal miniature painting from the sixteenth or seventeenth century that shows Muslim and Hindu court astronomers casting a horoscope for a newborn heir to the imperial throne.) At least one Hindu monarch reciprocated this practice by patronizing Muslim astronomers at his own court.[16]

The earliest record we have of Indian-Islamic encounters dealing with mathematical sciences dates to the very beginning of the Indo-Muslim period, namely the Arabic *Kitāb fī taḥqīq mā li'l-Hind* or *Book of inquiry concerning India* (henceforth *India*).[17] This broad survey of northern Indian religion, literature, and customs was written in the early eleventh century by the renowned Muslim astronomer Abū Rayḥan Muḥammad ibn Aḥmad al-Bīrūnī while he was simultaneously a captive and a protégé of Maḥmūd of Ghazni. Brought along by Maḥmūd on some of his raiding expeditions into the Panjab, al-Bīrūnī had frequent opportunities to converse with Indian scholars and examine Indian texts. He included in the *India* much of what he learned from them about Sanskrit mathematics, astronomy, and astrology.

8.2.1 Indian mathematics and astronomy as observed by al-Bīrūnī

A brilliant scientist and an assiduous reporter, al-Bīrūnī was nonetheless somewhat handicapped by his imperfect knowledge of Sanskrit and his frequent confusion about the chronology and content of Sanskrit texts.[18] De-

[15]For a detailed discussion of temple destruction, protection, and patronage under Indo-Muslim rulers, see [Eat2000], pp. 94–132. An example of an Indo-Muslim dynasty originating in a convert from a Hindu Brāhmaṇa family is recounted in [Knu2008], p. 21.

[16]Concerning Hindu court astronomers under the Mughals, and the abovementioned miniature painting, see [SarS1992]. Muslim court astronomers under the Hindu maharaja of Jaipur are discussed in section 8.2.2.

[17]It has been argued (see [Pin1978b], pp. 316–318) that a few quasi-Ptolemaic modifications of Sanskrit lunar and solar theory appearing in the work of Muñjāla and Śrīpati in the tenth and eleventh centuries respectively are derived from Islamic sources. However, so little is known about these hypothesized transmissions that it is difficult to draw any firm conclusions concerning them.

[18]The claim in the *India* that al-Bīrūnī was himself reading and translating Sanskrit works seems to be at best an exaggeration or misinterpretation; probably he meant that he would write down an Arabic paraphrase of a local pandit's verbal rendering of a Sanskrit

spite this, his *India* is invaluable as a contemporary record of how Indian mathematical science, or at least parts of it, looked to a non-Indian mathematician. Not surprisingly, as a product of the Greco-Islamic Euclidean tradition, al-Bīrūnī regarded the unfamiliar mathematical methodology of Sanskrit science with both interest and arrogance, a reaction that seems to have been mutual:

> According to their belief ... no created beings besides them have any knowledge or science whatsoever.... At first I stood to their astronomers in the relation of a pupil to his master, being a stranger among them and not acquainted with their peculiar national and traditional methods of science. On having made some progress, I began to show them the elements on which this science rests, to point out to them some rules of logical deduction and the scientific methods of all mathematics, and then they flocked together round me from all parts, wondering, and most eager to learn from me, asking me at the same time from what Hindu master I had learnt those things, whilst in reality I showed them what they were worth, and thought myself a great deal superior to them....
>
> [Y]ou mostly find that even the so-called scientific theorems of the Hindus are in a state of utter confusion, devoid of any logical order, and in the last instance always mixed up with the silly notions of the crowd, e.g., immense numbers, enormous spaces of time, and all kinds of religious dogmas.... Therefore it is a prevailing practice among the Hindus *jurare in verba magistri* [to appeal to the word of the master, i.e., to argue from authority]; and I can only compare their mathematical and astronomical literature, as far as I know it, to a mixture of pearl shells and sour dates, or of pearls and dung, or of costly crystals and common pebbles. Both kinds of things are equal in their eyes, since they cannot raise themselves to the methods of a strictly scientific deduction.[19]

The disdain al-Bīrūnī felt may have been partly inspired by his difficulties in grappling with the verse format of scientific Sanskrit:

> Grammar is followed by another science, called *chandas*, i.e. the metrical form of poetry, corresponding to our metrics—a science

text into a vernacular language. All that has survived of these paraphrases among al-Bīrūnī's voluminous extant writings is an Arabic version of a Sanskrit *karaṇa* work, now lost, in addition to one (in a single incomplete manuscript) of the *Yoga-sūtra* of Patañjali. See [Ken1978], [Pin1975], and [Pin1983].

[19] [Sac1992], vol. 1, pp. 23, 25; the Latin phrase is the translator's. So is the consistent use of "Hindus" where al-Bīrūnī's *al-hind* in the sense of "Indian people" may mean either Hindus as opposed to Buddhists or Indians in general. It is not clear how fully al-Bīrūnī grasped Indian sectarian divisions; for one thing, he seems to conflate Buddhist and Jain beliefs (e.g., in vol. 1, p. 119).

> indispensable to them, since all their books are in verse. By composing their books in metres they intend to facilitate their being learned by heart, and to prevent people in all questions of science ever recurring to a *written* text, save in a case of bare necessity.... They do not want prose compositions, although it is much easier to understand them.
>
> Most of their books are composed in *śloka*, in which I am now exercising myself, being occupied in composing for the Hindus a translation of the books of Euclid and of the Almagest, and dictating to them a treatise on the construction of the astrolabe, being simply guided herein by the desire of spreading science. If the Hindus happen to get some book which does not yet exist among them, they set at work to change it into *ślokas*.[20]

Subsequently al-Bīrūnī's discussion goes into some detail about Sanskrit terms for large decimal powers and for numbers in the concrete number notation.[21] However, since he could expect his readers already to be familiar with "Indian calculation" or decimal arithmetic, he skimped on description of the actual graphical use of numbers among the Indians, confining himself to a reference to their "reckoning in the sand" and a few other comments:

> The Hindus do not use the letters of their alphabet for numerical notation, as we use the Arabic letters in the order of the Hebrew alphabet.... The numeral signs which *we* use are derived from the finest forms of the Hindu signs....
>
> They use black tablets for the children in the schools, and write upon them along the long side, not the broad side, writing with a white material from the left to the right.[22]

Most of al-Bīrūnī's exposition of Sanskrit science—in fact, most of the content of the *India*—is devoted to its astronomical and astrological texts and its calendrical methods, and how they compare with those of other traditions familiar to him. As he notes,

> The science of astronomy is the most famous among them, since the affairs of their religion are in various ways connected with it. If a man wants to gain the title of an astronomer, he must not only know scientific or mathematical astronomy, but also astrology.[23]

---

[20] [Sac1992], vol. 1, pp. 136–137; see section A.2 for an explanation of the *śloka* meter. There is no textual record of any such Sanskrit translations of Euclid or the *Almagest* or any astrolabe text at this period, although there is an Arabic treatise by al-Bīrūnī on the astrolabe ([Ken1978], p. 156).

[21] [Sac1992], vol. 1, pp. 174–179.

[22] [Sac1992], vol. 1, pp. 175, 182. The remark about the lack of alphanumeric notation suggests that al-Bīrūnī's northern Indian informants were unfamiliar with the southern *kaṭapayādi* system.

[23] [Sac1992], vol. 1, pp. 152–153.

The *siddhāntas* and *karaṇas* that al-Bīrūnī either obtained or heard of include the *Pañca-siddhāntikā*, the *Paitāmaha-siddhānta*, the *Brāhma-sphuṭa-siddhānta*, the *Āryabhaṭīya*, and the *Khaṇḍa-khādyaka*, as well as some more obscure texts and commentaries. However, what he says about the content and authorship of these treatises often does not match up with the texts as we know them; in particular, he seems often to have mistaken statements in a commentary for part of the base text.[24]

This added to his difficulties in trying to figure out which theories were being advocated by which astronomers, particularly when cosmological models are in question. He was aware that the traditional flat-earth cosmology of the Purāṇas was different from the spherical universe of the astronomers, and that some astronomers invoked features from both systems. With his own grounding in Greco-Islamic science, where mathematical astronomy is at least superficially independent of scriptural cosmology, al-Bīrūnī found such syncretic arguments quite perplexing, and expended a good deal of ink in efforts to understand the psychological reasons that Indian authors advocated them. He was particularly vexed with Brahmagupta for his support of the traditional *smṛti* rejection of the moon as the cause of solar eclipses, discussed in section 4.3.6:

> Look, for instance, at Brahmagupta, who is certainly the most distinguished of their astronomers. For as he was one of the Brahmans who read in their Purāṇas that the sun is lower than the moon, and who therefore require a head biting the sun in order that he should be eclipsed, he shirks the truth and lends his support to imposture, if he did not—and this we think by no means impossible—from intense disgust at them, speak as he spoke simply to mock them, or under the compulsion of some mental derangement, like a man whom death is about to rob of his consciousness. . . .
>
> If Brahmagupta, in this respect, is one of those of whom God says (*Koran*, Sūra xxvii.14), "They have denied our signs, although their hearts knew them clearly, from wickedness and haughtiness," we shall not argue with him, but only whisper into his ear: If people must under circumstances give up opposing the religious codes (as seems to be your case), why then do you order people to be pious if you forget to be so yourself? Why do you, after having spoken such words, then begin to calculate the diameter of the moon in order to explain her eclipsing the sun, and the diameter of the shadow of the earth in order to explain its eclipsing the moon? Why do you compute both eclipses in agreement with the theory of those heretics, and not according to the views of those with whom you think it proper to agree?. . .
>
> I, for my part, am inclined to the belief that what made Brahma-

[24]See [Pin1983].

> gupta speak the abovementioned words (which involve a sin against conscience) was something of a calamitous fate, like that of Socrates, which had befallen him, notwithstanding the abundance of his knowledge and the sharpness of his intellect, and notwithstanding his extreme youth at the time. For he wrote the *Brahmasiddhanta* when he was only thirty years of age. If this indeed is his excuse, we accept it, and herewith drop the matter.[25]

For the most part, though, when al-Bīrūnī discussed Indian science in the *India* and in other works, he contented himself with supplying the names of Sanskrit technical terms for astronomical and astrological concepts, and explaining some of the Indian computational methods used for them. Among the practices he described was what he called the "Indian circle" procedure for determining the cardinal directions by means of two shadows touching the circumference of a circle, an ancient technique that we encountered in the *Kātyāyana-śulba-sūtra* (see section 2.2).[26]

A subject in Indian mathematics that especially attracted al-Bīrūnī's interest was the standard proportion rules beginning with the Rule of Three Quantities, discussed in sections 5.1.1, 5.1.3, and 6.2.1. He wrote a comprehensive Arabic treatise, the *Maqāla fī rāshīkāt al-Hind*, or "Treatise on the *rāśikas* (proportion rules) of the Indians," comparing these techniques to Euclidean notions of ratio. The following excerpt from the beginning of the work shows al-Bīrūnī's familiarity with both approaches to this subject:[27]

> A magnitude has a ratio only to all magnitudes of the same genus, regardless of whether one obtains it or does not arrive at it; either because of its form, as in the case of the diameter and the circumference owing to straightness and roundness, or because of its irrationality, as in the case of the diagonal and the side owing to incommensurability. Therefore a ratio always occurs between any two magnitudes of the same genus, under all circumstances. It is not expressible independently until it is defined or known, unless it is combined with another. Therefore it is dependent upon two ratios, and a proportion is among at least three magnitudes.
>
> For that reason Euclid said [*Elements*, book 5, def. 8]: A proportion consists in at least three terms.... Other kinds of ratios reduce to four proportional magnitudes. Upon this depend the calculations used in account books and commercial transactions, and [those] that are current in astronomy and mensuration.
>
> Euclid demonstrated in the sixteenth [proposition] from the sixth [book] that the plane [rectangle contained] by the first of them

[25] [Sac1992], vol. 2, pp. 110–112; see also [Plo2005a].

[26] See [Yan1986], pp. 17–18. The name "Indian circle" is still used for this and similar methods.

[27] [Bir1948], pp. 1–3. This translation is based in part upon an unpublished translation by Takanori Kusuba.

> [and] by the fourth is equal to the plane of the second by the third. So the first therefore is the counterpart of the fourth in multiplication, and the second is the counterpart of the third....
>
> The Indians call it "tray rāshīka," namely "that which has three places."... Moreover, they describe these [as] "three" because the known [magnitudes] in its data are three.
>
> These people follow in their computations a numerical approach as their custom concerning it. They rely, in verifying it, upon experiment and investigation of examples, without occupying [themselves] with explanation by means of geometric proof. They draw for that [purpose] two intersecting lines [which have] four regions in this form, and they say: if five are for fifteen, then three are for how much? Then they move the fifteen to the empty place and multiply it by [the number] above it, which is three, and forty-five is produced, and they divide it by the five. There results nine, which is what is required to be put in the empty place, so that three are for nine....

5	3
15	

The *trai-rāśika* methods used by al-Bīrūnī's informants evidently corresponded to the "two-column" formats established already in Brahmagupta's proportion rules in *Brāhma-sphuṭa-siddhānta* 12.10–13 (see section 5.1.3).

8.2.2 Translations of Islamic works into Sanskrit

Although al-Bīrūnī's promised Sanskrit translations of Euclid, Ptolemy, and his own text on the astrolabe either never materialized or left no traces in Indian literature, some Arabic and Persian scientific texts in subsequent centuries met with better success. The first known Sanskrit mathematical work to be based on Islamic ones was in fact a treatise on the construction and use of the plane astrolabe, whose title became the standard Sanskrit name of that device: *Yantra-rāja*, or "King of instruments." This translation or adaptation was made in 1370 by a Jain astronomer named Mahendra Sūri at the court of a monarch of the Tughluq dynasty during the Delhi Sultanate, Fīrūz Shāh. Several other Sanskrit works or chapters of works expounding the astrolabe were composed in succeeding centuries, including one that bestowed on the instrument the (transliterated) alternative name *Usturalāva-yantra*.[28] More details about the astrolabe and its Indian incarnations are supplied in section 8.3.1.

[28] See [SarS1999b]. No specific Islamic texts are named as sources in most of these works, so we cannot tell whether al-Bīrūnī's desire for a Sanskrit version of his own astrolabe treatise was partly realized more than three centuries after he wrote. We do know, however, that a direct translation of a different Islamic book on the astrolabe was made in Jaipur in the early eighteenth century (see the end of this section).

Several other Islamic astronomy texts as well were incorporated into Sanskrit *jyotiṣa*. A fifteenth-century Persian work on *hay'a* or the geometry of planetary orbs was translated into the Sanskrit *Hayata-grantha*, "Book on *hay'a*," probably in the seventeenth century. This translation, like the comparison of proportion rules in al-Bīrūnī's work, shows some interesting contrasts between the consciously Euclidean style adopted by most Islamic mathematical works and the exposition of Sanskrit mathematical prose. Texts on *hay'a* generally begin with definitions from basic geometry and Aristotelian physics; the following excerpts show what the Indian translator made of a few of them.[29]

> [Persian:] In the name of God, the Compassionate, the Merciful ... A sensible object [literally "indication"], if it is not divisible in any way, is called a point. If it is divisible in one way, it is called a line. If it is divisible in two ways, that is, divisible in length and width but not divisible in height, it is called a surface.... A surface is either plane or not plane. A plane [surface] is [such] that if any two points that may be supposed on it are connected by a straight line, that line is on that surface, in no way falling outside [it]. And when a curved line contains a surface in such a way that a point may be supposed on that surface [where] straight lines passing from that point to that line are all equal, that surface is called a circle, and that line the circumference of the circle, and a circular line; and that point [is called] its center.... Every straight line that divides the circle in two is called a chord.... And if it passes through the center, it is called a diameter.

> [Sanskrit:] Homage to Lord Gaṇeśa.... A point is in [the category of] visible indication. Because of its smallness, it is not in any way divisible.... And now, [an object] divisible in one way [and] not divisible in one way is expressed by the word "line."... And now, [an object] able to be divided in both ways is expressed by the word "surface."... And a surface is of two kinds, plane and not plane [literally "even" and "uneven"].... If, when one has marked two points, a straight line extended from one point to the other point stays on the surface,[30] that is a plane surface, otherwise, a non-plane surface. One curved line in the form of a circle encloses a plane surface. When one has made a point in the middle of that, [and] many lines extending from that [point] all around to the curved line, those lines are all equal.... Any straight line that

[29]The Persian text is from an unpublished manuscript; the Sanskrit version is published in [Bha1967], pp. 1–5. The extra gaps in the translation of the Sanskrit represent places where the Indian translator included transliterated Persian technical terms. For a fuller discussion of these two texts, see [Pin1996c] and [Pin1978b]; for a study of the mathematics and astronomy in Islamic *hay'a* texts, see [Rag1993].

[30]Reading *saṃlagnā* for *saṃjñakā*, [Bha1967], p. 3.

> makes two unequal parts of a circle is ... expressed by the word "full chord" [*pūrṇa-jyā*]. If that straight line goes through the center, then if it makes two equal parts of the circle,... it is called a diameter.

Even in a direct translation of a work in the Euclidean style, the Indian author's primary aim was to explain the terms in a clear and comprehensible way relying on the reader's understanding of geometric concepts, rather than to build up a logically watertight hierarchy of definitions constructing those concepts from scratch.

A Delhi Brāhmaṇa named Nityānanda in the early seventeenth century translated into Sanskrit the Indo-Persian *zīj* composed in 1628 for the emperor Shāh Jahān, and in 1639 produced a second Sanskrit version of it called the *Sarva-siddhānta-rāja*, or "King of all *siddhāntas*." This later version was deliberately more "Indianized," composed in verses claiming to describe the astronomy of a "Romaka," or Westerner, who was actually an incarnation of the Sun-god.[31] Such a consciously traditional presentation of new material suggests an attempt to overcome Indian skepticism or rejection of foreign sources, a conservative attitude that coexisted and to some extent competed with the more receptive views of Nityānanda and others. (This controversy is discussed in more detail in section 8.3.3.)

Forms of Islamic astrology too were largely naturalized in Sanskrit, beginning at least by the thirteenth or fourteenth century. Under the name *tājika*, meaning "Westerner" (from the Middle Persian *tāzīg*, or "Arab"),[32] an amalgam of Indian and Islamic astrological concepts found a secure niche beside traditional Sanskrit horoscopy, spreading from its birthplace in western India to the rest of the subcontinent by the seventeenth century. There was even a verse treatise in mixed Sanskrit and Persian composed on *tājika* around 1600.[33] In addition to these direct Sanskrit adaptations of Arabic and Persian sources on astral sciences, the Indian interest they inspired is also attested by the existence of Persian-Sanskrit dictionaries incorporating technical terms from mathematics and astronomy.[34]

The most far-reaching translation project for Islamic mathematical texts was that of the early eighteenth-century Hindu ruler of Amber and Jaipur in Rajasthan, Jayasiṃha or Jai Singh II. During most of his reign from 1700 to 1743, Jayasiṃha was a tributary ally of the Mughal emperor Aurangzeb and his successors, particularly Muḥammad Shāh. In fact, it was at Jayasiṃha's court in the 1730s that the imperial Persian *zīj* of Muḥammad Shāh was prepared by a team of astronomers representing both the Sanskrit and Islamic traditions. Jayasiṃha's own lifelong interest in the science of astronomy was manifested there in an ambitious project to reform it, by analyzing both Islamic and Indian theories and checking them against new observational

[31] [Pin2003b], and [Pin1996c], pp. 476–480.
[32] [Pin1997], p. 79.
[33] [Pin1997], pp. 79–90.
[34] [Pin1996c], pp. 474–475, and [SarSBD].

data.

To carry out this project, during the 1720s Jayasiṃha built in five north Indian cities observatories consisting of massive masonry instruments inspired by those of Islamic observatories in Central Asia, and set his scholars to work collecting data and translating texts. Subsequently, Jayasiṃha broadened the scope of his program to include investigation of contemporary European astronomy as well. He obtained from Portuguese and French Jesuits in Europe and India Latin astronomical works and some new instruments, including telescopes. He also maintained at his court a few Jesuit scholars to explain to his court astronomers the theories of Tychonic, and perhaps Copernican, astronomy; these interactions are described further in section 9.2.

In addition to obtaining copies of previously translated books, such as the *Hayata-grantha* and the works of Nityānanda, Jayasiṃha's multilingual research team churned out translations of Arabic recensions of Euclid's *Elements*, Ptolemy's *Almagest*, and the *Spherics* of Theodosius. Original Islamic works (or parts of them) on more recent developments in Greco-Islamic astronomy were also translated, including some works by the thirteenth-century Central Asian astronomer Naṣīr al-Dīn al-Ṭūsī, such as his Persian treatise on the plane astrolabe.[35]

8.2.3 Translations of Sanskrit works into Persian

Some Muslims in India regarded the study or appreciation of non-Muslim literature with puritanical horror, as a sacrilegious act. (This attitude was doubtless sometimes exaggerated or exploited for political advantage, as when the Mughal prince Aurangzeb in the mid-seventeenth century cleared his path to the Peacock Throne by encouraging the trial and execution of his elder brother Dārā Shukōh on charges of heresy, for having written admiringly of Hindu scriptures.) On the other hand, many Indo-Muslim rulers and scholars, like al-Bīrūnī before them, were more interested in foreign scientific knowledge than in doctrinal rigidity. Several of them sponsored or produced translations and adaptations of Sanskrit texts.

Astrology, as usual, was one of the first genres to catch the eye of translators. Around the same time that Mahendra Sūri was working on his *Yantra-rāja*, Muslim colleagues of his at the court of Fīrūz Shāh were translating into Persian a Sanskrit text on divination by the sixth-century astronomer Varāhamihira, and another, unnamed astrological work. Varāhamihira's text was subsequently re-translated at least twice at other courts. In the astronomy corpus, Bhāskara II's *Karaṇa-kutūhala* made its way into Persian via several translations from the fifteenth century onward.[36]

Although Sanskrit-derived works appear never to have seriously threat-

---

[35]A detailed description of Jayasiṃha's observatories and the records of the scholars who worked there is provided in [Shar1995]. See [Pin1999], [Pin1987b], and particularly [Pin2003a] for more information on the translations.

[36][Pin1970a], vol. 2, p. 13, and [Ans2004], pp. 592–593.

ened the dominance of the Greco-Islamic *zīj* genre in Indo-Muslim astronomy, there seems to have been a largely unexplained upsurge of interest among Muslim astronomers near the end of the eighteenth century in various Sanskrit treatises, including some representatives of the table text or *koṣṭhaka* genre that was itself originally inspired by *zīj* tables (see section 8.3.2). Several new *zījes* were composed as translations or adaptations of such Sanskrit sources.[37]

The mathematics textbooks of Bhāskara II, which were especially popular in the Muslim-controlled northern and western regions of India, also attracted the interest of a Persian-literate audience: the *Līlāvatī* was translated for the Mughal emperor Akbar in 1587 and the *Bīja-gaṇita* for his grandson Shāh Jahān in 1634–1635. New Persian translations of the *Līlāvatī* and perhaps the *Bīja-gaṇita* too were made in later years.[38] The translation of the *Līlāvatī*, by Akbar's renowned court scholar Abū al-Fayḍ Fayḍī, contains the first known appearance of a widespread story about Bhāskara II and a malfunctioning water clock that is frequently repeated even today. Fayḍī says in his preface:

> It is said that the composing the Lilawati was occasioned by the following circumstance. Lilawati was the name of the author's (Bhascara's) daughter, concerning whom it appeared, from the qualities of the Ascendant at her birth, that she was destined to pass her life unmarried, and to remain without children. The father ascertained a lucky hour for contracting her in marriage, that she might be firmly connected, and have children. It is said that when that hour approached, he brought his daughter and his intended son near him. He left the hour cup on the vessel of water, and kept in attendance a time-knowing astrologer, in order that when the cup should subside in the water, those two precious jewels should be united. But, as the intended arrangement was not according to destiny, it happened that the girl, from a curiosity natural to children, looked into the cup, to observe the water coming in at the hole; when by chance a pearl separated from her bridal dress, fell into the cup, and, rolling down to the hole, stopped the influx of the water. So the astrologer waited in expectation of the promised hour. When the operation of the cup had thus been delayed beyond all moderate time, the father was in consternation, and examining, he found that a small pearl had stopped the course of the water, and that the long-expected hour was passed. In short, the father, thus disappointed, said to his unfortunate daughter, "I will write a book of your name, which shall remain to the latest times—for a good name is a second life, and the ground-work of eternal existence."[39]

---

[37] [Ans2004], pp. 593–602.

[38] The Persian translations of the *Bīja-gaṇita* and *Līlāvatī*, respectively, are noted in [Pin1970a], vol. 1, pp. 39 and 44; vol. 4, pp. 300 and 308; and in [Ans2004], p. 592.

[39] The English translation is by the nineteenth-century Indologist Edward Strachey,

This account may well be a generic Sanskrit anecdote that became attached to the name of the illustrious Bhāskara without any factual basis. For example, the events described reappear with slight modifications in a Sanskrit story composed in Ahmedabad in 1600 (in this version, it is a grain of rice from the anonymous astrologer's ritual forehead mark rather than a pearl from his daughter's clothing that clogs up the water clock).[40]

8.3 INFLUENCE AND SYNTHESIS

General conclusions about the results of all the scientific cross-fertilization described in the previous section must remain tentative, since most of the known texts reflecting them are still unpublished and little studied, while there probably exist many others as yet unknown. Translated texts appear to have been sometimes little more than a sort of coffee-table book, a handsome volume of exotic learning destined to be briefly admired and seldom or never read.[41] On the other hand, a few aspects of Islamic mathematical sciences (in addition to astrology) were much more fully assimilated into the Sanskrit tradition. Thus the overall impact of Islamic works on Indian ones seems to have been significant, although ultimately not revolutionary.

8.3.1 Instruments

The advantages of various new devices to simplify observations and calculations were rapidly recognized by Indian scholars. Chief among these devices, as its Sanskrit name suggests, was the "king of instruments" or plane astrolabe. Not just Sanskrit expositions of the astrolabe but many examples of the instrument itself were crafted in India around the middle of the second millennium. (We have seen in section 7.5 the indications that one such instrument was used by Parameśvara in fifteenth-century Kerala.)

Many of the surviving astrolabes, being inscribed in Arabic or Persian characters, were evidently made by Indo-Muslim artisans for astronomers who were either Muslim themselves or at least sufficiently familiar with the Islamic tradition to decipher the inscriptions. Others, however, are marked additionally or solely in *nāgarī* script, for the use of astronomers working in Sanskrit. In the early eighteenth century, Jayasiṃha established in Jaipur a sort of astrolabe atelier that turned out numerous Sanskrit versions of the instrument, mostly for the use of his court astronomers. There is even one known Sanskrit astrolabe from northern India inscribed with *kaṭapayādi* numbers.[42]

---

quoted in [Hut1812], pp. 177–178.

[40][SarS2004], pp. 314–315.

[41]This seems to have been the fate of, for example, Nityānanda's first Sanskrit translation of Shāh Jahān's *zīj*, of which few or no manuscripts are known ever to have been made, other than the original presentation copies bestowed mostly on Muslim dignitaries, who probably did not read Sanskrit. See [Pin2003b], pp. 269–270.

[42]See [SarS1999b], and for the last-mentioned instrument, [SarS1999a].

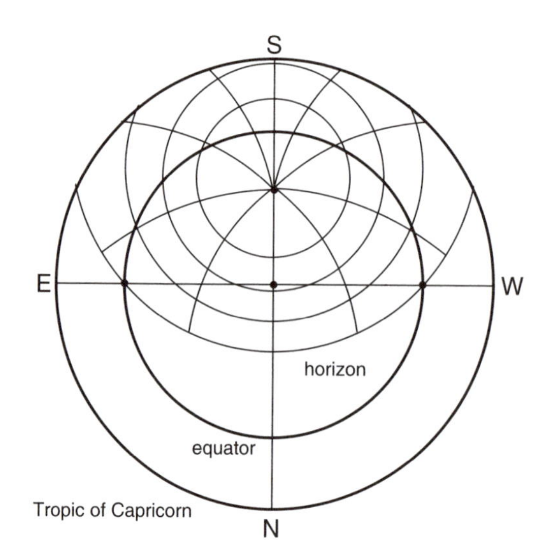

Figure 8.1 The astrolabe plate containing projections of celestial and terrestrial reference circles.

The mathematics underlying the design of the standard astrolabe was discussed in various Islamic astronomical treatises.[43] Its primary feature is a stereographic projection of the visible part of the celestial sphere onto the plane of the celestial equator, just as in the astrolabe's modern descendant, the planisphere. The plane of projection, called the "plate," is centered on the north celestial pole and marked with the projection of the horizon and related reference circles (circles of equal altitude and of equal azimuth) for a particular terrestrial latitude (see figure 8.1). Scales indicating time-degrees are inscribed about the plate's circular edge. Then the projected stars and reference circles are marked on a freely rotating lattice called the "rete." When the rete is placed on the plate and turned, the markings on the plate are visible through the spaces in the rete, so the projected stars appear to be rising and setting with respect to the projected horizon. The combination thus forms a sort of analog computer that displays graphically the appearance of the sky at any desired time at the given latitude, as illustrated in figure 8.2.

The star positions in the rete are represented by what we can think of as polar coordinates based on the stars' equatorial coordinates. If a radial line is drawn from the projected north pole in the center to a projected star, the angle between that line and the projected celestial equator's zero point is the star's right ascension, while the length of the line is proportional to

[43] A general explanation of the geometry of the astrolabe is provided in [Nor1974].

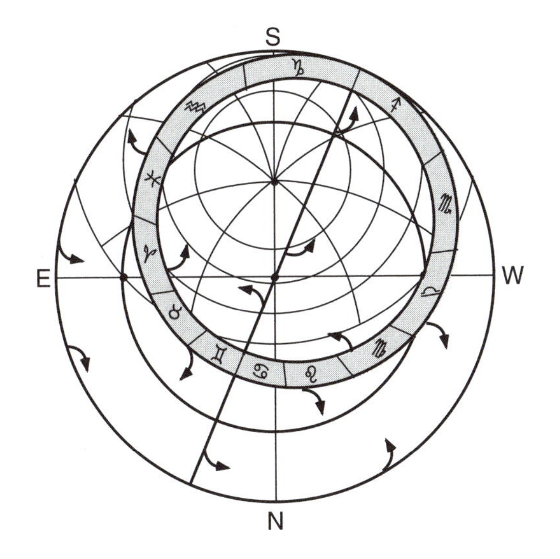

Figure 8.2 The rete with projection of the ecliptic and star-pointers overlaid on the astrolabe plate.

the sine of the star's declination. However, since premodern star catalogues typically give positions in ecliptic rather than equatorial coordinates, the astrolabe designer has to employ coordinate conversions based on spherical trigonometry to turn the catalogued values into the desired ones.

In general, Sanskrit astrolabe texts seem to have glossed over the spherical trigonometry of coordinate conversions, concentrating instead on explaining the use of the device. For example, the standard Islamic formulas for computing stellar declinations were transformed by Mahendra Sūri in his original *Yantra-rāja*, either deliberately or accidentally, into approximate versions. One of these approximations was repeated in a subsequent Sanskrit astrolabe treatise as late as the eighteenth century.[44]

The celestial globe, another import from Islamic cultures, apparently aroused less interest than the astrolabe among Indian astronomers. However, there do exist some surviving examples of this instrument with Sanskrit inscriptions, as well as a description of it in a seventeenth-century Sanskrit commentary on the *Siddhānta-śiromaṇi* of Bhāskara II.[45]

---

[44]See [Plo2000] for a fuller discussion of these coordinate conversion formulas in and after the *Yantra-rāja*.

[45]See [SarS1992], pp. 247–248, and [SarSAK1993].

8.3.2 Table texts

We have seen in section 8.1 how certain features of Sanskrit texts influenced the early development of Islamic mathematical astronomy. During the Indo-Muslim period, a key feature of the Islamic *zīj* in its turn apparently influenced the structure of Sanskrit astronomy, namely, its reliance on elaborate numerical tables. Such tables, ultimately inspired by those in the works of Ptolemy, formed the chief part of a typical Greco-Islamic *zīj*. Planetary mean motions were tabulated for various units of time ranging from several minutes to several centuries; planetary equations, or corrections to mean positions, were tabulated for every minute or even every second of anomaly; and so forth. Thus, to compute a planet's true position, for example, an astronomer would not go through the laborious calculations we examined in sections 4.3.2 and 4.3.3. Instead, he would simply compute the interval between his desired time and the earlier epoch time for which known mean positions were recorded. Then he would add up from the tables the additional amount of mean longitude accumulated by the planet during that interval, add it to the epoch mean longitude, reduce the sum to the appropriate value of anomaly, and find in the equation tables the size of the correction to be applied for that amount of anomaly.

In this system, all the work of carrying out complicated trigonometric calculations for mathematical astronomy functions fell to the lot of the compiler of the *zīj* tables, so that all the user needed to do was to interpolate between the tabulated values at the appropriate places. The format of such texts is illustrated in figure 8.3, which shows a *zīj* table reproduced from the first Arabic *zīj* known to have been composed in India, dating to around the middle of the thirteenth century. (Note that the tabulated values, as usual in *zījes*, are in Arabic alphanumeric notation rather than in Indo-Arabic decimal place value numerals.)[46]

Of course, the concept of preserving in a treatise useful lists of data, such as Sine values and planetary parameters, had always been integral to the medieval *siddhānta* and *karaṇa*. But such lists incorporated in verse texts were generally kept as concise as possible, to ease the tasks of composition and memorization. The strong preference remarked by al-Bīrūnī for calculation based on memorized verses rather than on written documents dominated the style of Indian technical works up to the early second millennium.

The Sanskrit bias in favor of orality, however, evidently did not withstand the attractions of the labor-saving *zīj* tables, a quintessentially graphical and visual style of text. From about the twelfth century on, extensive numerical tables or *koṣṭhakas* in Sanskrit such as the one illustrated in figure 8.4 became an increasingly important part of Indian astronomy.[47] Some of these

---

[46]My thanks to Benno van Dalen, who described this *zīj* in [Dal2002], for kindly providing this illustration from a manuscript of it.

[47][Pin1981a], pp. 41–46. Figure 8.4 shows f.2r of a manuscript purchased in Jaipur in 2004 and now in the collection of the Jaina Vidya Sansthan, Jaipur. The listed year range is from Śaka 1640–1720 or 1718–1798 CE.

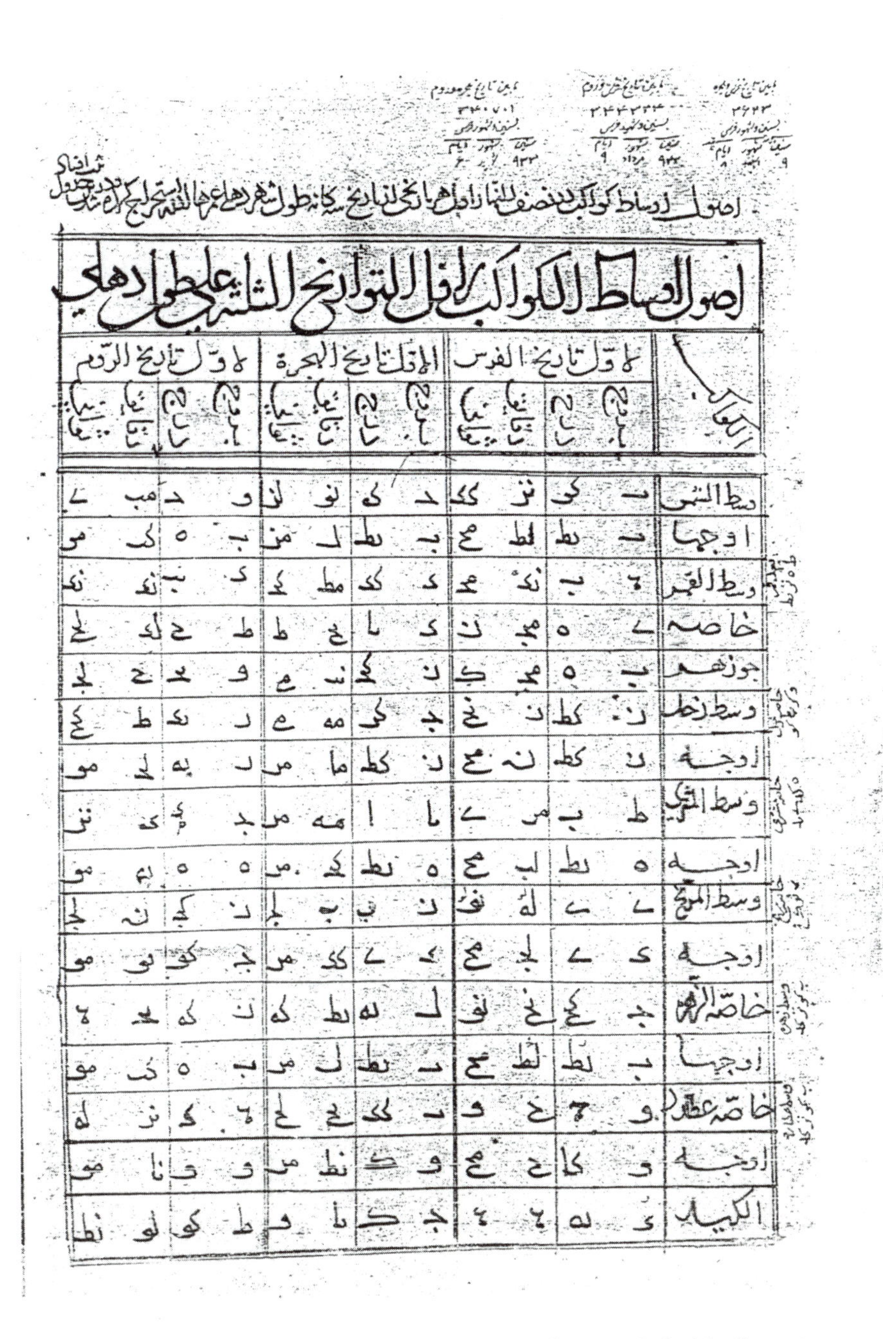

Figure 8.3 A table of epoch mean positions from an Indo-Muslim *zīj*.

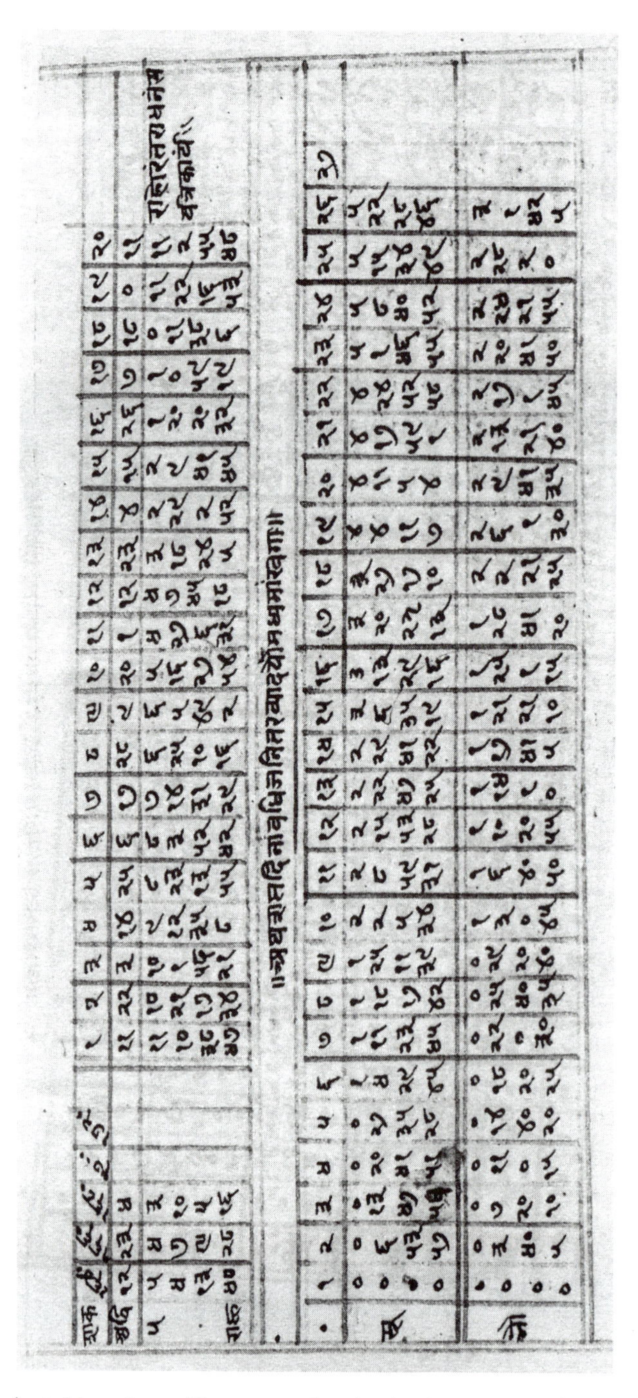

Figure 8.4 A table of weekly mean longitude increments from a Sanskrit manuscript.

Indian tables have a structure resembling that of Greco-Islamic ones, with mean longitude increments tabulated for time units of varying sizes and equation increments tabulated for varying amounts of anomaly. However, others reflect a range of innovations and ingenious computational adjustments evidently devised by astronomers working within the Sanskrit tradition.

For example, some planetary tables pre-mix the calculation of mean positions and their corrections to produce tabulated values of true longitudes at equal time-intervals, a structure somewhat reminiscent of the *vākya* astronomy of southern India described in section 7.4. (However, there seems to have been no attempt among northern Indian calculators to adapt extensive tabular data to an oral format, the way that *vākya* astronomers did with *kaṭapayādi* sentences.) Other tables are designed for the needs of calendar makers, setting out the necessary data for finding the consecutive endpoints of time units such as days and *tithis* (see section 4.4.2) in any desired year. The variety and ingenuity of these *koṣṭhakas* and their role in later Indian astronomy are discussed further in section 9.3.2.

8.3.3 Assimilation and rejection of Islamic theories

While astrolabes and *zīj*-style tables had a significant impact on the practice of second-millennium Sanskrit astronomy, other aspects of Islamic mathematical science apparently produced no detectable effects. For example, the axiomatic deductive geometry that so many Islamic mathematicians pursued with enthusiasm, on the model of Hellenistic Euclidean texts, seems to have sparked no similar imitations in Sanskrit. Nor was there (outside of Jayasiṃha's court, at least) a strong interest in debating different versions of geocentric astronomy to identify a single self-consistent geometric model with optimal predictive power. Islamic methods, unlike the later impact of modern science under colonial European rule, did not fundamentally reshape or eclipse Sanskrit mathematical science.

The efforts at a more comprehensive assimilation of Islamic mathematical science that were launched in the courts of Delhi and Jaipur apparently did not make much impression on Indians outside of those cosmopolitan elite settings. In fact, some Sanskrit astronomers viewed foreign technical innovations with strong distrust, as potentially damaging to the purity of traditional knowledge. The concerns on this subject are illustrated by the controversy between representatives of two rival families of Benares astronomers in the eighteenth century, one of whom opposed Islamic opinions on concepts such as precession and the composition of the orbs of the planets, while the other accepted not only the disputed opinions but also other Islamic learning, such as the theories of geometric optics. Much of this controversy turned on the question of whether this foreign knowledge was contrary to the teachings of *smṛti* and therefore unacceptable.[48]

Suspicion of foreign learning did not prevent a few Islamic mathematical

[48] [Pin1978b], pp. 320–323.

and astronomical concepts from working their way even into conservative Sanskrit texts. For example, Islamic-style spherical trigonometry rules for stellar declination, similar to the one that had been somewhat garbled or modified in the fourteenth-century *Yantra-rāja*, appear without any caveats in the work of one of the more traditionalist Benares astronomers of the eighteenth century. Such rules had by this time evidently ceased to count as "foreign learning" and become naturalized as simply part of *gaṇita*.[49]

Other Sanskrit technical texts that have not yet been thoroughly studied also indicate more familiarity with Islamic ideas than was previously suspected.[50] Thus the overall influence on Indian mathematical science of Islamic texts or of reactions against them cannot yet be conclusively assessed. But it is worth bearing these issues in mind as we explore in our final chapter various developments near the end of the Sanskrit mathematical tradition, before it was supplanted in the nineteenth and twentieth centuries by modern mathematics under European colonialism.

[49] [Plo2002a], pp. 85–87.
[50] [Pin1996c], pp. 480–481, 484.

Chapter Nine

Continuity and Changes in the Modern Period

The interactions with Islamic mathematics and astronomy discussed in the previous chapter were by no means the only major developments affecting Sanskrit science during the second millennium. Until recently, though, this period was largely overshadowed in the historiography of India by the eras preceding and following it, namely, pre-Islamic India, on the one hand, and European colonization on the other. In the nineteenth and twentieth centuries, the previous half-millennium or so was often dismissed as a moribund interval of intellectual "stagnation" or "decay" for India, particularly in scientific thought.[1] This opinion had its conveniences for some European imperialists, who wanted to argue that foreign domination was helping India advance out of its "Dark Ages."[2] In the words of a recent historian,

> Although introducing some scientific and technical skills of its own, Islam was largely seen to have been destructive of the remnants of the old Indian civilisation.... This Orientalist triptych—contrasting the achievements of ancient Hindu civilisation with the destruction and stagnation of the Muslim Middle Ages and the enlightened rule and scientific progress of the colonial modern age—has had a remarkably tenacious hold over thinking about the science of the sub-continent. It was a schema deployed not only by British scholars, officials and polemicists but also by many Indians, for whom it formed the basis for their own understanding of the past and the place of science in Indian tradition and modernity.[3]

[1]Consider, for example, these remarks by two seminal historians of Indian mathematics: "In abnormal times when there were foreign invasions, internal warfares [sic] or bad government and consequent insecurity, the study of mathematics and, in fact, of all sciences and arts languished.... It is certain ... that after the 12th century very little original work was done in India. Commentaries on older works were written and some new works brought out, but none of these had sufficient merit as regards exposition or subject matter, so as to displace the works of Bhāskara II, which have held undisputed sway for nine centuries (as standard text books)." [DatSi1962], vol. 1, pp. 127–128.

[2]For example, in 1834 a British official and enthusiastic amateur of Sanskrit astronomy named Lancelot Wilkinson wrote: "[T]o give a command and powerful influence over the native mind, we have only to revive that knowledge of the system therein [in *siddhāntas*] taught, which ... has ... been allowed to fall into a state of utter oblivion ..." ([Wilk1834], p. 507; see the more detailed discussion in [Plo2007b], section 6.2).

[3][Arn2000], pp. 3–4.

In consequence, most post-Bhāskara II works on mathematics and mathematical astronomy were neglected by historians until quite recently, and many of them still remain unpublished.

Current scholarship, however, has begun to investigate evidence for the late medieval and early modern periods, comprising approximately the fourteenth to eighteenth centuries, as a time of great activity and innovation rather than decline in Sanskrit intellectual traditions.[4] Especially after around 1500, scholars writing in Sanskrit began to promulgate self-consciously new approaches in a range of disciplines, including logic, epistemology, textual exegesis, and grammar. It is still far from clear exactly what inspired this trend toward new directions in Sanskrit learning. It certainly cannot be explained as a simple parallel to the contemporary rise of modernity in Europe, many of whose manifestations—among them printing, global exploration, "mixed" or applied mathematics in fields such as ballistics and mechanics, and heliocentric cosmology—were not present in South Asian cultures of the time. Without attempting to analyze the larger intellectual currents in late medieval and early modern Indian thought, this chapter explores some developments in Sanskrit mathematical sciences of this period that seem to represent similarly novel perspectives.

9.1 INDIVIDUALS, FAMILIES, AND SCHOOLS

Thanks to the greater (although still not as lavish as we would like) abundance of the documentary record in the second millennium as compared to earlier times, we have a fuller historical picture of some Indian mathematicians after about 1500 than we have been able to obtain of most of their predecessors. The first thing to note about Indian scholars in this later era of the Sanskrit tradition is that there are many more of them. It has been estimated that the total population of the subcontinent, after having grown only slowly since the end of the first millennium, nearly doubled between about 1550 and 1800.[5] This increase is reflected in the increased numbers of known authors of Sanskrit works or commentaries on astronomy, astrology, and mathematics. Of course, the numbers in this period also appear greater simply because more recent works had a better chance of surviving to the present. In addition, they had the advantage of the growing availability and affordability of paper. (Manuscripts were still copied by hand, however; the printing press, although introduced to the subcontinent in the mid-sixteenth century by European missionaries, became widespread only in the nineteenth.)[6]

A few examples of scientific "schools" or family traditions in particular localities during this period illustrate the kind of contextual information

---

[4] [Pol2001b], [Min2002].

[5] See [Ric1997], p. 207.

[6] [Arn2000], p. 6.

available (at least in some cases).[7]

Pārthapura in the fifteenth/sixteenth century. Pārthapura, a town on the Godavari River identified with modern Pathri perhaps 110 kilometers southeast of Aurangabad in Maharashtra, was a regional center of astronomical and mathematical activity, for reasons that are still unknown. The earliest known ancestor in this tradition is the fourteenth-century great-great-great-grandfather of the astronomer Jñānarāja around the turn of the sixteenth century, whose two sons, nephew, and great-nephew all produced mathematical, astronomical, or astrological works. In a village in the neighborhood of Pārthapura lived another family of *jyotiṣa* specialists, one of whom, named Divākara, in the mid-sixteenth century went to another regional center of science, Nandigrāma (modern Nandod) in Gujarat. There he studied with the scholar who became perhaps the most famous mathematician of the age, Gaṇeśa (see below).

Nandigrāma in the sixteenth century. Gaṇeśa, born in 1507, was the author of probably the most popular astronomical works of the second millennium, a *karaṇa* and a *koṣṭhaka*, discussed in section 9.3.2. His father, his brother, his nephew, and his great-grandson are also known to have written *jyotiṣa* works. The family seems to have stayed in Nandigrāma for at least five generations, but the above example of the visiting student from the Pārthapura region indicates that they were well-known enough to attract pupils from elsewhere.

Varanasi in the sixteenth and seventeenth centuries. The greatest metropolis of Sanskrit learning at this time was the city of Varanasi, also known as Benares or Kāśī, on the Ganga River in Uttar Pradesh. The city was a magnet for scholars in other regions, some of whom seem to have received funding from their local rulers to study and work there.[8] Gaṇeśa's student Divākara from the Pārthapura region moved to Varanasi after completing his studies in Nandigrāma, and there he had at least five sons, two grandsons, and three great-grandsons, most of whom were active in astronomical studies. In the late sixteenth century a scion of another Maharashtra family living slightly northwest of Aurangabad also moved to Varanasi, where some of his many descendants carried on a sort of scholarly feud with the earlier immigrants over the proper use of elements of Islamic astronomy in *jyotiṣa*. Varanasi continued to be a center of astronomical learning up through the eighteenth century.

Courts at Delhi and elsewhere. Section 8.2 mentioned the activities of some scholars who apparently served as court astronomers under the Tughluq emperor Fīrūz Shāh the Mughal emperor Shāh Jahān, and the Rajasthani ruler Jayasiṃha. Other scholars are known to have worked on aspects of

[7]See [Pin1981a], pp. 119–130, for fuller details, including the names of all the individuals involved. The Pārthapura school in particular is discussed at length in [Knu2008], pp. 9–22.

[8]See [MinNil].

jyotiṣa at the court of Shāh Jahān's grandfather, the emperor Akbar, and other Mughal rulers.[9]

The level of polymathia or multidisciplinary scholarship perceptibly increased among authors of mathematical works in this period. Whereas most earlier mathematicians and astronomers are not known for compositions in other Sanskrit genres, several later scientific authors were renowned poets or philosophers as well. Section 7.2 noted that in the later Kerala school, Nārāyaṇa and Acyuta were famed for their scholarship in poetics and grammar, respectively. And for some other astronomers of this period also, mathematical interests went hand in hand with mastery of literary, religious, or philosophical disciplines. For instance, a son of the Pārthapura astronomer Jñānarāja, one Sūryadāsa, was simultaneously a mathematician (among other works, he composed commentaries on Bhāskara II's *Līlāvatī* and *Bīja-gaṇita*) and a specialist in the high art of Sanskrit poetic narrative. He is best known as an early exponent of the genre of Sanskrit bidirectional poetry, where the syllables of a poem convey different meanings depending on whether they are read from beginning to end or backwards from the end to the beginning: one of his works is a bidirectional poem that simultaneously recounts the lives of the two divine heroes Rāma and Kṛṣṇa, in opposite directions.[10]

9.2 CONTACTS WITH EUROPE

The well-known story of Christopher Columbus's voyage in 1492 in search of a westward route to "the Indies" illustrates the importance attached by early modern European powers to breaking the Arab-Venetian monopoly on the global spice trade. Six years later, the Portuguese explorer Vasco da Gama made his way from Africa across the Indian Ocean to the commercial center of Calicut on the Kerala coast, opening the door to Portuguese domination along much of the west coast of India and in Sri Lanka.

In the early seventeenth century, the British East India Company obtained permission from the Mughal emperor Jehangir to establish trading centers on the subcontinent's eastern coast, at Calcutta and Madras. The Portuguese and British were followed by the French, the Dutch, and the Danes, mostly in southern and eastern India. During the seventeenth and eighteenth centuries, several of the colonial powers maneuvered and fought with one another for territory and influence. Proxy wars and political deals eventually consolidated the dominance of the East India Company over most of the subcontinent in the nineteenth century. During the course of these developments, Indians and Europeans perforce became acquainted to some extent with each other's languages and cultures, and in the process transmitted knowledge about their respective systems of mathematical science.

[9] [Pin1981a], pp. 116–117.
[10] See [Min2004b].

9.2.1 European discovery of Indian mathematical sciences

Foreign traders and missionaries residing in India had pressing practical motives for figuring out the computational, calendric, and geographic systems of the local population. Some of them, as well as some colonial officials stationed in India, began to pursue these subjects further. The founding of the Asiatic Society of Bengal by the Calcutta jurist and scholar Sir William Jones in 1784 provided one of the earliest venues for their scholarship on a variety of subjects, ranging from "the laws of the Hindus and Mahomedans; the history of the ancient world; proofs and illustrations of scripture; traditions concerning the deluge" to "Arithmetic and Geometry and mixed sciences of Asiaticks."[11] A number of their compatriots back home in European countries also took an interest in what the emigrés were able to decipher concerning Indian sciences.

For instance, in the 1730s the young Leonhard Euler, at the behest of a historian colleague at the St. Petersburg Academy of Sciences who corresponded with Danish Protestant missionaries in south India, took a stab at explaining some of the mathematical and astronomical techniques in Indian texts that they had translated.[12] And a former assistant to the British Astronomer Royal Nevil Maskelyne, a mathematician named Reuben Burrow, investigated Indian mathematics during his work in Bengal as an instructor of the engineers' corps.[13] He published in 1790 a paper entitled "A Proof that the Hindoos had the Binomial Theorem," inspired by a rule from an unnamed Sanskrit text that is reminiscent of ones treated as part of the "net of numbers" or combinatorics by Mahāvīra and later authors:

> With respect to the Binomial Theorem ... the following question and its solution evidently shew that the Hindoos understood it in whole numbers to the full as well as Briggs, and much better than Pascal....
>
> "A Raja's palace had eight doors; now these doors may either be opened by one at a time; or by two at a time; or by three at a time; and so on through the whole, till at last all are opened together: it is required to tell the numbers of times that this can be done?
>
> "Set down the number of the doors, and proceed in order gradually decreasing by one to unity and then in a contrary order as follows:
>
> 8 7 6 5 4 3 2 1
> 1 2 3 4 5 6 7 8

[11]From a memorandum "Objects of Enquiry during my residence in Asia," written by Jones en route to Calcutta; see [JonW1807], vol. 2, pp. 3–4. The admiration expressed by several early European visitors for the mental calculation skills of Indians is described in [SarS1997], p. 191.

[12]See [Plo2007a].

[13][Dha1971], p. 279.

> "Divide the first number eight by the unit beneath it, and the quotient eight shows the number of times that the doors can be opened by one at a time: multiply this last eight by the next term seven, and divide the product by the two beneath it, and the result twenty-eight is the number of times that two different doors may be opened . . . and lastly, one is the number of times the whole may be opened together, and the sum of all the different times is 255."[14]

Burrow also expessed an intention to translate what he called "the *Leelavotty* and *Beej Geneta*, or the Arithmetic and Algebra of the Hindoos," but apparently did not do so. The first English translations of Bhāskara's mathematics were made by other researchers in the early nineteenth century.[15]

Indian astronomy enjoyed a brief vogue in late eighteenth-century Europe as a possible source of ancient observations that could be used to test modern hypotheses in celestial mechanics. As the theories of Newtonian mathematical astronomy became more detailed, they outstripped the capabilities of contemporary instruments to provide adequate data for testing them within the span of a few years or decades. The curiosity of Europeans about Indian data had originally been sparked by some astronomical tables acquired from Thailand and India in the seventeenth and eighteenth centuries, respectively, and analyzed by Domenic Cassini in 1693 and by Guillaume Le Gentil in 1776.[16] At this time, most European scholars were still so ignorant of even recent Indian chronology that they could accept, say, a 1772 description of Jayasiṃha's early eighteenth-century masonry observing instruments in Benares (see section 8.2.2) as "built by Mawnsing, the son of Jysing, about 200 years ago."[17] They knew practically nothing of the history of ancient and medieval Indian astronomy, or of the great historical and textual gaps separating surviving Sanskrit treatises from the eras that they treated as reference points.

In particular, it seemed plausible to several European scientists that the mean conjunction assigned by Indian texts to the start of the Kaliyuga in 3102 BCE might represent part of a continuous observational record going back thousands of years. Such early data could provide a usable observational baseline extending over nearly five millennia—a tempting prospect. The French astronomer Jean-Sylvain Bailly in 1787 published a treatise on "Indian and Oriental astronomy" that assessed very positively their potential usefulness to contemporary astronomers:

> Astronomy, which supplies dates, assists history to throw some light on the chronology on the nations of Asia.. . . But this Astron-

[14][BurR1790], pp. 101–102.

[15][BurR1790], p. 96 (Burrow's statement); [Pin1970a], vol. 4, pp. 300, 308, 311.

[16][Lal1693], vol. 2, pp. 186–227, and [Leg1776].

[17][Dha1971], p. 7. The confusion may have arisen from the fact that the Benares building itself, on which the eighteenth-century Jayasiṃha's (or Jai Singh's) instruments stood, may have been used as an observing platform in the time of Jayasiṃha's great-grandfather Māna Singh: [Shar1995], pp. 1, 191–192.

> omy can above all be useful to our modern sciences, in providing us with ancient data, which serve us as a point of comparison.... [Only] centuries can teach us [about] what changes imperceptibly; it is there that time is worth more than genius....
>
> When M. de la Grange published in 1782 the result of his new and profound researches, he appealed to the future to confirm them: a future which neither he, nor we, nor several generations more will ever see; he had not hoped that what remains to us of antiquity could furnish data sufficiently exact and sufficiently ancient to verify his theory. These remains of antiquity are [found] among the Indians....[18]

The Scottish mathematician John Playfair followed suit in 1790 with the publication of "Remarks on the astronomy of the Brahmins" that mostly recapitulated Bailly's findings.[19] Bailly's and Playfair's arguments depended principally on the identification of several Indian astronomical parameters that corresponded to what modern celestial mechanics retrodicted for conditions several millennia earlier. Some of the correspondences, they claimed, were close enough to rule out the possibility of their having been back-calculated for the desired epoch by Indian methods at any time in or near the common era, because the imprecision of the methods would have produced less accurate values.[20]

However, these conclusions involved some rather selective reasoning to justify accepting the Indian parameters that agreed with modern calculations while discarding or explaining away the ones that did not. If a parameter matched the modern retrodicted value for the beginning of the Kaliyuga in 3102 BCE, that was taken as evidence that it was founded on actual observational records for that date. And if it did not match, that was taken as evidence that it was founded on observational records for a date even earlier:

> The year in the tables of Tirvalore is therefore too great by $1'$, $5''\frac{1}{2}$.... And hence it is natural to conclude, that this determination of the solar year is as ancient as the year 1200 before the Calyougham, or 4300 before the Christian era....
>
> The obliquity of the ecliptic is another element in which the Indian astronomy and the European do not agree.... This gives us $23°$, $51'$, $13''$, which is $8'$, $47''$, short of the determination of the Indian astronomers. But if we suppose, as in the case of the sun's equation, that the observations on which this determination is founded, were made 1200 years before the Calyougham, we shall find ... that the error of the tables did not much exceed $2'$.[21]

[18] [Bai1787], pp. ii, lxvi. See [Rai2003] for a fuller discussion of Bailly's work.

[19] [Pla1790], pp. 11–12.

[20] [Pla1790], pp. 27–28, 45–46, 62–63.

[21] [Pla1790], pp. 37–39. Note however that Playfair expresses some doubts about this

The Indian assumption that the Kaliyuga conjunction was a mean conjunction of all the planets, ignoring their considerable longitudinal spread (see section 4.6.2), also required some special pleading:

> These tables ... have an obvious reference to the great epoch of the Calyougham. For if we calculate the places of the planets from them, for the beginning of the astronomical year, at that epoch, we find them all in conjunction with the sun in the beginning of the moveable zodiac.... According to our tables, there was, at that time, a conjunction of all the planets, except Venus, with the sun; but they were, by no means, so near to one another as the Indian astronomy represents.... [T]here is reason to suspect, that some superstitious notions, concerning the beginning of the Calyougham, and the signs by which nature must have distinguished so great an epoch, has, in this instance at least, perverted the astronomy of the Brahmins.[22]

Such speculations did not conduce to a systematic rigorous scrutiny of the Indian parameters as a whole. Consequently, pioneers such as Bailly and Playfair ultimately failed to reconstruct a convincingly consistent data set that practicing astronomers around the turn of the nineteenth century could rely on. Harsh critiques based less on analysis of the evidence than on Christian chauvinism and other prejudices also reduced the perceived credibility of arguments for ancient Indian data.[23] At the same time, improvements in instrumental precision continued to deflect astronomers' attention away from the quest for ancient observations in general, and the scrutiny of early Indian texts by modern scientists was eventually abandoned.

9.2.2 European mathematical science in India

At the same time that the achievements of Indian sciences were being explored in Europe, those of European sciences were coming under scrutiny in India. Jesuit missionaries from Portugal and France undertook partial cartographic surveys and other astronomical observations in their countries' Indian colonies, in the process introducing the first telescopes into the subcontinent.[24] Beginning in the late eighteenth century, some Indo-Muslim scholars wrote in Persian or Urdu on European theories.[25] But the most comprehensive effort to assimilate European mathematical science seems to have been that of Jayasiṃha, the early eighteenth-century ruler of Jaipur.

---

hypothesis in the first case quoted, although he seems to have no qualms about accepting it in the second.

[22] [Pla1790], pp. 42–43. After acknowledging the "loose conjunction" problem, Bailly comments that planetary observations were probably considered less important than lunar and solar ones and hence less accurate; [Bai1787], pp. xxviii, lxvii.

[23] See, for example, the ravings of John Bentley in [Ben1970], briefly discussed in section 4.6.2.

[24] [Ans1985], pp. 10–13.

[25] [Ans1985], pp. 51–52, and [Ans2002a].

We have seen in section 8.2.2 that Jayasiṃha undertook a program of astronomical reform intended to draw on both Sanskrit and Islamic mathematical models. A few years later, when he heard accounts of European astronomy from visiting Jesuits from the Portuguese colony of Goa, he sent a delegation to Portugal to find out more about it. The travellers returned to Jayasiṃha's court with books, instruments, and a young Portuguese astronomer whose pedagogical efforts were soon supplemented by those of a French adventurer from Delhi.[26]

The chief task of the Jaipur astronomers' investigations into European theories was to read and comprehend their copy of the Latin treatise *Tabulae astronomicae* or *Astronomical Tables* of Philippe de la Hire. This was far from easy, as de la Hire's text involved a variety of unfamiliar geometrical models relying on heliocentric assumptions (although the discreet de la Hire officially disclaimed any "hypotheses," claiming to base his system solely on observations).[27] The French author also assumed that his readers would be familiar with the use of logarithms, both of ordinary numbers and of trigonometric functions. Jayasiṃha's scholars made a Sanskrit version of the logarithm tables, which they apparently attributed to one "Don Juan Napier," but struggled to ascertain their meaning and use.[28]

Evidently on account of these difficulties, the Maharaja requested in 1732 additional expertise from French Jesuits in southern India. The two missionaries subsequently sent to join the Jaipur project appear to have been more competent astronomers and teachers than the other two Europeans. For the Indian astronomers were subsequently able to use de la Hire's text well enough to produce a Sanskrit summary of it and to incorporate some of its data into the *zīj* that they composed, in addition to writing another short Sanskrit treatise called "Aid to representing the Phiraṅgis' (Europeans') [theory] of the moon." This latter work appears to have been informed (despite the Church's official ban on the heliocentric theory at this time) by some explicitly Copernican-Keplerian explanations from the new European advisers that were absent from the published work of de la Hire. In its discussion of lunar theory, it shows a diagram of a cone cut by a plane to form an ellipse and refers to the lunar orbit as "fish-figured," apparently alluding to the (cusped) oval figures created from intersecting circular arcs to construct perpendicular bisectors.[29] Clearly, the hypothesis of elliptical orbits was by this time known to the author(s) of this work. On another of its diagrams appears the comment "there the earth is assumed to move, [and] the sun is assumed to be fixed."[30]

However, this introduction to heliocentric theory does not seem to have

---

[26]See [Pin1999], pp. 75–80, [Pin2002b], p. 123.

[27][Pin1999], p. 81; [Pin2002b], p. 124; [Pin2002c], p. 436.

[28][Pin2002b], p. 125.

[29]Of course, an elliptical orbit does not have cusps, so the term "fish-figure" for it is a bit misleading. Was the writer unfamiliar with the term "long circle" used, for example, in the *Gaṇita-sāra-saṅgraha*, or did he think that it referred to a shape other than the ellipse?

[30][Pin2002b], pp. 129–130; [Pin2002c], pp. 436, 442.

penetrated any further into the Sanskrit mathematical astronomy tradition. In fact, despite all the impressive scientific activity at his court, Jayasiṃha's ambitious scheme of producing a revised and improved astronomical system ultimately more or less fizzled out. New observations and calculations were used to modify some parameters in the production of the new *zīj* dedicated to Muḥammad Shāh, and a number of its tables were adapted from those of de la Hire, but overall this work followed the form and content of traditional *zīj* texts. The telescopes brought to Jaipur by foreign visitors were not used in the collection of its data, although it does describe the use of a telescope to view hitherto unknown phenomena like the phases of Mercury and Venus and the satellites of Jupiter.[31]

Most importantly, astronomical models were not explicitly revised, nor were the theories of Sanskrit, Islamic, and European texts systematically compared or evaluated. Considering the sheer scale of such an undertaking—even more massive and ramified than the comparison of Latin Ptolemaic and Copernican systems that took Tycho Brahe and Kepler some four decades to work through—this is hardly to be wondered at. The linguistic, philosophical, doctrinal, technical, and practical issues involved in a radical comparative analysis of these systems with all their diverse variants are so complex that one wonders if any of Jayasiṃha's astronomers (including the European ones) ever fully understood just what they were attempting. In any case, after Jayasiṃha's death, although some of his successors kept up his observatories and his library, his larger project of comprehensive astronomical reform was abandoned.

9.3 SANSKRIT MATHEMATICS AND ASTRONOMY, 1500–1800

The centuries from about 1500 to about 1800 saw a blend of innovative and conservative approaches in Indian mathematical science. The texts of Bhāskara II continued to be regarded as standard, as indicated by their predominance among Sanskrit technical works as candidates not only for translation, but subsequently for publication in Sanskrit print as well.[32] At the same time, new texts on mathematics and astronomy appeared, along with a plethora of commentaries on earlier ones.

9.3.1 New works on astronomy and mathematics

To do justice to the mathematical and astronomical treatises of the early modern period would require another book. The difficulties are exacerbated by the fact that none of the works in question has been fully translated or studied in detail, and several of them have never yet been published. A brief

[31] [Shar1995], p. 243. The dependence of some of the Jaipur tables on those of de la Hire has been examined in [Mer1984] and [Dal2000].

[32] [Pin1970a], vol. 4, pp. 308, 311, 318.

listing of a few significant works, along with a comment here and there on some of their innovative or notable features, gives an idea of the scope of the material:

- In 1503 Jñānarāja of Pārthapura composed a *Siddhānta-sundara* or "Beautiful *siddhānta*" using Saura-pakṣa parameters. Jñānarāja seems to have invented the idea of using the concrete number system with a literary and mathematical double meaning, as some south Indian authors used the *kaṭapayādi* system. His sample problems employ concrete numbers in ambiguous ways: if read with their standard numerical interpretations, they supply the given quantities needed for solving the problem; if read with their literal verbal meanings, they transform the problem into a mini-narrative with no mathematical significance. Jñānarāja also wrote an algebra treatise (still unpublished), the *Bījādhyāya* ("Algebra lesson"). His great-nephew wrote a *Gaṇita-mañjarī* ("Flowers of computation"), also unpublished.[33]

- In 1658 Kamalākara of Varanasi completed the *Siddhānta-tattva-viveka* ("Investigation of the truth of *siddhāntas*") that integrated traditional *siddhānta* computations with Greco-Islamic cosmological notions. He also included a summary of some results in geometric optics (he appears to have been the only Sanskrit author to discuss this subject) and a comprehensive survey of algebra.[34]

- Munīśvara, a contemporary of Kamalākara belonging to a rival family of Varanasi astronomers, wrote the *Siddhānta-sarva-bhauma* ("Ground of all *siddhāntas*") in 1646. He was much more resistant to incorporating foreign cosmology into *jyotiṣa* than his opponent Kamalākara, but he included without comment in his treatise a number of results on celestial coordinate conversions derived from Islamic spherical trigonometry. Some of these results had been incorrectly interpreted in Mahendra Sūri's astrolabe treatise of the late fourteenth century, so Munīśvara's work shows that they must have been retransmitted and fully assimilated by Sanskrit authors in the meantime.[35]

- The Islamic cosmological notions accepted by Kamalākara's family or school included the concept of transparent "crystalline" celestial spheres for the physical support and impulsion of the orbiting bodies. Munīśvara's side objected to this idea on the grounds that such a material could not be strong enough for this purpose. If any such spheres existed, they would have to be made of metal, conforming to traditional Indian descriptions of the blue sky as a round metal surface. This debate and related controversies were continued between

[33]A translation and discussion of part of the *Siddhānta-sundara*, including these problems, is provided in [Knu2008]. Jñānarāja's other works and that of his great-nephew are also noted in [Pin1981a], p. 64.

[34][Pin1981a], p. 31; for some of Kamalākara's algebra rules, see [DatSi1962] 2, passim.

[35]See [Pin1981a], pp. 30–31, and [Plo2002a], pp. 86–87.

Kamalākara's brother and Munīśvara's cousin in two treatises entitled *Loha-gola-khaṇḍana* ("Smashing of the iron spheres") and *Loha-gola-samarthana* ("Vindication of the iron spheres"), respectively.[36]

- In 1782 another Varanasi astronomer, Mathurānātha Śukla, composed the *Jyotiḥ-siddhānta-sāra* ("Essence of astronomical *siddhāntas*"), as well as a new treatise on the astrolabe containing diagrams in the style of Greco-Islamic geometry with labeled points, as in the translations of Western works produced by Jayasiṃha's scholars. Neither of these works has been published.[37]

Even this sketchy survey should be enough to show that claims of stagnation and decline in Indian mathematical science in the second millennium are at best greatly exaggerated and at worst wholly inaccurate. Fuller publication and study of the mathematical texts of this era will doubtless reveal many more interesting developments, even if none of the sources seriously challenged the canonical status of Bhāskara II's mathematical works from the twelfth century.

9.3.2 Computation and tables

The area in which some texts of the early modern period did decisively displace older ones was that of practical astronomical computation. Despite the continued interest in mathematical astronomy theory indicated by the abovementioned works, *karaṇas* or handbooks and particularly *koṣṭhakas* or table texts seem to have attracted even more attention. A *karaṇa* composed in about 1520, the *Graha-lāghava* ("Brevity [of calculation] for the planets") of Gaṇeśa of Nandigrāma, achieved not only widespread popularity but even the status of a new *pakṣa* or astronomical school. The establishment of this "Gaṇeśa-pakṣa" was due to Gaṇeśa's adopting a new set of computational parameters, which according to his commentator Mallāri, writing a few decades later, were inspired by his own observations, prompting him to select some parameters from the existing *pakṣas* and modify others:

> The differences in [the positions of] the planets [were] determined by that teacher [i.e., Gaṇeśa], observing the planets with a [sighting]-tube. As follows: [the positions] of the sun and the moon's apogee [were] as in the Saura-pakṣa, the moon as in the Saura-pakṣa less nine arcminutes, Mars, Jupiter and the lunar node as in the Ārya-pakṣa.[38]

The new parameters were not Gaṇeśa's only innovation in the *Graha-lāghava*. Although, as we have seen, *karaṇa* authors in general prided themselves on ingeniously minimizing computational effort by various means, including the construction of shorter and simpler Sine tables, Gaṇeśa seems

[36] See [Pin1978b], pp. 321–322.

[37] See [Pin1970a] 4, pp. 349–350. I used MS. Benares (1963) 35245 of Mathurānātha's astrolabe text.

[38] Commentary of Mallāri on *Graha-lāghava* 1.10; [JosK1994], p. 26.

to have been the first to decide to throw out the Sine table altogether, as he announces in the third verse of his text:

> Although intelligent and great [authors] composed *karaṇas*, there is no use in them if [their tables of] Sines and arcs are left out. Therefore, I am undertaking to make an accurate treatise on the planets of a very easy kind, entirely without computations of Sines and arcs.[39]

This of course requires him to provide algebraic approximations for every desired quantity whose computation would usually involve trigonometry. The following instance, a Three Questions rule for computing the degrees of terrestrial latitude from the noon equinoctial shadow, invites comparison with the corresponding rule of Bhāskara II that we examined in section 6.2.5:

> The latitude-shadow [i.e., noon equinoctial shadow] multiplied by five is decreased by a tenth part of the square of the latitude-shadow. The degrees of latitude towards the south [of the location are the result].[40]

In other words, the noon equinoctial shadow s, expressed in digits, of which there are twelve in the standard gnomon, is related to the terrestrial latitude ϕ in degrees by

$$\phi \approx 5s - \frac{s^2}{10}.$$

Actually, of course, since $\frac{\text{Sin}\,\phi}{R} = \frac{s}{\sqrt{s^2 + 12^2}}$, where R is the trigonometric radius,

$$\phi = \text{arcSin}\left(R \cdot \frac{s}{\sqrt{s^2 + 12^2}}\right).$$

Gaṇeśa's simple rule is in fact quite accurate, deviating from the exact result by less than 10 arcminutes for any ϕ less than about 37° (a higher latitude than any Indian astronomer would need to worry about). Its values are at least as good as the ones that linear interpolation in a small crude Sine table would produce.

Literally dozens of Sanskrit table texts or *koṣṭhakas* were produced between the late fifteenth and late eighteenth centuries,[41] suggesting that a significant portion of the mathematical interest and energy of early modern Indian astronomers went into devising methods for computing tables and for using them to generate yearly calendars and predictions of planetary positions at desired times. Hardly any of these texts have been published or thoroughly studied, and their mathematically interesting features lie buried

---

[39] *Graha-lāghava* 1.3; [JosK1994], p. 7.
[40] *Graha-lāghava* 4.6; [JosK1994], p. 119.
[41] [Pin1981a], pp. 41–46.

within the unspecified algorithms used to compute their entries, so it will take a good deal of analysis to reconstruct the details of what their authors were actually doing in the process of creating them.

The most famous *koṣṭhaka* is probably the *Tithi-cintāmaṇi* ("Thought-jewel of *tithis*"), composed by the same Gaṇeśa who wrote the *Graha-lāghava*. His concise tables enabled *jyotiṣa* specialists to compute all the information required for the elaborate Indian calendar with its five separate sequences of time units (see section 4.4.2). The user needed to know nothing except the present year in the Śaka era, the distance of his location from the Indian prime meridian passing through Ujjain, and the number of *tithis* from the first new moon of the year up to the current date, as well as the techniques of basic decimal and sexagesimal arithmetic. To make the tables easy for the user, though, Gaṇeśa had to set up an impressively complicated structure of calculation for himself, involving, for example, mean time units, true time units, intermediate time units computed with respect to the true motion of the sun but the mean motion of the moon, and procedures for converting between them. He left no record of the algorithms he devised for creating his tables, and historians are still attempting to reconstruct exactly what his mathematical procedures were.[42]

9.3.3 Commentaries and mathematical proof

Did the role of demonstration in mathematics significantly change during the general ferment of early modern Sanskrit learning? We have seen in section 6.4 that the work of medieval mathematical authors apparently embraced a wide range of views on proof. Sometimes demonstrations or rationales for mathematical statements were provided in commentaries, but sometimes they were omitted in expositions consisting only of verbal glosses and examples. Sometimes, as we saw in the remarks of the seventh-century commentator Bhāskara, the use of a figure in an explanatory demonstration might even be disparaged as something needed "to convince the slow-witted." The usefulness of persuasive rationales was certainly recognized, but they were not necessarily considered mandatory.

The idea of detailed demonstration seems to have gripped more and more mathematical imaginations in the sixteenth century and later. The same Gaṇeśa who wrote such concise *karaṇa* and *koṣṭhaka* texts also composed in 1545 a commentary on Bhāskara II's *Līlāvatī* that was lavish in explanatory detail. He echoes at the beginning of the commentary Bhāskara's own sentiments from the start of the *gola* section of the *Siddhānta-śiromaṇi*, as follows:

> In arithmetic or in the knowledge of algebra, what is stated without demonstration is certainly not free from uncertainty, and it will not become established in truth among good mathemati-

[42] A translation and detailed study of the *Tithi-cintāmaṇi*, which however does not reproduce all its tables, is in [IkPl2001].

> cians. That clear [demonstration] is perceived directly like a mirror in the palm of the hand. Therefore, to increase understanding, I undertake to explain completely the best demonstration.[43]

In such commentaries, detailed justifications are by no means disdained as a crutch for the "slow-witted." Here, for instance, is the commentator Kṛṣṇa (ca. 1600) on the *Bīja-gaṇita* of Bhāskara II, discussing the comparatively straightforward concept that we would express notationally as $a^2+b^2\pm 2ab = (a \pm b)^2$:

> How can we state without demonstration that twice the product of two quantities when added or subtracted from the sum of their squares is equal to the square of the sum or difference of those quantities? That it is seen to be so in a few instances is indeed of no consequence. Otherwise, even the statement that four times the product of two quantities is equal to the square of their sum, would have to be accepted as valid. For, that is also seen to be true in some cases. For instance, take the numbers 2, 2. Their product is 4, four times which will be 16, which is also the square of their sum 4. Or take the numbers 3, 3. Four times their product is 36, which is also the square of their sum 6.... Hence the fact that a result is seen to be true in some cases is of no consequence, as it is possible that one would come across contrary instances also. Hence it is necessary that one would have to provide a rationale for the rule that twice the product of two quantities when added or subtracted from the sum of their squares results in the square of the sum or difference of those quantities.[44]

A thorough comparative study of all contemporary commentaries will have to be undertaken before we can know how far they represented new developments in Indian ideas of mathematical proof in the early modern period. At present we can tentatively say that they do seem to indicate, as we saw earlier in the rationales of the Kerala school of mathematicians, a conviction that demonstrations are not only important for verifying statements but also mathematically and philosophically interesting in their own right, even in the case of results that are well-known or trivial.

If this is an accurate assessment, then what prompted the development of this conviction? Was it just a natural amplification and refinement of ideas about demonstration that had been expressed earlier by, for example, Bhāskara II? Was it inspired by new analytical approaches and revivals of commentarial traditions in nonmathematical fields such as philosophy, logic, and grammar? Was it influenced by exposure to demonstration-rich Islamic

---

[43] *Buddhi-vilāsinī* 4, [Apt1937], p. 1; see also [Srin2005], p. 229.

[44] Quoted in [Srin2004b], pp. 9–10, and in [SarK2008], vol. 1, Epilogue; I have omitted some of the passage and retranslated a couple of the terms.

mathematical texts? Or was it some mixture of all three? In any case, it is clear that we cannot speak of "the" Indian attitude toward mathematical proof; the concept appears to have had many different interpretations at different times and according to different individuals.

9.3.4 Mathematical models and sacred cosmology

In what appears to be a new development from about the start of the sixteenth century onward, the geometric models of earlier astronomers came under direct attack from orthodox supporters of the cosmology of the Purāṇas. We have seen in section 3.2.3 that some scriptural texts described the activities of mathematical astronomers and astrologers as suspect or sinful. And medieval astronomers in their turn openly criticized Purāṇic ideas of the size and shape of the earth, the configuration of celestial bodies, and so forth, on the grounds that they failed to provide quantitatively predictive models for astronomy.

At this time, however, there seems to have been a resurgence of orthodox scrupulousness toward nonscriptural notions about the universe. Perhaps it was related to what might be called the "new interdisciplinarity" spreading among Sanskrit scholars and the increased interaction between the exact sciences and other fields: the idea of explicitly comparing scriptural perspectives with various "secular" ones was in the air.[45] In consequence, the conflicts between astronomical and Purāṇic assumptions were amplified, leading some astronomers as well as some other scholars to undertake the task of promoting what they called "noncontradiction" or "removal of contradiction" to reconcile the two world systems, although never very successfully.

The arguments of Purāṇa supporters were based not only on scriptural authority but on commonsense critiques of the *jyotiṣa* models' ability to explain ordinary experience of the world, such as its perceived flatness and the impossibility of heavy bodies' hovering unsupported.[46] Astronomers countered with assertions of their own scriptural authority, emphasizing the foundation of their *pakṣas* on direct revelations by gods or seers.[47]

A number of disputants on both sides also offered new cosmological interpretations that could accomodate the truth claims of both systems. These were the "noncontradiction" arguments, involving various compromises such as putting the Purāṇas' earth-supporting tortoise or serpent inside the shell of a spherical earth, to support it from within. Or a text might simply combine both cosmologies, either metaphysically (claiming, for example, that the Purāṇic universe represents transcendent truth while the spherical universe of the *siddhāntas* is adapted to degraded human perceptions) or physically (for example, situating the spherical earth above the larger flat earth, through which the planets pass as they orbit the spherical one).[48] There

[45] See the suggestions to this effect in, e.g., [MinNil].

[46] [Min2004a], pp. 352–353.

[47] [Pin1996c], pp. 471–472.

[48] See [Min2004a], pp. 356–357 and pp. 378–379 for examples of such compromises.

was apparently no strong polarization of this dispute into two uniformly opposed camps: as in earlier periods, an astronomer might object to one detail of scriptural cosmology and simultaneously rebuke another astronomer for objecting to a different one.

It is tempting to speculate that the controversy over issues of *gola* or the cosmic sphere could have been related to astronomers' increasing interest in producing computational texts as opposed to theoretical ones. Perhaps the purely algorithmic structure of handbooks and tables seemed attractive partly because it bypassed the quarrels over cosmology. Conversely, perhaps reliance on computational texts made the geometrical models underlying them seem less important, so astronomers were more willing to accommodate Purāṇic ideas about the universe. On the other hand, as we have seen, the authors of commentaries continued to expound the models of earlier texts. Gaṇeśa, for example, may have written the nontheoretical and hence noncontroversial *Graha-lāghava* and *Tithi-cintāmaṇi* in preference to composing a new work dealing with *gola*, but he also wrote a commentary on Bhāskara II's *gola*-intensive *Siddhānta-śiromaṇi*, so concern about cosmological controversy cannot have been a significant deterrent.

9.4 CONCLUSION

The developments described above suggest that if there was a systemic change in the Indian pursuit of mathematical science in the early modern period, it was far from a simple surrender to decay or stagnation. It is true that the twelfth-century works of Bhāskara II retained their canonical status in India for centuries; but so did the writings of Ptolemy and Euclid in the Islamic world and in Europe, even after the European twelfth-century renaissance. The repudiation of older texts is at best a crude yardstick by which to measure the vitality of a scientific tradition.

On the contrary, at the end of the period surveyed here, Sanskrit mathematics and astronomy were following many interesting paths of discovery and controversy. But the days of all of them were numbered. As a result of a combination of vernacularization, colonial rule, and modern globalization, the academic tradition of Sanskrit mathematical science was essentially extinct by the early twentieth century, despite some nineteenth-century efforts to assimilate new scientific theories into it. Some modern imitations or hybridized forms of it still flourish in popular venues, such as the so-called "Vedic Mathematics," or traditional *jyotiṣa* astrology blending *siddhānta* computations with astronomical ephemeris data produced by modern observatories or software applications. And longstanding traditions of folk mathematics, as practiced in vernacular languages among, for example, agricultural workers, can still shed light on the practical and social needs that inspired and preserved some of the content of premodern Sanskrit mathematics. Intriguingly, there may even be traces in the work of some Indian mathematicians in the modern period of ideas and approaches at least distantly inspired by traditional *gaṇita*. But the Sanskrit texts themselves by then had ceased to affect the production of contemporary mathematics in India or elsewhere.

Perhaps this historical distance has prevented us from fully realizing the complexity of the Indian mathematical tradition, and the extent to which its most basic issues are still very incompletely understood. Indian mathematics is still presented in most histories of mathematics simply as a collection of sample results showcasing ingenuity or foreshadowing modern mathematical concepts. The question of what Indian mathematics was fundamentally about, what made it a coherent intellectual discipline in the universe of Sanskrit learning and inspired particular paths of development, has been only partially explored. A few examples of fundamental questions that remain largely unanswered will give an idea of how much mystery still surrounds the subject:

- What determined the basic building-blocks or subjects of mathematics? For example, why did Mahāvīra consider it possible in the *Gaṇita-sāra-saṅgraha* to dispense with addition and subtraction of numbers as canonical arithmetic operations? How did the operations and procedures of medieval arithmetic texts originate, and how did a particular problem get assigned to a particular category?

- How did the distinction between arithmetic and algebra develop? What caused, say, the pulverizer to be identified as an arithmetic topic as well as an algebra one? What effect, if any, did this division have on the distinction between arithmetic and algebra in Western mathematical traditions?

- How did ideas of proof develop over time? How much were they influenced by epistemological considerations in other branches of learning? How did they interact with the conventions of the commentarial genre?

- What was the role of foreign sources in Indian mathematics? How was it affected by "Sanskrit purism," or the goal of keeping Sanskrit *śāstra*s as part of sacred learning, separate from foreign knowledge outside the world of *dharma*?[49] Did this attitude have the effect of masking transmission from other cultures?

Imagine not being able to answer such questions in the case of, say, Hellenistic or Islamic or early modern Latin mathematics. We tend to take for granted that in those scientific traditions, we know to a large extent what mathematicians thought they were doing and why. Perhaps we owe this awareness ultimately to classical philosophers' interest in mathematics as

[49]This phenomenon is seldom analyzed in histories of Indian science but often assumed; for example, the same historians quoted at the beginning of this chapter, remarking on al-Bīrūnī's difficulties in finding an Indian scholar who could explain mathematical texts to him, casually assert that "no respectable *paṇḍit* would agree to help a foreigner" ([DatSi1962], vol. 1, p. 128).

a uniquely certain form of knowledge, which provoked explicit discussions of the nature of mathematical thought, the connection between later results and earlier ones, and the relationship of mathematical models to physical reality. In the Indian "ocean of calculation," on the other hand, we encounter a very different set of assumptions, and we are still in the early stages of trying to integrate them into Sanskrit learning as a whole.

Appendix A

Some Basic Features of Sanskrit Language and Literature

Classical Sanskrit literature in all genres tends to be very conscious of the language's phonetic and grammatical structure. Mathematical texts in particular may make use of the Sanskrit alphabet for verse-friendly forms of number representation, or construct syllabic abbreviations for mathematical quantities, or adapt grammatical technical terms as mathematical ones. For these reasons, and also simply to be able to read transliterated Sanskrit terms correctly, students of the history of Indian mathematics will find it helpful to have at least an inkling of the linguistic and literary structure of Sanskrit.[1] A brief introduction to the characteristics of Sanskrit manuscripts and a glossary of transliterated technical terms used in this book are also provided in this chapter.

A.1 ELEMENTS OF SPOKEN AND WRITTEN SANSKRIT

The phonetically determined order of the Sanskrit alphabet, as codified by Classical grammarians, is shown in table A.1. The written characters of the script typically used for Sanskrit, a script known as *nāgarī* and developed in medieval northern India, are represented here and throughout this book by a standard roman transliteration system (the International Alphabet of Sanskrit Transliteration, or IAST).

First, in the sequence shown, come the simple vowels in their short and long forms, such as *a*, *ā*, *i*, *ī*, and so forth. The long forms of the vocalic *r* and *l* are very rare, and are included mostly for phonetic consistency. They are followed by the four diphthongs and the two vowel modifications, namely, *anusvāra* or nasalization (*ṃ*) and *visarga* or aspiration (*ḥ*). Then the sequence of twenty-five *sparśa* or "contact" consonants is divided into five categories or *vargas* in a 5×5 square. The consonants are distinguished according to their places of production in the mouth (identified in the table by their modern phonetic designations), from the throat forward to the lips. Within each *varga* the phonetic order of the consonants is voiceless, voiceless aspirated, voiced, voiced aspirated, nasal. After the five *vargas* come the four semivowels (also in order of their oral location), the three sibilant consonants,

[1]The basic linguistic information given here is more fully discussed in introductory Sanskrit textbooks such as [Cou1976].

and finally the aspiration *h*.

Table A.1 The Sanskrit alphabet.

Vowels										
Simple vowels:	a	ā	i	ī	u	ū	ṛ	ṝ	ḷ	ḹ
Diphthongs:	e	ai	o	au						
Vowel modifications:	ṃ	ḥ								
Consonants										
Velar:	k	kh	g	gh	ṅ					
Palatal:	c	ch	j	jh	ñ					
Retroflex:	ṭ	ṭh	ḍ	ḍh	ṇ					
Dental:	t	th	d	dh	n					
Labial:	p	ph	b	bh	m					
Semivowels:	y	r	l	v						
Sibilants:	ś	ṣ	s							
Aspiration:	h									

The basic phonetic unit of Sanskrit is the syllable consisting of zero or more consonants followed by a vowel. In the Sanskrit *nāgarī* script, the short *a* is the understood "default" vowel that follows any consonant not marked with a different vowel sign. Thus, Sanskrit consonants that are actually not followed by any vowel at all, as when they are combined with other consonants or placed at the end of a word, need special letter-forms and diacriticals in *nāgarī* to indicate the absence of the short *a*. Modern Hindi and many other North Indian languages also use the *nāgarī* script or variant forms of it, but they have mostly dropped Sanskrit's "default short *a*" convention. Consequently, a Sanskrit name such as Gopāla written in *nāgarī* will generally be read by a Hindi speaker as "Gopāl." Many Dravidian languages of South India, on the other hand, add nasalization to the "default short *a*" rather than dropping it, so the usual modern South Indian equivalent of Sanskrit Gopāla is "Gopālam" or "Gopālan." Such influences from modern Indian languages lead to many superficial discrepancies in nonstandard transliterations of Sanskrit terms, but they are generally trivial and should not interfere too much with comprehension.

Sanskrit is a heavily compounded language, resulting in what English speakers often find to be dauntingly long words, particularly in titles of works (e.g., the *Śiṣya-dhī-vṛddhida-tantra* of Lalla). The spelling of compound words is modified in accordance with the phonetic changes produced by the resulting combinations of phonemes. In English, such phonetic changes are not generally reflected in spelling: for instance, our compound word "cup" + "board" continues to be spelled "cupboard," although it is invariably pronounced more like "cubboard," where the unvoiced consonant *p* is assimilated to its voiced neighbor *b*. In Sanskrit, however, orthography follows phonology. Thus the compound *bṛhat* "great" + *jātaka* "horoscopy" produces *Bṛhaj-jātaka*, not *Bṛhat-jātaka*. Similarly, *parama* "supreme" + *īśvara* "lord" combine their adjacent vowels to form the name Parameśvara, an epithet of Śiva.

Vowels within a word can also change depending on the word's meaning, a form of what linguists call vowel gradation. A derivative form of a word frequently displays the lengthened or diphthong form of the word's root vowel: so *sūra* or *sūrya*, "sun," gives rise to Saura-pakṣa, "the school *of* the sun." And the short *a* in the name of the god Brahman is lengthened in *Brāhma-sphuṭa-siddhānta*, "the corrected *siddhānta of* Brahman."

The end of a word, on the other hand, changes in accordance with its grammatical case, or function in the sentence. In English, Sanskrit names are generally transliterated in their basic or "stem" forms, disregarding grammatical case endings. For the majority of masculine and neuter nouns, these stem forms end in short *a*, as in Bhāskara, Brahmagupta, and *Sūrya-siddhānta*. Feminine nouns, often used as the titles of treatises, mostly end in long *ā* or long *ī*, as *Tri-śatikā* and *Līlāvatī*. Again, nonstandard or phonetically incomplete transliterations, especially ones influenced by modern Indian languages, can produce minor inconsistencies in the results.

When whole Sanskrit phrases and sentences are transcribed into roman letters, however, the case endings of their words are retained. The divisions between words may not always be apparent, because written Sanskrit in *nāgarī* combines adjacent sounds phonetically even when they belong to different words. Thus there are few if any interword spaces or dividers within a sentence, and some transliterators imitate this practice by running words together in transcription. The majority, though, will insert spaces between words in the transcribed text, even if this requires breaking apart conjunct consonants in some of the syllables. The end of a sentence or a line of verse is marked with the traditional *nāgarī* punctuation mark, the vertical bar or *daṇḍa* "stick," appearing singly (|) for minor breaks or pauses and doubled (||) for major ones.

The pronunciation of correctly transliterated Sanskrit words, at least well enough to pass muster in the ears of English speakers, is actually quite easy. The simple vowels and diphthongs are pronounced more or less as they are in Italian, except for vocalic *r* and *l*, which are approximately equivalent to the final sounds of the English words "singer" and "mumble." The differences between the short and long forms of the simple vowels can be neglected except in the case of *a*, whose short form is similar to the schwa, while its long form is like the *a* in "father."

The Sanskrit consonants *c* and *r* also sound like their Italian counterparts. Aspirated consonants (transliterated as consonant + *h*) are pronounced a little "harder" or more breathily than their unaspirated forms rather than as fricatives, so Sanskrit *maṭha* "school" is spoken approximately as "mut-ta," not with a fricative *th*, as in English "mother" or "math." Otherwise, Sanskrit consonants can be acceptably imitated by their English counterparts, without worrying about the differences among the various *n*-nasals or between retroflex and dental forms of *t* and *d*. Similarly, the first two sibilants can both be pronounced like English *sh*. There is only one tricky conjunct consonant: the combination *jñ*, as in Jñānarāja, which is pronounced more like *ny-*: "Nyāna-rāja." (Hindi speakers often use the sound *gy-* instead, so that the Sanskrit name Jñānarāja corresponds to modern Hindi "Gyan Raj.") And if we're really being fussy, a Sanskrit *v* following another consonant ought to be pronounced more like English *w*.

Sanskrit syllables are classified as either *laghu*, "light" (pronounced without stress) or *guru*, "heavy" (stressed). A syllable that ends in a short simple vowel is light if and only if the syllable following it begins with at most one consonant. (Bear in mind that in transliteration, the combination of consonant + *h* counts as one aspirated consonant.) A heavy syllable, on the other hand, ends in a long vowel, diphthong, or modified vowel (or in a consonant at the end of a sentence), and/or precedes two or more combined consonants at the start of the next syllable. Thus the stress sequence of the name Jñānarāja is heavy-light-heavy-light, while that of Mādhava is heavy-light-light. The reader may like to practice on more complicated examples such as the following, where heavy syllables are marked with ′ and light ones with ˘ :

′	˘	′	˘	′	˘	′	′
Nī-	la-	ka-	ṇṭha	So-	ma-	yā-	jin
˘	′	˘	˘	′	′	˘	
Va-	*ṭe-*	*śva-*	*ra-*	*si-*	*ddhā-*	*nta*	

A.2 THE STRUCTURE OF SANSKRIT VERSE

All Sanskrit prosody is based on the distinction between heavy and light syllables. A metrical verse or *śloka* pattern is typically divided into four identical parts or *pādas* ("quarters"). Each of these four *pādas*, which are generally identified in modern editions and translations by the roman letters a through d, has a given number of syllables in a particular sequence of light ones and heavy ones. (In some verse formats the first and third *pādas* share a syllabic pattern which is slightly different from that of the second and fourth *pādas*.) Thus there are in theory 2^n distinct syllabic patterns for a *pāda* of n syllables.[2]

Prosody/metrics, or *chandas*, is the discipline in which the different verse meters are identified and classified. Of course, there are not really 2^n accepted meters for every possible value of n, owing to the difficulty of con-

[2] There is also a second variety of Sanskrit metrics, particularly common in sung poetry and derived partly from vernacular verse genres. In this type of meter, the metrical form is determined not by a fixed sequence of light and heavy syllables but by the total number of what Western prosodists call *morae* (units of syllabic weight) in a line. For example, if a light syllable is assigned one *mora* and a heavy syllable two, then a line in an eight-*mora* meter could contain eight light syllables, or four heavy ones, or two light and three heavy, and so forth. For brevity's sake the current discussion is restricted to the fixed syllabic meters, but more information about the so-called "moric meters" can be found in, for example, [Ger1989], pp. 535–538.

structing correct Sanskrit phrases in some syllabic patterns (for example, ones with no heavy syllables). Only values of n ranging from about 6 into the 20's have a significant number of distinct meters associated with them, producing a total of over 300 widely recognized metrical formats.

To distinguish and organize these formats, Indian prosodists evolved an elegant system of classifying syllabic patterns. As discussed in section 3.3, at least as far back as the *Chandaḥ-sūtra* of Piṅgala around the start of the Classical period, the 2^3 distinct patterns of syllabic triples were associated with letters of the Sanskrit alphabet:

m	*y*	*r*	*s*	*t*	*j*	*bh*	*n*
/ / /	˘ / /	/ ˘ /	˘ ˘ /	/ / ˘	˘ / ˘	/ ˘ ˘	˘ ˘ ˘

At some point in the development of *chandas*, these letters were assigned definite vocalizations with either short or long *a*, and ordered into the following single mnemonic phrase:[3]

ya-mā-tā-rā-ja-bhā-na-sa-la-gā

The extra two syllables at the end stand for *laghu* "light" and *guru* "heavy" respectively, and are vocalized accordingly. Each successive triple of syllables within this phrase then displays the syllabic pattern associated with its initial consonant in the *Chandaḥ-sūtra*: *ya-mā-tā* is the *y*-pattern or light-heavy-heavy, *mā-tā-rā* is the *m*-pattern or heavy-heavy-heavy, and so forth. In short, this mnemonic resembles a modern de Bruijn sequence, a minimal sequence containing all possible strings of length n of characters from an alphabet of size k; in this case, $k = 2$ (light or heavy) and $n = 3$. It has the advantage of allowing all eight patterns for syllabic triples to be compactly represented in one "word" of only ten syllables.

These syllabic abbreviations for patterns of syllabic triples are used in medieval metrics texts to construct sample *pādas* illustrating the form of each meter.[4] For example, the eleven-syllable meter known as *Indra-vajrā* or "Indra's thunderbolt" with syllabic pattern / / ˘ / / ˘ ˘ / ˘ / / is simultaneously defined and exemplified in the following *pāda*:

/ / ˘ / / ˘ ˘ / ˘ / /

syād indravajrā yadi tau jagau gaḥ |

When [there are] two *ta* [triples], a *ja* and a *ga*, [and another] *ga*, [that] is the *Indra-vajrā* [meter].

Metrical examples also frequently employ the Sanskrit concrete number notation so ubiquitous in astronomical and mathematical works (see section 3.1), as in the canonical *pāda* for the nineteen-syllable meter *Śārdūla-vikrīḍita*, "tiger's sport":

[3]It is still very unclear when or how this mnemonic *sūtra* originated; it might have been known to Piṅgala himself or it might have been invented in the medieval or early modern period. See [Kak2000b].

[4]Such metrical examples and analyses appear, e.g., in the *Vṛtta-ratnākara* or "Verse-treasury" of Kedāra around 1000; [Cou1976], p. 310, [Pol1998], p. 14.

> / / / ◡ ◡ / ◡ / ◡ ◡ ◡ / / / ◡ / / ◡ /
>
> *sūryāśvair yadi mātsajau satatagāḥ śārdūlavikrīḍitam |*
>
> The *Śārdūla-vikrīḍita* [meter] is by means of [syllables numbering] "sun" [twelve] and "horse" [seven], when [there are], after a *ma*, a *sa* and *ja*, a *sa*, a *ta*, [another] *ta* and a *ga*.

The excerpts in section 7.3.2 above from Mādhava's verses encoding Sine and Versine series coefficients in the *kaṭapayādi* alphanumeric system are a nice example of this challenging meter.

However, the real workhorse of Classical Sanskrit *śloka*s or verses is a more indefinite metrical pattern called *anuṣṭubh*, or just *śloka*. This meter has eight syllables per *pāda*, and most of them may be either light or heavy as the poet requires. However, in the first and third *pādas* the fifth, sixth and seventh syllables should follow the sequence light-heavy-heavy, while in the second and fourth *pādas* the same syllables should run light-heavy-light. There are other rules and variations in this pattern, but these are its main defining characteristics.

The *anuṣṭubh* or *śloka* meter was employed at least as far back as late Vedic times. It is also the primary format of the great epics *Mahābhārata* and *Rāmāyaṇa*, as well as carrying most of the content of Classical literature in every genre. The following example of the text of Mādhava's *anuṣṭubh* verse on the arc-difference (see section 7.3.1) shows the metrical flexibility of this verse format:

> ◡ / / / ◡ / / ◡
>
> *jyayor āsannayor bheda-*
>
> / / / / ◡ / ◡ /
>
> *bhaktastatkoṭiyogataḥ |*
>
> / / / ◡ ◡ / / /
>
> *chedastena hṛtā dvighnā*
>
> / / / ◡ ◡ / ◡ /
>
> *trijyā taddhanurantaram ||*

The six *kaṭapayādi* verses of Mādhava's Sine table in section 7.3.2 also use the *śloka* meter. In general, it is fairly typical for authors in the *śāstra* or didactic genres to compose invocatory, rhetorically ornate, or otherwise special verses in one of the fancier poetic meters, falling back on the more forgiving *śloka* format or the shorter regular meters for most of the ordinary material of the treatise.

A.3 THE DOCUMENTARY SOURCES OF TEXTS

Until well into the modern period, Sanskrit texts were perpetuated by a succession of handwritten manuscripts. We have seen in section 6.1 that manuscripts on mathematics were often copied by students and scholars for their own use or that of their pupils. Sometimes members of scribal castes were paid to do the copying.

The primary writing materials in the first millennium seem to have been palm leaves, particularly in southern India, and birch bark in the north. Palm leaves were usually inscribed with a stylus and the resulting grooves filled in with ink powder or paste to create the visible letters. Birch bark was usually written on with pen and ink. The traditional format of a manuscript book was usually rectangular with the long edge horizontal (probably originally designed to maximize the writing surface of long narrow palm leaves). They were not bound, although they sometimes had one or two holes in the center of the leaves through which a string could be threaded to keep the leaves together and in order. The loose leaves were often sandwiched between thin boards and wrapped in a cloth for their protection and preservation.

The Chinese use of paper as a writing material was apparently known but not widely imitated in medieval India. Only after their introduction from the Islamic world, sometime near the beginning of the second millennium, did paper manuscripts become popular. In fact, by the fourteenth century Indian paper was apparently so ubiquitous and cheap that vendors in Delhi used it to wrap sweets for their customers![5] But the Sanskrit manuscript, whether paper or palm leaf, generally retained its traditional format: a collection of loose leaves with width greater than their height. Sometimes, particularly in later centuries and in places with Islamic influence such as sultans' courts, Sanskrit manuscripts were produced in a format imitating that of Western books, higher than wide and bound in fabric. (Not usually, though; for example, note the difference in orientation between the Indo-Muslim table in figure 8.3 and the Sanskrit one in figure 8.4.) Materials such as parchment, vellum, and leather bindings were less popular due to the anti-animal-slaughter principles of Hindu, Jain and Buddhist scholars.

Indian manuscripts generally had a lifespan of at most a few centuries, being vulnerable to a wide variety of destructive influences in the subtropical climate of most South Asian regions: mildew, insects, moisture, mice, and so forth. Unbound leaves often became separated or disordered; for technical works in particular, since students would start out studying the beginning of a text and possibly not make it through to the end, the beginnings of manuscripts are more likely to be soiled and damaged while their ends are more likely to be lost. Owing to the need for constant recopying, almost all surviving manuscripts are distant descendants of original documents, copied within the past three hundred years, and many of them are incomplete. (The Bakhshālī Manuscript described in section 5.2, a birch-bark manuscript apparently copied somewhere between the eighth and twelfth centuries CE, is an extremely rare exception among mathematical texts.)

A Sanskrit manuscript is typically written on both sides of each leaf (folio), except for the recto (front) of the first folio and sometimes the verso (back) of

---

[5]See [Hab1985], pp. 14–15.

the last, to avoid soiling. Scribes, whether amateur or professional, generally identified the texts they copied by colophons at the end rather than by title pages at the beginning. Sometimes, though, the title of the work appears in the center of the first recto, title-page style, occasionally accompanied by identifying information about the owner and the collection to which it belonged. Folios are numbered in sequence in one or more corners of the verso, and occasionally an abbreviation of the title accompanies each folio number.

The written content generally appears as a centered rectangular block of justified lines surrounded by uniform margins (and sometimes with blank spaces in the center for the string hole). Since there are no capital letters or hyphens in *nāgarī* script and few interword spaces, the lines mostly look like uniform streams of syllables punctuated by vertical *daṇḍas*. Paragraphing, indentation, page breaks between chapters, and line breaks between verse lines are not used, although often the beginning phrases of chapters or other textual units are written in red ink or "highlighted" by being smeared with red or yellow ocher. Overall, there is little experimentation with the visual aspects of writing, such as elaborate calligraphy or page formatting. This probably reflects the traditional perception of Sanskrit as primarily a spoken rather than a written language. In fact, many manuscripts appear (judging from their phonetically inspired errors of spelling) to have been "mass-produced" by simultaneous multiple transcriptions from an oral dictation rather than visually copied from an earlier manuscript.

The individual verses of a text are generally followed by verse numbers delineated by *daṇḍa* lines, but not infrequently a scribe would lose track of the current number and introduce off-by-one errors. Apparently it was also common to leave spaces between the *daṇḍas* and fill in the verse numbers afterward; many scribes evidently omitted to complete this task, or stopped partway through the manuscript. Errors in copying, when sought and detected (which not every scribe bothered to do), would be fixed by inserting the correct form of the erroneous or omitted text within a margin, with marks to indicate where it belonged in the body of the page.

A commentary is usually written in the same text stream as the verses of the base text, without any page formatting to distinguish the two. Often the base text verses are not even quoted in full; only their opening phrases are cited at the appropriate places in the commentary, and the reader is presumably supposed to recall the rest of the verse from memory. Occasionally, though, the base text is presented in one or two central text blocks, surrounded by blocks of commentary.

In mathematical and astronomical manuscripts, small sections of numeric or symbolic text such as Sine tables or equations are displayed in bordered boxes or arrays within a text block, interrupting the lines of text (see, for example, the sample leaf from the Bakhshālī Manuscript shown in figure 5.7). Diagrams or figures are infrequent, and usually no more complex than a rough sketch of a simple geometric shape, perhaps labeled with the names of its component lines. Except in copies of Sanskrit works influenced by Islamic or European ones, there are no Euclidean-type diagrams with labeled points illustrating geometric rationales or demonstrations. Even descriptions of visual procedures such as the construction of an armillary sphere or the picturing of the moon's crescent are unlikely to be accompanied by an illustrative figure. Occasionally, rectangular blanks in the text blocks reveal the intended presence of figures or tables that the scribe never got around to inserting.

To produce a printed text of a previously unpublished Sanskrit treatise, a researcher ("textual editor") must collect photocopies or microfilms of a selection of its surviving manuscripts. (Despite conservation efforts, the number of surviving usable manuscripts is decreasing with the passage of time. The practice of replacing decayed manuscripts with new copies started to decline significantly in India with the general use of printing in the nineteenth century, and had essentially stopped by the late twentieth century.) Then the manuscripts must be read and compared with one another, for the purpose of sifting out the accumulated scribal errors and reconstructing an approximate version of the original text. If the variant readings of the manuscripts are published along with the reconstructed text, the publication is called a critical edition.

The Sanskrit manuscript tradition thus presents a number of challenges for the textual editor. It is often difficult to tell which work or works a particular manuscript contains, especially if the end of the manuscript with its identifying colophon has been lost. Many manuscripts are incomplete, due to loss of pages or to a scribe's inadvertently skipping sections of the text, or both. The continuous format of the typical written text can make textual omissions or lacunae by a scribe hard to spot. "Displayed" data such as small tables or figures are often displaced from their proper position in the text stream or omitted altogether. And all errors of transmission are multiplied by the need for frequent re-copying.

A.4 MEANING AND INTERPRETATION: CAVEAT LECTOR!

The vocabulary of Sanskrit is very fluid, with most words having multiple meanings and connotations, as well as abundant synonyms and paraphrases—a useful feature in a literary tradition where so much of the corpus consists of metrical verse. Just about any concept can be successfully expressed in any type of Sanskrit verse by means of an appropriate synonym or metaphor. Semantic flexibility is thus more important than avoidance of ambiguity. Even technical terms in the various *śāstras* may have more than one definition, and their meaning in a particular instance often has to be determined from context, if possible.

For example, the mathematical technical term *jyā* (literally synonymous with *jīvā*, "bowstring") appears originally to have meant "Chord," later giving rise to the derived term *ardhajyā* or *jyārdha*, "half-Chord," for the Sine. But the original term *jyā* eventually became the primary word for Sine, without entirely losing its original meaning of Chord. Sometimes if Chords and Sines both appear in a rule or explanation, they will be deliberately distinguished by the use of *ardhajyā* for Sine or a more descriptive term like *samastajyā*, "whole Sine," used for Chord—but not always. So Sanskrit discussions of trigonometry must be read alertly, with constant awareness of the potential alternative meanings of the terms; and the same applies to other mathematical and astronomical topics.

The semantic uncertainty produced by such a multivalent vocabulary is magnified by the ubiquitous practice of word compounding, with its accompanying phonetic changes in the compounded words. It is often unclear exactly which words are present in the compound and what their intended grouping should be. For instance, a compound formed of the words *cāpa* ("bow," "arc"), *ardha* ("half"), and *jyā* ("Sine," "Chord") coalesces into *cāpārdhajyā*. Is that meant to signify *cāpa* + *ardhajyā*, "the Sine of the arc," or *cāpārdha* + *jyā*, "the Sine of half the arc" or perhaps even "the Chord of half the arc"? All these interpretations are plausible, and again, only context can determine unequivocally which is the true one.

Consequently, there is hardly any such thing as a true literal translation of a Sanskrit text. All translations require choices among various possible meanings, sometimes equally applicable. Most translators try to maximize transparency by using typographical cues to distinguish their own additions and corrections from the directly translated material. (Throughout the present book, I have used square brackets for this purpose.) But much ambiguity of meaning is silently suppressed in the "direct translation" itself.[6]

These characteristics may seem somewhat startling to practitioners or students of modern mathematics, who have been schooled for years to believe that the supreme aim of mathematical exposition is precision and clarity. It is important to keep in mind that in the Sanskrit tradition, multiplicity of meanings is not a bug but a feature. Semantic duality and paranomasia were highly prized by Indian poets, most notably in the elaborately ornamented genres of belles-lettres, for their richness of expression and connotation. The technical verses of *śāstra* texts tend to be less rhetorically complex than those of literature and drama, but they seldom entirely avoid ambiguity.

In fact, it may be that linguistic ambiguity was deliberately exploited by Sanskrit mathematicians for pedagogical purposes. When a mathematical statement has more than one possible meaning, mediocre students who simply want to apply it by rote in "cookbook" fashion will be unable to do so. To know clearly what the rule actually says, they will need to think about the mathematics involved. Consider the following example of a geometrical

---

[6] Translators disagree about how best to convey the varied meanings of Sanskrit texts. Some of the broader issues are addressed in [Pol1996], while [Pin1996d] discusses technical questions of translating Sanskrit scientific texts. I have tried to follow the basic principle set forth in [Wuj1993], vol. 2, p. xxix, that a translated sentence should be meaningful and syntactically complete even if the editorial material in square brackets is omitted, but I have not always been able to accomplish this, particularly in very concise and cryptic *sūtra* passages.

rule from Bhāskara's *Līlāvatī*:[7]

> *karṇasya vargād dviguṇād viśodhyo*
> *doḥkoṭiyogaḥ svaguṇo 'sya mūlam |*
> *yogo dvidhā mūlavihīnayuktaḥ*
> *syātāṃ tadardhe bhujakoṭimāne ||*
>
> The sum of the arm and the upright [of a right triangle], subtracted from the doubled square of the hypotenuse, self-multiplied, [is found]. The square root of that [is taken]. The sum, [set down] twice, is [separately] diminished and increased by the square root. The halves of those [results] are the amounts of the arm and the upright [respectively].

Here, the hypotenuse c of a right triangle and also the sum of its other two sides, the "arm" a and the "upright" b, are known, and the separate quantities a and b are sought. The sum $(a+b)$ is supposed to be subtracted (*viśodhya*) from $2c^2$ and "self-multiplied" or squared (*sva-guṇa*). But should the squaring take place before or after the subtraction? The construction of the sentence could bear either interpretation: the words for "subtracted" and "squared" are separate modifiers agreeing with the word for "sum," and in Sanskrit (as in other highly inflected languages), meaning is not dependent on word order in the verse. Thus the rule could be implying either that

$$a = \frac{(a+b) - \sqrt{2c^2 - (a+b)^2}}{2}, \qquad b = \frac{(a+b) + \sqrt{2c^2 - (a+b)^2}}{2},$$

or that

$$a = \frac{(a+b) - \sqrt{(2c^2 - (a+b))^2}}{2}, \qquad b = \frac{(a+b) + \sqrt{(2c^2 - (a+b))^2}}{2}.$$

Only working through the equations will establish with certainty that the former interpretation is the correct one. Bhāskara, like all Sanskrit authors familiar with the conventions of double meaning, must have been aware of the possibility of misinterpretation. It seems plausible that he and other mathematical authors consciously considered it beneficial not to coddle their students by supplying too much precision and clarity in their texts. (Another factor may have been the desire to defend against plagiarism by devising formulas whose derivation is difficult to understand, as illustrated in section 4.4.)

Unfortunately, modern translators are not always smarter than medieval students when it comes to disambiguating the versified formulas of the base texts. Usually the translator can seek help from a commentary that expounds or at least glosses the meaning of the verses (and doubtless the students did the same). The commentaries, however, can introduce some uncertainties of their own. Consider the helpful explanation of the *Līlāvatī*

---

[7] *Līlāvatī* 158, [Apt1937], pp. 147–148. See section 6.2.1.

rule above furnished by the prose commentary of Gaṇeśa, reproduced in its entirety below:

> Now, when the hypotenuse and the sum of the arm and upright, or [their] difference, are known, for the purpose of determining [them] separately he states a rule in *Indra-vajrā* [meter]: "*karṇasya vargād dviguṇād viśodhya.*"
>
> From the doubled square of the hypotenuse the self-multiplied, [i.e.,] squared, sum of the arm and upright is subtracted. The square-root of this is taken. The sum of arm and upright is set down twice. [It] is diminished and increased by that square-root. The two halves of that, in order, are the two amounts of the arm and upright.
>
> It is to be considered in the same way also when the difference of the arm and upright is known. That is as follows: From the doubled square of the hypotenuse the self-multiplied difference of the arm and upright is subtracted. The square-root of this is taken. That, [set down] twice, [is] diminished and increased by the difference of the arm and upright; so the two halves of that are the two amounts of the arm and upright.
>
> Demonstration of this: The square of the hypotenuse *is* the sum of the squares of the arm and upright. But [if] the sum of the squares [is separately] increased and diminished by twice the product, [the results] are the squares of the sum and difference. And therefore Śrī Keśava[8] said: "The sum of the squares of two quantities [is] increased and decreased by two times the product; those two [results] in order are the squares of the sum and difference." Hence the sum of the squares of two quantities increased by two times the product must be the square of the sum; decreased, the square of the difference. And the sum of the two [results] must be the sum of the square of the sum and the square of the difference. Hence, when there is elimination of the doubled product due to [its] being equally-positive-and-negative, the doubled square[9] of the hypotenuse itself must be the sum of the squares of the sum and difference of arm and upright. Hence when the square of the sum of arm and upright is subtracted from the doubled square of the hypotenuse, the square of [their] difference remains. So when the square of the difference is subtracted, the square of the sum remains. Its square-root will be the sum or difference. With those two, by means of a rule of concurrence, the arm and upright [are] found; thus the demonstration.

---

[8] Gaṇeśa's father, the astronomer. See appendix B.

[9] Edition has *karṇayoga* for *karṇavarga*, [Apt1937], p. 148.

To paraphrase this rationale in modern mathematical notation,

$$
\begin{aligned}
c^2 &= a^2 + b^2, \\
a^2 + b^2 + 2ab &= (a+b)^2, \\
a^2 + b^2 - 2ab &= (a-b)^2, \\
(a^2 + b^2 + 2ab) + (a^2 + b^2 - 2ab) &= (a+b)^2 + (a-b)^2, \\
2a^2 + 2b^2 = 2c^2 &= (a+b)^2 + (a-b)^2, \\
2c^2 - (a+b)^2 &= (a-b)^2,
\end{aligned}
$$

and consequently

$$\sqrt{2c^2 - (a+b)^2} = a - b.$$

So Gaṇeśa confirms that the squaring must precede the subtraction in Bhāskara's formula.

The footnote in the quotation reveals that I have emended a word in the published text of Gaṇeśa's commentary, changing *karṇa-yoga*, "the sum of the hypotenuse," to *karṇa-varga*, "the square of the hypotenuse," since the latter reading makes much more mathematical sense than the former in this context. Such modifications are routinely required in translating published Sanskrit texts. However, since the Sanskrit edition of Gaṇeśa's commentary that I used is not a critical one, it is impossible to know whether I was fixing a minor error by the editor, by a scribe somewhere in the textual lineage of the manuscripts used by the editor, or by Gaṇeśa himself. Moreover, it is always possible that the published reading actually has a more subtle but correct meaning that I simply failed to understand.

To sum up, the path from what a medieval mathematician composing in Sanskrit intended to say to what a modern translator claims that he said is frequently long and perilous. *If* the author's words were successfully preserved and transcribed by a lengthy succession of scribes, and *if* the commentator(s) who later encountered manuscripts of the text correctly understood and explained it, and *if* the editor(s) who published the text selected suitable manuscript readings in cases where one or another scribe slipped up, and *if* the translator had enough linguistic and technical knowledge, plus enough critical imagination, to pick the most appropriate of the possible interpretations of the resulting version of the text, *then* what the reader encounters will most likely be a pretty fair approximation of the original. But there are significant uncertainties present at every stage of this process, so caution is always warranted in dealing with the end result.

A.5 GLOSSARY OF TRANSLITERATED TECHNICAL TERMS

Most of the following terms are Sanskrit; other languages are indicated in parentheses.

ahar-gaṇa The number of days accumulated between a given epoch date and some desired date.

aṅgula Digit, as a unit of linear measure (finger-breadth).

anuṣṭubh A Sanskrit verse meter consisting of four quarter-verses of eight syllables each.

ardha Half.

bīja 1. Algebra. 2. A conversion constant to adjust astronomical parameters. (Literally, "seed.")

chandas Prosody or poetic metrics. One of the six Vedāṅgas.

daṇḍa A punctuation mark in the form of a vertical stroke.

dharma The Hindu doctrine of sacred law or righteousness; the moral and religious duties prescribed for humans.

gaccha In Jainism, a lineage of teachers and pupils.

gaṇita 1. Computation, calculation; mathematics in general. 2. The mathematical component of astronomy and astrology.

ghaṭikā A time unit equal to one-sixtieth of a day.

gola Spherics, a sphere.

gotra Among Brāhmaṇas, a lineage tracing its descent to one of the legendary sages of Hinduism.

hay'a (Arabic) Islamic astronomy, particularly its geometric models.

illam (Malayalam) Among Brāhmaṇas in Kerala, a family residence or estate.

jātaka Genethlialogy; the practice or study of casting nativity horoscopes.

jāti Hereditary social and occupational group, often translated "caste."

jīvā Sine (or chord; literally, "bowstring").

jñāti Hereditary social and occupational group, sometimes a subdivision of *jāti* and sometimes synonymous with it.

jyā Sine (or chord; literally, "bowstring").

jyotiṣa Astronomy and astrology. One of the six Vedāṅgas.

kalā 1. A time unit equal to one-sixtieth of a *muhūrta*. 2. In the *Jyotiṣa-vedāṅga*, a time unit equal to 201/20 of a *ghaṭikā*. 3. Arcminute.

Kaliyuga The last of the four ages in a *mahāyuga*, equal to 432,000 years. Indian cosmology holds that the current Kaliyuga began on 18 February 3102 BCE.

kalpa 1. A lifetime of the universe, conventionally equal to 4,320,000,000 years. 2. The Vedāṅga of ritual practice.

karaṇa 1. A handbook for astronomical computation. 2. A time unit equal to half a *tithi*.

karaṇī Literally, "making"; a surd or non-square number.

kaṭapayādi An alphanumeric notational system for verbally representing numerals.

kha Void; zero.

koṣṭhaka A set of tables for astronomical computation.

mahāyuga A cosmological time span equal to 4,320,000 years.

manda The inequality in a planet's orbital motion that depends on its position with respect to the stars, analogous to zodiacal anomaly. (Literally, "slow.")

manvantara A cosmological time span equal to seventy-one *mahāyugas* or 306,720,000 years.

mīmāṃsā The Vedāṅga of philosophical investigation of scripture; a division of Indian philosophy.

muhūrta A time unit equal to one-thirtieth of a day.

mūla 1. Square root. 2. Base text (on which an expository commentary is written).

nāgarī (also called *deva-nāgarī*.) The name of the script most commonly used to write Sanskrit.

nakṣatra 1. One of the twenty-seven or twenty-eight constellations that lie in the orbital path of the moon. 2. An arc of the ecliptic with length $13°20'$, associated with one of the abovementioned lunar constellations. 3. An interval of time during which the moon traverses an arc of $13°20'$.

pāda A quarter-verse. (Literally, "foot.")

padabhā Literally, "foot-shadow"; a gnomon shadow measured in units of feet.

pakṣa 1. A school or faction. 2. One side of an equation. (Literally, "wing" or "side.")

pañcāṅga A yearly calendar tracking the succession of various civil, liturgical, and astronomical time units.

pāta An astrologically significant event when the longitudes of the sun and moon add up to 180° or 360°.

pāṭī Arithmetic (literally, "board," for the dust board on which computations were written out).

Purāṇa A sacred work on cosmology, mythology, and historical legend.

pūrṇa Full, complete.

ṛṇa Subtracted, negative.

rūpa Known number; unity. Abbreviated "*rū*" in algebra.

śāstra Learned discipline, science.

siddhānta Technical treatise; in particular, a comprehensive treatise on astronomy.

śīghra The inequality in a planet's orbital motion that depends on its position with respect to the sun, analogous to synodic anomaly. (Literally, "fast.")

śloka Verse, especially a verse in *anuṣṭubh* meter.

smṛti Sacred texts ascribed to human authorship. (Literally, "remembering.")

soma A ritual beverage used in some ancient Vedic sacrifices.

śruti Sacred texts ascribed to divine revelation, such as the Vedas. (Literally, "hearing.")

śulba A measuring cord; ritual geometry for altar construction.

śūnya Void; zero.

sūtra A rule or algorithm (literally, "thread").

tājika An Islamic form of horoscopic astrology.

tātkālika Literally, "at that time"; momentary, instantaneous.

tithi A "lunar day," or one-thirtieth of a synodic month.

trai-rāśika Rule of Three Quantities, simple proportion.

vākya Literally, "sentence"; a celestial longitude value encoded in alphanumeric notation.

varga 1. Square, in arithmetic or geometry. Abbreviated "*va*" in algebra. 2. The first twenty-five consonants in the Sanskrit alphabet.

varṇa Literally, "color." 1. One of the four major divisions of humanity in Hinduism. 2. A way of designating unknown quantities in algebra; different unknowns are referred to by the names of different colors.

Vedāṅga One of the six "limbs," or supporting disciplines, of the sacred Vedas.

yāvattāvat Literally, "as much as so much"; in algebra, an unknown quantity. Abbreviated "*yā*."

yoga Literally, "sum"; an interval of time during which the combined motions of the sun and moon add up to $13°20'$.

yojana A unit of distance, probably somewhere between five and ten kilometers.

yuga 1. A calendar intercalation cycle. 2. A cosmological time interval, especially a *mahāyuga*.

yuta Added. Abbreviated "*yu*" in algebra.

zīj (Arabic, Persian) A set of tables and algorithms for astronomical computation in Islamic science.

Appendix B

Biographical Data on Indian Mathematicians

The following very brief sketches summarize available biographical information about most of the mathematical authors named in this book, and in some cases describe the sources from which the information is known or inferred. In addition, some attempts are made to clear up confusion between reasonably well-established facts or inferences and unsubstantiated legends. If no source is stated for a particular assertion, that means that it is derived from statements in the author's own work or a related original source deemed reliable. All authors are known or presumed to be Hindu Brāhmaṇas unless otherwise noted.

The citations at the end of each entry reference English-language sources with more detailed information about the author in question. (Note that many of the sources contain Sanskrit quotations that are not directly translated, although most of the information in them is summarized in English.) Author names marked with an asterisk are the subjects of articles in the *Dictionary of Scientific Biography* [Gille1978]. Cross-references to names of other authors in this list are indicated in boldface type. Names of treatises are not translated unless they have not previously appeared in this book.

***Acyuta Piṣāraṭi.** Born around 1550, died 7 July 1621. Author of several astronomical works. Native or resident of Tṛkkaṇṭiyūr (Sanskrit Kuṇḍapura) near Tirur in Kerala. A student of **Jyeṣṭhadeva** in the **Mādhava** or Kerala school of mathematics and astronomy, a protégé of King Ravivarman (ruled 1595–1607), and an authority on Sanskrit poetics and grammar. Member of an Ambalavāsi non-Brāhmaṇa caste called Piṣāraṭi in Malayalam. He was the first known Indian astronomer to specify the conversion of planetary motion on an inclined orbit to motion on the ecliptic. [Pin1970a], vol. 1, pp. 36–38; [SarK1972], pp. 64–65.

Āpastamba. See ***Śulba-sūtra* authors.**

***Āryabhaṭa (I).** Author of the *Āryabhaṭīya* and a lost work using a midnight epoch, which became respectively the primary texts of the Ārya-pakṣa and the Ardha-rātrika-pakṣa. Aged twenty-three after 3600 years of the Kaliyuga, corresponding to 499 CE (hence born 476), which is not necessarily the date of composition of either of his works. Refers to "the knowledge honored in Kusumapura," which his seventh-century commentator **Bhāskara I** identifies with Pāṭaliputra, now identified with modern Patna in Bihar. Bhāskara I also calls Āryabhaṭa "the one from Aśmaka," apparently referring to a region between the Godavari and Narmada rivers in central India, and the sixteenth-century commentator **Nīlakaṇṭha** asserted that he was born there. Thus his native area and his later place of residence may have been different.

The myths and misunderstandings about Āryabhaṭa's life and work are legion, and some of them are many centuries old. Allegations that he was a leader in the renowned center of Buddhist learning ("university") of Nalanda in medieval Bihar all seem to be ultimately extrapolated from his reference to Kusumapura and from an unattributed verse calling him a *kulapa*, "chief" or "head." He evidently advocated an astronomical model with a rotating earth, but not a heliocentric system. [Pin1970a], vol. 1, pp. 50–53; [ShuSa1976], pp. xvii–xxvii.

***Āryabhaṭa (II).** Author of the *Mahā-siddhānta*, also known as the *Ārya-siddhānta*. This work has some similarities with that of **Śrīdhara**. A peculiar feature of it—the division of the ecliptic into "decans" or arcs of 10°—is attributed to "Āryabhaṭa" by **Bhāskara II** in the twelfth century but does not appear in the *Āryabhaṭīya* of **Āryabhaṭa I**. Āryabhaṭa II is thus inferred to have lived after **Śrīdhara** and before **Bhāskara II**, and usually estimated at ca. 950–1100. [Pin1970a], vol. 1, pp. 53–54; [DviS1910], pp. 1–23.

Baudhāyana. See ***Śulba-sūtra* authors**.

***Bhāskara (I).** Author of the astronomical works *Mahā-bhāskarīya* and *Laghu-bhāskarīya*, and a commentary on the *Āryabhaṭīya* which he wrote in (elapsed) year 1986123730 of the *kalpa*; according the Ārya-pakṣa this corresponds to (elapsed) Kaliyuga year 3730 or 629 CE. References in some of his sample problems to the Aśmaka region (see **Āryabhaṭa I**) and to the city of Valabhī, now identified with Vala in the Saurashtra/Kathiawar peninsula of Gujarat, have led to the inference that he lived in one or both of those areas. [Pin1970a], vol. 4, pp. 297–299; [Shu1976], pp. xvii–xxxv; [Kup1957], pp. xi–xxiv.

***Bhāskara (II).** Also called "Bhāskara Ācārya," meaning "teacher" or "learned one." Author of the *Līlāvatī*, the *Bīja-gaṇita*, the *Siddhānta-śiromaṇi* with auto-commentary (in Śaka 1072 or 1150 CE at age thirty-six), and the *Karaṇa-kutūhala* (epoch date Thursday 24 February 1183), as well as a commentary on the *Śiṣya-dhī-vṛddhida-tantra* of **Lalla**. The first two works are sometimes considered part of the third. From the above data, he must have been born in 1114 CE. Member of the Śāṇḍilya *gotra*. An inscription commemorating the establishment of a school in 1207 under the leadership of his grandson for the study of his works lists several generations of his ancestors and descendants. Refers to his father and teacher, Maheśvara, as a resident of a city called Vijjaḍaviḍa in the Sahyādri mountains, that is, the Western Ghats between the Tapti and Godavari rivers. Some of his ancestors and descendants are associated with the courts of various rulers in central and western India, so Bhāskara too may have been a court astronomer. His alleged connection with an alleged observatory at Ujjain is undocumented. So is the existence of his daughter after whom the *Līlāvatī* was allegedly named; see section 6.2.1. [Pin1970a], vol. 4, pp. 299–326.

***Brahmagupta.** Author of the *Brāhma-sphuṭa-siddhānta* (at the age of thirty in Śaka 550 or 628 CE; hence born 598), the earliest surviving treatise of the Brāhma-pakṣa; the last chapter is a separate *karaṇa* work, the *Dhyāna-grahopādhyāya* ("Thought-grasping teacher," a pun on *graha*, "planet"), with epoch date at the vernal equinox in Śaka 550, or 21 March 628. In the text he mentions his father by name, Jiṣṇugupta. He also composed a *karaṇa* called *Khaṇḍa-khādyaka* following the Ardha-rātrika-pakṣa, with epoch date falling in 665 CE. The ninth-century commentator **Pṛthūdakasvāmin** calls him "the teacher from Bhillamāla," now identified with Bhinmal near Mount Abu in Rajasthan. As he makes a general reference in one of the *Brāhma-sphuṭa-siddhānta*'s verses to showing celestial phenomena to "the people or the king," it is conjectured that he may have been a court astronomer under the ruler Vyāghramukha of that place and period. There is no known reason to connect him with any observatory at Ujjain. The alternative translation of the first two words of the title *Brāhma-sphuṭa-siddhānta* ("Corrected [*sphuṭa*] *siddhānta* of Brahman") as "Opening of the universe" seems to be based on a modern misreading of "Brahma" for "Brāhma." [Pin1970a], vol. 4, pp. 254–257; [Cha1970] 1, pp. 1–6.

Brahman. Deity credited with the authorship of some astronomical works including the *Paitāmaha-siddhānta.* [Pin1970a], vol. 4, pp. 259–260.

Chajaka's son. The scribe of the Bakhshālī Manuscript or an author of the commentary in it, or both; the only named author or scribe associated with the Bakhshālī Manuscript. Self-described as a Brāhmaṇa and "king of mathematicians," writing or copying the text for the use of "Hasika, son of Vasiṣṭha" and his descendants. See section 5.2 for more on the Bakhshālī Manuscript. [Hay1995], pp. 84–86, 148–150.

Citrabhānu. Author of a *karaṇa* text (dated to Kaliyuga day 1691513 or about 30 March 1530, but with epoch date Kaliyuga year 4608 or 1507 CE), a treatise on algebraic problems, and also a commentary on a poetic text. One of the pupils of **Nīlakaṇṭha** and one of the teachers of **Śaṅkara** in the Kerala school. A Nampūtiri Brāhmaṇa and member of the Gautama *gotra*, native of Covvaram (Sanskrit Śivapura) near Trichur. [Pin1970a], vol. 3, p. 47; [SarK1972], p. 57; [Hay1998].

Dāmodara. Son of **Parameśvara** and teacher of both **Nilakantha** and **Jyeṣṭhadeva** in the Kerala school; thus probably flourished in the second half of the fifteenth century and the early sixteenth. None of his own astronomical works survive even by name, but he is occasionally quoted by **Nīlakaṇṭha** and other authors. [SarK1972], pp. 54–55.

***Gaṇeśa.** Author of the highly popular *Graha-lāghava* (with epoch date Śaka 1442 or 1520 CE), the *Tithi-cintāmaṇi*, a commentary on the *Līlāvatī* of **Bhāskara II**, and several other works on astronomy and astrology. A tradition recounts that he wrote the *Graha-lāghava* at the age of thirteen; if so, he was born in 1507 CE. The *Tithi-cintāmaṇi* was composed in Śaka 1447 or 1525 CE, and the *Līlāvatī* commentary in Śaka 1467 or 1545 CE. His father was Keśava and his mother Lakṣmī, and he belonged to the Kauśika *gotra*. He and his family for several generations, most of whom seem to have been *jyotiṣa* specialists, resided in Nandigrāma (modern Nandod) in Gujarat. [Pin1970a], vol. 2, pp. 94–106.

Govindasvāmin. Author of a commentary on the *Mahā-bhāskarīya* composed by **Bhāskara I** and of three other works, two of which are lost. He was the teacher of one Śaṅkaranārāyaṇa, who was a court astronomer under King Ravivarma (born 844 CE) of Kerala and mentioned in his own work dates corresponding to 866 and 869 CE. Thus Govindasvāmin is inferred to have lived and worked in Kerala near the middle of the ninth century. [Pin1970a], vol. 2, pp. 143–44; [SarK1972], pp. 44–45; [Kup1957], pp. xlvi–xlix.

***Jayasiṃha (Jai Singh II).** Royal sponsor of astronomical and mathematical research and author of a couple of astronomical works. Born 1686 CE, died 2 October 1743. A Kṣatriya of the Kachwāha Rājput family, he was the ruler of Amber in Rajasthan under the sultan Muḥammad Shāh and founded the city of Jaipur in 1728 CE. Sponsored an ambitious comparative astronomy program involving the building of observatories with large masonry instruments in five cities and the translation of astronomical and mathematical works from Arabic, Persian, and Latin by teams of Hindu, Muslim, and Jesuit scholars. [Pin1970a], vol. 3, pp. 63–64; [Pin1999]; [Shar1995], esp. pp. 1–18.

Jñānarāja. Author in Śaka 1425 or 1503 CE of the *Siddhānta-sundara*, the first known major *siddhānta* after the *Siddhānta-śiromaṇi* of **Bhāskara II**; also wrote a *Bījādhyāya*. Resident of Pārthapura, modern Pathri in Maharashtra. Father of **Sūryadāsa** and Cintāmaṇi. A nineteenth-century lineage assigns him to the Bhāradvāja *gotra*. [Pin1970a], vol. 3, pp. 75–76; [Knu2008].

Jyeṣṭhadeva. A student of **Dāmodara** and a teacher of **Acyuta Piṣāraṭi** in the Kerala school, hence probably working around the middle of the sixteenth century. Author of the *Yukti-bhāṣa* in Malayalam, containing detailed demonstrations for many mathematical results including infinite series in trigonometry. A traditional chronology says he was a native of Ālattūr in Kerala. [Pin1970a], vol. 3, pp. 76–77; [SarK1972], pp. 59–60.

***Kamalākara.** Author in Śaka 1580 or 1658 CE of the *Siddhānta-tattva-viveka*, a *siddhānta* combining traditional Sanskrit and Islamic astronomical and mathematical concepts, and other astronomical and astrological works. Member of a prominent family of astronomers of the Bhāradvāja *gotra* who had migrated to Varanasi (Benares) from a village near Pārthapura (modern Pathri) in Maharashtra in the sixteenth century. Contemporary and scholarly rival of **Munīśvara**. [Pin1970a], vol. 2, pp. 21–23; [Pin1981a], pp. 118–130.

Kātyāyana. See ***Śulba-sūtra* authors**.

Kṛṣṇa. Uncle of **Munīśvara** in the family of Varanasi astronomers whose chief professional rivals were the family of **Kamalākara**. This family of the Devarāta *gotra* moved from Maharashtra to Varanasi in the late sixteenth century. Kṛṣṇa was a protégé of the Mughal emperor Jahāngir (ruled 1605–1627), and is best known for his commentary on the *Bīja-gaṇita* of **Bhāskara II**. [Pin1970a], vol. 2, pp. 53–54; [Pin1981a], pp. 118–130.

Lagadha. Authority to whom the ancient *Jyotiṣa-vedāṅga* or *Vedāṅga-jyotiṣa* is ascribed, sometimes said to have been actually written by Śūci. Probably around the fifth century BCE, but astrochronological arguments have also been use to claim a date in the late second millennium BCE. [Pin1970a], vol. 5, pp. 538–543; [SarKu1984].

***Lalla.** Author of the *Śiṣya-dhī-vṛddhida-tantra*, of an astrology text, and of other astronomical and mathematical works now lost. He is first mentioned by **Śrīpati** in the mid-eleventh century and appears to follow in some respects the style of the *Brāhma-sphuṭa-siddhānta* of **Brahmagupta**, which would place him between the seventh and eleventh centuries. He gives a set of planetary corrections that seem to indicate the year Śaka 670 or 748 CE, on the strength of which he is generally dated to the eighth or ninth century. His occasional references to the Lāṭa region, in modern south Gujarat, may imply that he lived there. His father's name was Trivikrama. [Pin1970a], vol. 5, pp. 545–546; [Cha1981] 2, pp. xi–xxi.

Mādhava. Founder of the so-called "Mādhava school" in second-millennium Kerala. A few of his astronomical works survive, but they do not contain the achievements he is most famous for, namely the derivation of the Mādhava-Leibniz infinite series for π, the Mādhava-Newton power series for sine and cosine, and related results known from quotations in the works of his followers. An Emprāntiri Brāhmaṇa, he lived in Irinjālakkuḍa (Sanskrit Saṅgamagrāma) near modern Kochi or Cochin. He was the teacher of **Parameśvara**, so he probably flourished around the turn of the fifteenth century. [Pin1970a], vol. 4, pp. 414–415; [SarK1972], pp. 51–52.

***Mahāvīra.** A Digambara Jain and the author of the *Gaṇita-sāra-saṅgraha* on mathematics. As the introductory verses of this work eulogize the Rashtrakuta king Amoghavarṣa (ruled mid-ninth century) of the region corresponding to modern Karnataka and Maharashtra, it is generally assumed that Mahāvīra worked at his court. [Pin1970a], vol. 4, p. 388; [Ran1912].

***Mahendra Sūri.** A Jain who composed in Śaka 1292 or 1370 CE the first known Sanskrit treatise on the Islamic astrolabe (*Yantra-rāja* or "King of instruments," which became the standard Sanskrit technical term for the instrument) for the Tughluq sultan Fīrūz Shāh. The only clue to his place of origin is the fact that his teacher came from Bhṛgupura (modern Broach in Gujarat). Mahendra was described by a student and commentator of his as the foremost astronomer at the sultan's court in Delhi. [Pin1970a], vol. 4, pp. 393–395; [SarS2000].

Mallāri. The son of a pupil of **Gaṇeśa**, he wrote a commentary on **Gaṇeśa**'s *Tithi-cintāmaṇi* sometime in the late sixteenth century, as well as another astronomical work. His father moved to Varanasi (Benares), so Mallāri probably lived and worked there; his nephew's son was **Kamalākara**. [Pin1970a], vol. 4, pp. 365–367; [Pin1970a], vol. 5, pp. 284–285; [Pin1981a], pp. 118–130.

Mallikārjuna Sūri. Composed a commentary on the *Śiṣya-dhī-vṛddhida-tantra* of **Lalla** which mentions a date in Śaka 1100 corresponding to 21 March 1178, and one on the *Sūrya-siddhānta*. He mentioned having written another commentary in Telugu and referred to his birth in the Veṅgi region, an area south of the Godavari River in what is now Andhra Pradesh. However, in his commentary on the *Śiṣya-dhī-vṛddhida-tantra* he recorded the calculation of an ominous celestial phenomenon (on a date corresponding to Tuesday 5 March 1185) at a place called Prakāśa-pattana located at coordinates approximately equivalent to those of modern Patna in Bihar, so he may have resided in the north as well. He belonged to the Kauṇḍinya *gotra*. [Pin1970a], vol. 4, p. 368; [Pin1970a], vol. 5, pp. 285–286; [Cha1981] 2, pp. xxiii–xxvi; [Shu1990], p. 3.

Mānava. See ***Śulba-sūtra* authors**.

Mañjula. See **Muñjāla**.

***Munīśvara Viśvarūpa.** Born 17 March 1603, a member of the Varanasi family of astronomers (in the Devarāta *gotra*) who were contemporary with, and often disputing with, that of **Kamalakara**. Author of the *Siddhānta-sarva-bhauma* (completed Monday 8 Bhādrapada Śaka 1568 or 7 September 1646), commentaries on works of **Bhāskara II**, and other mathematical and astronomical texts. Nephew of **Kṛṣṇa**. [Pin1970a], vol. 4, pp. 436–441.

***Muñjāla.** This author's name was recorded by his first known commentator as Mañjula, but he was uniformly known among later writers as Muñjāla. Author of the *karaṇa Laghu-mānasa* with epoch date Saturday 10 March 932, and of another *karaṇa* now lost. Member of the Bhāradvāja *gotra*, resident of a place called Prakāśa, possibly the same as the Prakāśa-pattana mentioned by **Mallikārjuna**. [Pin1970a], vol. 4, pp. 435–436; [Shu1990], pp. 1–4.

Nārāyaṇa. A pupil of **Citrabhānu** in the Kerala school and a member of a Kerala Nampūtiri family named Mahiṣamaṅgala, Nārāyaṇa was renowned as a poet and scholar on nonmathematical subjects as well as on mathematics and astronomy. Completed at the age of eighteen the unfinished *Kriyā-kramakarī* commentary of **Śaṅkara** on the *Līlāvatī* of **Bhāskara II**, at the request of his father, also named Śaṅkara. Later wrote another commentary on the *Līlāvatī* as well as astronomical works. A protégé of the kings of Kochi (Cochin) Vīrakerala Varman (ruled 1537–1565) and Rāma Varman (ruled 1565–1601). Mentions in one of his works Kaliyuga day 1,719,937, which fell in 1607 CE. [Pin1970a], vol. 4, p. 137; [SarK1975], pp. xvi–xxiv. The earlier [SarK1972], pp. 57–58, mistakenly attributes the entire *Kriyā-kramakarī* to Nārāyaṇa.

***Nārāyaṇa Paṇḍita.** Completed his *Gaṇita-kaumudī* on the date Thursday 10 November 1356; also wrote the *Bīja-gaṇitāvataṃsa*. His father's name was Nṛsiṃha or Narasiṃha, and the distribution of the manuscripts of his works suggests that he may have lived and worked in the northern half of India. [Pin1970a], vol. 3, pp. 156–157; [Kus1993], pp. 1–7; [Hay2004], pp. 386–390.

***Nīlakaṇṭha Somayājin.** Author of the astronomical work *Tantra-saṅgraha* (completed on Kaliyuga day 1680553, falling in 1500 CE), the *Jyotir-mīmāṃsā* on the philosophical analysis of astronomy, and a commentary on the *Āryabhaṭīya* of **Āryabhaṭa I**, as well as several other works. Pupil of **Damodara** and teacher of **Citrabhānu** and **Śaṅkara** in the Kerala school. Born on Kaliyuga day 1660181 or 14 June 1444, and may have lived as long as a hundred years. Native of Tṛkkaṇṭiyūr (Sanskrit Kuṇḍapura) near Tirur; a Nampūtiri Brāhmaṇa belonging to the Garga *gotra*, and a protégé of the Nampūtiri caste leader Netranārāyaṇa. The name Somayājin indicates that he (or his family?)

performed the Vedic *soma* sacrifice. [Pin1970a], vol. 3, pp. 175–177; [SarK1977b], pp. xxiv–xxxviii. The earlier [SarK1972], pp. 55–57, puts his birthdate in December 1444.

Nityānanda. Resident of Delhi in the early seventeenth century, apparently a court astronomer of the Mughal emperor Shāh Jahān (ruled 1627–1658). Commissioned by a minister of the emperor to translate into Sanskrit the Indo-Persian *zīj* composed for Shāh Jahān in 1629; the translation is dated Saṃvat 1685 or 1628 CE but was completed in the early 1630s. Wrote in Saṃvat 1696 or 1639 CE the *Sarva-siddhānta-rāja* treating the same material in a more traditional *jyotiṣa* context. He was a Brāhmaṇa of the Gauḍa family and the Mudgala *gotra*, and lists in the *Sarva-siddhānta-rāja* a few generations of his ancestors. [Pin1970a], vol. 3, pp. 173–174; [Pin2003b].

***Parameśvara.** Pupil of **Mādhava** in the Kerala school. Author of the *Dṛg-gaṇita* dated Śaka 1353 or 1431 CE and of several other astronomical and astrological works and commentaries, including one dated Śaka 1365 or 1443 CE. He recorded a number of eclipse observations, and his other treatises are mostly about the computation of eclipses and about spherical astronomy. Remarks by his son's student **Nīlakaṇṭha** assert that he completed the *Dṛg-gaṇita* after fifty-five years of observing the sky and that he personally explained some of his teachings to **Nīlakaṇṭha**; if both statements are true, he would have lived from at least the 1360s to the 1450s. A Nampūtiri Brāhmaṇa of the Bhṛgu *gotra*, and a resident of the Vaṭaśśeri (Sanskrit Vaṭaśreṇi) estate on the banks of the Niḷā river. [Pin1970a], vol. 4, pp. 187–192; [Kup1957], pp. l–liv; [SarK1972], pp. 52–54.

Pṛthūdakasvāmin. Wrote commentaries on the *Brāhma-sphuṭa-siddhānta* and *Khaṇḍa-khādyaka* of **Brahmagupta**; the latter commentary uses the date Śaka 786 or 864 CE in several of its worked examples. He is known as Caturveda, "[the one knowing] four Vedas," and gives his father's name as Madhusūdhana Bhaṭṭa. [Pin1970a], vol. 4, pp. 221–222; [Ike2002], pp. 1–14.

Śaṅkara Vāriyar. Student of **Nīlakaṇṭha** in the Kerala school and author of commentaries on his teacher's *Tantra-saṅgraha* and a partial one, *Yukti-dīpikā* (later completed by **Nārāyaṇa**), on the *Līlāvatī* of **Bhāskara II**. The *Yukti-dīpikā* is to a large extent a Sanskrit verse version of the Malayalam *Yukti-bhāṣa* of **Jyeṣṭhadeva**. Mentions in his works the Kaliyuga days 1691302, falling in 1529 CE, 1692972, falling in 1534 CE, and 1700000, falling in 1554 CE. Member of an Ambalavāsi non-Brāhmaṇa caste called Vāriyar in Malayalam (Sanskrit *pāraśava*), protégé of Nīlakaṇṭha's patron Netranārāyaṇa, and resident of Tṛkkaṇṭiyūr (Sanskrit Kuṇḍapura) near Tirur. [SarK1972], pp. 58–59; [SarK1975], pp. xiii–xxii; [SarK1977b], pp. xliii–lxviii.

***Sphujidhvaja.** An Indo-Greek king in western India under the Western Kshatrapa (Śaka) ruler Rudrasena II. (His name is more correctly spelled Sphujiddhvaja, literally "banner of Āsphujit," a Sanskritization of the Greek "Aphrodite.") Composed in Śaka 191 or 269 CE a Sanskrit verse version, the *Yavana-jātaka*, of a Sanskrit prose translation of a Greek work on horoscopy; this had been written in Śaka 71 or 149 CE during the reign of the Western Kshatrapa Rudradāman by an author known only as "Yavaneśvara," "lord of the Greeks," evidently another Indo-Greek king. [Pin1981a], pp. 81–82; [Pin1978c].

***Śrīdhara.** Author of the *Pāṭī-gaṇita* and *Tri-śatikā* on arithmetic, and a lost work on algebra. Nothing more is definitely known about him except the following: he is mentioned by **Bhāskara II**; he appears to refer to **Brahmagupta**, although not by name; **Govindasvāmin** quotes a verse identical to one in the *Tri-śatikā*; and either he borrowed from **Mahāvīra** or vice versa. This puts Śrīdhara most likely somewhere in the eighth or ninth century. [Pin1981a], p. 58; [Shu1959]; [Jai2001], p. 41.

***Śrīpati.** Author of the *Siddhānta-śekhara*, *Gaṇita-tilaka* ("Forehead-mark of calculation"), and other works on astronomy and astrology; one of these has a date corresponding to 1039 CE and another to 1056 CE. Member of the Kāśyapa *gotra* and resident of Rohiṇīkhaṇḍa in Maharashtra some 250 kilometers south of Ujjain. [Pin1981a], passim; [Misr1932].

Śūci. See **Lagadha**.

***Śulba-sūtra* authors.** The ancient ritual geometry texts on the construction of fire altars, called *Śulba-sūtras*, are found in some collections of texts on ritual procedures. The authorship of each of these collections, including its *Śulba-sūtra* text if it has one, is ascribed to a legendary sage. But nothing is known about these sages as historical figures, and the texts attributed to them can be dated only in relation to other texts, on the basis of their content and linguistic style. Of the four major *Śulba-sūtras* associated with the names of Baudhāyana, Mānava, Āpastamba, and Kātyāyana, the first is thought to date to the first half of the first millennium BCE and the last to the last few centuries BCE, with the others probably somewhere in between. [SenBa1983], pp. 2–5; [Pin1981a], pp. 3–7.

Sūrya. Deity credited with the authorship of the *Sūrya-siddhānta*. [Pin1981a], p. 3.

Sūryadāsa. Also known as Sūrya, Sūrya Paṇḍita, etc., he was a son of **Jñānarāja** of Pārthapura, in the Bhāradvāja *gotra*. Wrote a commentary on the *Bīja-gaṇita* of **Bhāskara II** in Śaka 1460 or 1538 CE at the age of thirty-one; hence born in 1507 CE. Also commented on **Bhāskara II**'s *Līlāvatī* and wrote several other mathematical and astrological works, but is better known as an author of poetic and philosophical works. [Jai2001], pp. 1–8; [Pin1981a], pp. 118–130.

Sūryadeva Yajvan. Commentator on several astronomical and astrological works, including the *Āryabhaṭīya* of **Āryabhaṭa I** and the *Laghu-mānasa* of **Muñjāla**. Born in Śaka 1113 on a date corresponding to Monday 3 February 1192. His *Laghu-mānasa* commentary uses an epoch date in Śaka 1170 or 1248 CE, and in it he gives latitude and longitude information for a locality called Gaṅgāpura that identifies it with modern Gangai Konda Cholapuram in Tamil Nadu. He belonged to the Nidhruva *gotra*; the name "Yajvan" indicates that he performed the Vedic *soma* sacrifice. [SarK1976a], pp. xxv–xxx; [Shu1990], pp. 23–30

***Varāhamihira.** Author of the *Pañca-siddhāntikā*, which is a summary of five earlier astronomical works, and several immensely popular astrological texts. Some of the works summarized in his *Pañca-siddhāntikā* use an epoch date corresponding to 505 CE, and **Brahmagupta** refers to him in the *Brāhma-sphuṭa-siddhānta* of 628 CE, so he is inferred to have lived sometime in or around the sixth century. Member of the Maga Brāhmaṇa group descended from Iranian Zoroastrians, a native of the Avantī region in central India, and a resident of Kāpitthaka (location unknown). He mentioned the city of Ujjain as lying on the prime meridian, but there is no explicit evidence linking him to any putative observatory or other astronomical institution there. [Pin1970a], vol. 5, pp. 563–595.

Vararuci. Author of a work in the *vākya* or "sentence" genre of south Indian astronomy, in which tables of true longitudes are encoded as *kaṭapa-yādi* sentences. He is traditionally assigned to the fourth century CE and credited with the origin of the *vākya* system and the *kaṭapayā-di* notation which it employs. [Pin1970a], vol. 5, p. 558; [SarK1972], p. 43.

***Vaṭeśvara.** Author of the *Vaṭeśvara-siddhānta* and of a *karaṇa* text. Born in Śaka 802 or 880 CE, and composed the *Vaṭeśvara-siddhānta* at the age of twenty-four or in 904 CE. He gives his father's name as Mahadatta and residence as Ānandapura with a latitude of 24°, possibly identical to modern Vaḍnagar in northern Gujarat. [Pin1970a], vol. 5, pp. 555–556; [Shu1986] 2, pp. xxiii–xxx.

Yavaneśvara. See **Sphujidhvaja**.

Bibliography

[Ach1997] Achar, B. N. Narahari. A note on the five-year yuga of the Vedāṅga Jyotiṣa. *Electronic Journal of Vedic Studies* 3 (4), 1997, 21–28 (http://www.ejvs.laurasianacademy.com/issues.html).

[Ans1985] Ansari, S. M. Razaullah. *Introduction of Modern Western Astronomy in India during 18–19 Centuries*. New Delhi: Institute of History of Medicine and Medical Research, 1985.

[Ans2002a] Ansari, S. M. Razaullah. European astronomy in Indo-Persian writings. In [Ans2002b], pp. 133–144.

[Ans2002b] Ansari, S. M. Razaullah, ed. *History of Oriental Astronomy*. Dordrecht: Kluwer, 2002.

[Ans2004] Ansari, S. M. Razaullah. Sanskrit scientific texts in Indo-Persian sources, with special emphasis on *siddhāntas* and *karaṇas*. In [BurC2004], pp. 587–605.

[Apt1937] Āpaṭe, V. G. *Līlāvatī* (Ānandāśrama Sanskrit Series 107), 2 vols. Puṇe: Ānandāśrama Press, 1937.

[Arn2000] Arnold, David. *Science, Technology and Medicine in Colonial India* (The New Cambridge History of India III, vol. 5). Cambridge: Cambridge University Press, 2000.

[BabuCT] Babu, Senthil. Counting in Tamil: Making texts from arithmetic practice in colonial Madras. Presented at the Association for Asian Studies Annual Meeting, San Francisco, April 2006.

[BagSa2003] Bag, A. K., and Sarma, S. R., eds. *The Concept of Śūnya*. Delhi: Indian National Science Academy, 2003.

[Bai1787] Bailly, Jean-Sylvain. *Traité de l'astronomie indienne et orientale*. Paris: Debure, 1787.

[Ben1970] Bentley, John. *A Historical View of the Hindu Astronomy*. Osnabrück: Biblio Verlag, repr. 1970.

[BerA1878] Bergaigne, A. *La religion védique d'après les hymnes du Rig-Veda*, 3 vols. Paris: F. Vieweg, 1878–83.

[BerJ1986] Berggren, J. L. *Episodes in the Mathematics of Medieval Islam.* New York: Springer, 1986, repr. 2003.

[BerJ2007] Berggren, J. L. Mathematics in medieval Islam. In [Kat2007], pp. 515–675.

[Bha1967] Bhattācārya, V. B. *Hayata.* Varanasi: Varanaseya Sanskrit Vishvavidyalaya, 1967.

[Bil1971] Billard, Roger. *L'astronomie indienne: investigation des textes sanskrits et des données numériques.* Paris: École Française d'extrême-orient, 1971.

[Bir1948] al-Bīrūnī, Abū Rayḥān Muḥammad ibn Aḥmad. *Rasā'il u'l-Bīrūnī.* Hyderabad: Osmania Oriental Publications Bureau, 1948.

[Bir1954] al-Bīrūnī, Abū Rayḥān Muḥammad ibn Aḥmad. *Al-Qānūn al-Mas'ūdī,* 2 vols. Hyderabad: Osmania Oriental Publications Bureau, 1954.

[Bre2002] Bressoud, David. Was calculus invented in India? *The College Mathematics Journal* 33, 2002, 2–13.

[Bro2001] Bronkhorst, Johannes. Pāṇini and Euclid: reflections on Indian geometry. *Journal of Indian Philosophy* 29, 2001, 43–80.

[Bro2006] Bronkhorst, Johannes. Commentaries and the history of science in India. *Asiatische Studien/Études Asiatiques* 60, 2006, 773–788.

[Bro2007] Bronkhorst, Johannes. *Greater Magadha: studies in the culture of early India* (Handbook of Oriental Studies, Section 2 South Asia, 19). Leiden-Boston: Brill, 2007.

[BroDe1999] Bronkhorst, Johannes, and Deshpande, Madhav M., eds. *Aryan and Non-Aryan in South Asia: Evidence, Interpretation, and Ideology* (Harvard Oriental Series, Opera Minora 3). Cambridge, MA: Harvard University Press, 1999.

[BroFaIAAS] Brown, D. R., and Falk, H. *The Interactions of Ancient Astral Science* (Vergleichende Studien zu Antike und Orient 3). Bremen: Hempen, forthcoming.

[BrPa2005] Bryant, Edwin F., and Patton, Laurie L., eds. *The Indo-Aryan Controversy: Evidence and Inference in Indian History.* London: Routledge, 2005.

[BurE1971] Burgess, Ebenezer. *Translation of the Sūrya-siddhānta.* Delhi: Indological Book House, repr. 1971.

[BurC2002] Burnett, Charles. Indian numerals in the Mediterranean Basin in the twelfth century, with special reference to the "eastern forms." In [Dol2002], pp. 237–288.

[BurC2004] Burnett, Charles, et al., eds. *Studies in the History of the Exact Sciences in Honour of David Pingree.* Leiden: Brill, 2004.

[BurR1790] Burrow, Reuben. A proof that the Hindoos had the binomial theorem. *Asiatic Researches* 2, 1790, 487–497. Repr. in [Dha1971], pp. 94–103.

[Cha1970] Chatterjee, Bina. *The Khaṇḍa-khādyaka of Brahmagupta*, 2 vols. New Delhi: Bina Chatterjee, 1970.

[Cha1981] Chatterjee, Bina. *Śiṣya-dhī-vṛddhida-tantra of Lalla*, 2 vols. New Delhi: Indian National Science Academy, 1981.

[ChKe2002] Chemla, Karine, and Keller, Agathe. The Sanskrit *karaṇīs* and the Chinese *mian.* In [Dol2002], pp. 87–132.

[Col1817] Colebrooke, H. T. *Algebra, with Arithmetic and Mensuration, from the Sanscrit of Brahmegupta and Bhascara.* London: John Murray, 1817.

[Cou1976] Coulson, Michael. *Teach Yourself Sanskrit.* New York: David McKay Co., 1976.

[Dal2000] van Dalen, Benno. Origin of the mean motion tables of Jai Singh. *Indian Journal of History of Science* 35, 2000, 41–66.

[Dal2002] van Dalen, Benno. The *Zīj-i Nāṣirī* by Maḥmūd ibn 'Umar. In [BurC2004], pp. 825–862.

[DanA1986] Dani, Ahmad Hasan. *Indian Palaeography*, 2nd ed. New Delhi: Munshiram Manoharlal, 1986.

[DanS1993] Dani, S. G. Myth and reality: On "Vedic mathematics." *Frontline*, vol. 10, 21 (22 October 1993), pp. 90–92, and 22 (5 November 1993), pp. 91–93. Revised version at http://www.math.tifr.res.in/~dani/vmissc.pdf.

[Dat1993] Datta, Bibhutibhusan. *Ancient Hindu Geometry: the Science of the Sulba.* New Delhi: Cosmo Publications, repr. 1993.

[DatSi1962] Datta, B., and Singh, A. N. *History of Hindu Mathematics: A Source Book*, 2 vols. Bombay: Asia Publishing House, repr. 1962.

[Dau2007] Dauben, Joseph W. Chinese mathematics. In [Kat2007], pp. 186–384.

[Del2005] Delire, Jean Michel. Quadratures, circulature and the approximation of $\sqrt{2}$ in the Indian Śulba-sútras. *Centaurus* 47, 2005, 60–71.

[Dha1971] Dharampal. *Indian Science and Technology in the Eighteenth Century: Some Contemporary European Accounts.* Delhi: Impex India, 1971.

[Dhu1957] Dhupakara, A. Y. *Taittirīyasaṃhitā,* 2nd ed. Pāraḍīnagara: Bhāratamudraṇālaya, 1957.

[Div2007] Divakaran, P. P. The first textbook of calculus: *Yukti-bhāṣa. Journal of Indian Philosophy* 35, 2007, 417–443.

[Dol2002] Dold-Samplonius, Yvonne, et al., eds. *From China to Paris: 2000 Years Transmission of Mathematical Ideas.* Stuttgart: Steiner Verlag, 2002.

[Duk2005a] Duke, Dennis. Comment on the origin of the equant papers by Evans, Swerdlow, and Jones. *Journal for the History of Astronomy* 36, 2005, 1–6.

[Duk2005b] Duke, Dennis. The equant in India: The mathematical basis of Indian planetary models. *Archive for History of Exact Sciences* 59, 2005, 563–576.

[Duk2005c] Duke, Dennis. Hipparchus' eclipse trios and early trigonometry. *Centaurus* 47, 2005, 163–177.

[Duk2008] Duke, Dennis. Mean motions and longitudes in Indian astronomy. *Archive for History of Exact Sciences* 62, 2008, 489–509.

[DukeIPT] Duke, Dennis. Indian planetary theories and Greek astronomy. (https://people.scs.fsu.edu/~dduke/Thursday_slides).

[DviK1987] Dvivedī, K. C. *Sūrya-siddhānta.* Varanasi: Sampurnanand Sanskrit University, 1987.

[DviP1936] Dvivedī, Padmākara. *Gaṇita-kaumudī of Nārāyaṇa* (Saraswati Bhavana Texts 57), 2 vols. Benares: Government Sanskrit College, 1936, 1942.

[DviS1899] Dvivedī, Sudhākara. *Tri-śatikā.* Benares: Pandit Jagannātha Śarmā Mehtā, 1899.

[DviS1901] Dvivedī, Sudhākara. *Brāhma-sphuṭa-siddhānta* (*The Pandit* NS 23–24). Benares: Government Sanskrit College, 1901–1902.

[DviS1908] Dvivedī, Sudhākara. *Yājuṣa-Jyautiṣa and Ārca-Jyautiṣa.* Benares: Medical Hall Press, 1908.

[DviS1910] Dvivedī, Sudhākara. *Mahā-siddhānta* (Benares Sanskrit Series 148–150). Benares: Braj Bhushan Das & Co., 1910.

[DviS1912] Dvivedī, Sudhākara. *Līlāvatī* (Benares Sanskrit Series 153). Benares: Braj Bhushan Das & Co., 1912.

[Eat2000] Eaton, Richard M. *Essays on Islam and Indian History*. New Delhi: Oxford University Press, 2000.

[Egg1897] Eggeling, J. *Satapatha-Brahmana According to the Text of the Madhyandina School*, vol. 4 (Sacred Books of the East 43). Oxford: Oxford University Press, 1897.

[Emc2005] Emch, Gerard, et al., eds. *Contributions to the History of Indian Mathematics* (Culture and History of Mathematics 3). New Delhi: Hindustan Book Agency, 2005.

[End2003] Endress, Gerhard. Mathematics and philosophy in medieval Islam. In [HoSa2003], pp. 121–176.

[Eva1998] Evans, James. *The History and Practice of Ancient Astronomy*. New York: Oxford University Press, 1998.

[Fal2000] Falk, Harry. Measuring time in Mesopotamia and ancient India. *Zeitschrift der Deutschen Morgenländischen Gesellschaft* 150 (1), 2000, 107–132.

[Far2004] Farmer, Steve, Sproat, Richard, and Witzel, Michael. The collapse of the Indus-script thesis: the myth of a literate Harappan civilization. *Electronic Journal of Vedic Studies* 11 (2), 2004, 19–57 (http://www.ejvs.laurasianacademy.com/issues.html).

[Fil1993] Filliozat, Pierre-Sylvain. Making something out of nothing. *UNESCO Courier*, November 1993.

[Fil2005] Filliozat, Pierre-Sylvain. Ancient Sanskrit mathematics: an oral tradition and a written literature. In Robert S. Cohen et al., eds., *History of Science, History of Text* (Boston Studies in Philosophy of Science 238). Dordrecht: Springer Netherlands, 2005, pp. 137–157.

[Gar1996] Garzilli, Enrica, ed. *Translating, Translations, Translators from India to the West* (Harvard Oriental Series Opera Minora 1). Cambridge, MA: Harvard University Press, 1996.

[GauCa1937] Gauḍa, V. S., and Caudhuri, C. S. *Śatapathabrāhmaṇa*, 3 vols. Kāśī: Acyutagranthamālākāryālaya, Saṃ. 1994–1997 (1937–1940 CE).

[Ger1989] Gerow, Edwin. Jayadeva's poetics and the classical style. *Journal of the American Oriental Society* 109 (4), 1989, 533–544.

[Gille1978] Gillespie, Charles C., ed. *Dictionary of Scientific Biography*, 16 vols. New York: Scribner's, 1978.

[Gillo2007] Gillon, Brendan S. Pāṇini's *Aṣṭādhyāyī* and linguistic theory. *Journal of Indian Philosophy* 35, 2007, 445–468.

[Gok1966] Gokhale, Shobhana Laxman. *Indian Numerals*. Poona: Deccan College, 1966.

[GolPi1991] Gold, David, and Pingree, David. A hitherto unknown Sanskrit work concerning Mādhava's derivation of the power series for sine and cosine. *Historia Scientiarum* 42, 1991, 49–65.

[Gom1998] Gomperts, Amrit. Sanskrit mathematical and astral sciences in ancient Java. *Newsletter of the International Institute for Asian Studies* 16, 1998, p. 23.

[Gom2001] Gomperts, Amrit. Sanskrit *jyotiṣa* terms and Indian astronomy in Old Javanese inscriptions. In Marijke J. Klokke and Karel R. van Kooij, eds., *Fruits of Inspiration: Studies in Honour of Prof. J. G. de Casparis*. Groningen: Egbert Forsten, 2001, pp. 93–134.

[Gon2002] González-Reimann, Luis. *The Mahabharata and the Yugas: India's Great Epic Poem and the Hindu System of World Ages* (Asian Thought and Culture, vol. 51). New York: Peter Lang, 2002.

[Gos2001] Gosvami, Bijoya. *Lalitavistara*. Kolkata: Asiatic Society, 2001.

[Goy2000] Goyal, S. R. *The Kauṭilīya Arthaśāstra: its author, date, and relevance for the Maurya period*. Jodhpur: Kusumanjali Book World, 2000.

[Gup1967] Gupta, R. C. Bhāskara I's approximation to sine. *Indian Journal of History of Science* 2, 1967, 121–136.

[Gup1969] Gupta, R. C. Second order interpolation in Indian mathematics up to the fifteenth century. *Indian Journal of History of Science* 4, 1969, 86–98.

[Gup1971] Gupta, R. C. Fractional parts of Āryabhaṭa's sines and certain rules found in Govindasvāmi's bhāṣya on the *Mahā-bhāskarīya*. *Indian Journal of History of Science* 6 (1), 1971, 51–59.

[Gup1974] Gupta, R. C. An Indian form of third order Taylor series approximation of the sine. *Historia Mathematica* 1, 1974, 287–289.

[Gup1986] Gupta, R. C. On derivation of Bhāskara I's formula for the sine. *Gaṇita Bhāratī* 8, 1986, 39–41.

[Gup1987] Gupta, R. C. South Indian achievements in medieval mathematics. *Gaṇita Bhāratī* 9, 1987, 15–40.

[Gup1992a] Gupta, R. C. Abū'l Wafā' and his Indian rule about regular polygons. *Gaṇita Bhāratī* 14, 1992, 57–61.

[Gup1992b] Gupta, R. C. Introduction. In L. C. Jain, *The Taō of Jaina Sciences*. Delhi: Arihant International, 1992, pp. viii–xvi.

[Gup1992c] Gupta, R. C. On the remainder term in the Mādhava-Leibniz series. *Gaṇita Bhāratī* 14, 1992, 68–71.

[Gup1993] Gupta, R. C. New researches in Jaina mathematics: The work of Prof. L. C. Jain. *Ṇāṇasāyara: The Ocean of Jaina Knowledge* 9, 1993, 22–27, 96.

[Gup1995] Gupta, R. C. Who invented the zero? *Gaṇita Bhāratī* 17, 1995, 45–61.

[Gup2002] Gupta, R. C. India. In Joseph W. Dauben and Christoph J. Scriba, eds., *Writing the History of Mathematics: Its Historical Development*. Basel: Birkhäuser, 2002, pp. 307–315.

[Gup2003] Gupta, R. C. Technology of using śūnya in India. In [BagSa2003], pp. 19–24.

[Gup2004a] Gupta, R. C. Area of a bow-figure in India. In [BurC2004], pp. 517–532.

[Gup2004b] Gupta, R. C. Vedic circle-square conversions: New texts and rules. *Gaṇita Bhāratī* 26, 2004, 27–39.

[GupAJM] Gupta, R. C. *Ancient Jain Mathematics*. Tempe, AZ: Jain Humanities Press, n.d.

[Hab1985] Habib, Irfan. *Medieval Technology Exchanges between India and the Islamic World* (Aligarh Oriental Series 6). Aligarh: Viveka Publications, 1985.

[Ham1987] Hamadanizadeh, Javad. A survey of medieval Islamic interpolation schemes. In [KiSa1987], pp. 143–152.

[Han1970] Hanaki, T. *Selections from Aṇuogaddārāīṃ*. Vaishali, Bihar: Prakrit Jain Research Institute, 1970.

[Hay1990] Hayashi, Takao. A new Indian rule for squaring a circle etc. *Gaṇita Bhāratī* 12, 1990, 75–82.

[Hay1991] Hayashi, Takao. A note on Bhāskara I's rational approximation to Sine. *Historia Scientiarum* 42, 1991, 45–48.

[Hay1995] Hayashi, Takao. *The Bakhshālī Manuscript: an Ancient Indian Mathematical Treatise*. Groningen: Egbert Forsten, 1995.

[Hay1997a] Hayashi, Takao. Āryabhaṭa's rule and table for sine-differences. *Historia Mathematica* 24, 1997, 396–406.

[Hay1997b] Hayashi, Takao. Calculations of the surface of a sphere in India. *Science and Engineering Review of Doshisha University* 37 (4), 1997, 194–238.

[Hay1998] Hayashi, Takao. Twenty-one algebraic normal forms of Citrabhānu. *Historia Mathematica* 25, 1998, 1–21.

[Hay2000] Hayashi, Takao. The *Caturacintāmaṇi* of Giridharabhaṭṭa: A sixteenth-century Sanskrit mathematical treatise. *SCIAMVS* 1, 2000, 133–208.

[Hay2003] Hayashi, Takao. Indian mathematics. In Helen Dunmore and Ivor Grattan-Guinness, eds., *Companion Encyclopedia of the History and Philosophy of the Mathematical Sciences*, vol. 1. Baltimore: JHU Press, 2003, pp. 118–130.

[Hay2004] Hayashi, Takao. Two Benares manuscripts of Nārāyaṇa Paṇḍita's *Bīja-gaṇitāvataṃsa*. In [BurC2004], pp. 386–516.

[HayKY1989] Hayashi, T., Kusuba, T., and Yano, M. Indian values for π derived from Āryabhaṭa's value. *Historia Scientiarum* 37, 1989, 1–16.

[HayKY1990] Hayashi, T., Kusuba, T., and Yano, M. The correction of the Mādhava series for the circumference of a circle. *Centaurus* 33 (2–3), 1990, 149–174.

[Hee2007] Heeffer, Albrecht. The tacit appropriation of Hindu algebra in Renaissance practical arithmetic. *Gaṇita Bhāratī* 29, 2007, 1–60.

[Hid2000] Hidayat, Bambang. Indo-Malay astronomy. In [Sel2000], pp. 371–384.

[Hoc1999a] Hock, Hans Henrich. Out of India? The linguistic evidence. In [BroDe1999], pp. 1–18.

[Hoc1999b] Hock, Hans Henrich. Through a glass darkly: Modern "racial" interpretations vs. textual and general prehistoric evidence on ārya and dāsa/dasyu in Vedic society. In [BroDe1999], pp. 145–174.

[Hoc2005] Hock, Hans Henrich. Philology and the historical interpretation of the Vedic texts. In [BrPa2005], pp. 282–308.

[HoSa2003] Hogendijk, Jan P., and Sabra, Abdelhamid I., eds. *The Enterprise of Science in Islam*. Cambridge, MA: MIT Press, 2003.

[Hoy2002] Høyrup, Jens. *Lengths, Widths, Surfaces: A Portrait of Old Babylonian Algebra and Its Kin*. New York: Springer-Verlag, 2002.

[HuPi1999] Hunger, Hermann, and Pingree, David. *Astral Sciences in Mesopotamia*. Leiden: Brill, 1999.

[Hut1812] Hutton, Charles. History of algebra. In Charles Hutton, *Tracts on Mathematical and Philosophical Subjects*, 3 vols. London: F.&C. Rivington, 1812, vol. 2, pp. 143–305.

[Ike2002] Ikeyama, Setsuro. *The Brāhma-sphuṭa-siddhānta Chapter 21 with the Commentary of Pṛthūdakasvāmin*. Ph.D. dissertation, Brown University, Providence, RI, 2002.

[Ike2004] Ikeyama, Setsuro. A survey of rules for computing the true daily motion of the planets in India. In [BurC2004], pp. 533–551.

[IkeSIM] Ikeyama, Setsuro, trans. T. Hayashi, T. Kusuba, and M. Yano, *Studies in Indian Mathematics: Series, Pi and Trigonometry* (in Japanese; Tokyo, 1997). Forthcoming.

[IkPl2001] Ikeyama, Setsuro, and Plofker, Kim. The *Tithicintāmaṇi* of Gaṇeśa, a medieval Indian treatise on astronomical tables. *SCIAMVS* 2, 2001, 251–289.

[Jai2001] Jain, Pushpa Kumari. *The Sūryaprakāśa of Sūryadāsa*, vol. I. Vadodara: Oriental Institute, 2001.

[Jha1949] Jhā, Acyutānanda. *The Bīja-gaṇita (Elements of Algebra) of Śrī Bhāskarāchārya* (Kashi Sanskrit Series 148). Banaras: Jaya Krishna Das Gupta, 1949.

[JonA2004] Jones, Alexander. An '*Almagest*' before Ptolemy's? In [BurC2004], pp. 129–136.

[JonAMA] Jones, Alexander. Merely an aberrant development? How much did knowledge of Babylonian astronomy affect Greek astronomy? In [PanaProc].

[JonW1807] Jones, William. *The Works of Sir William Jones*, ed. Lord Teignmouth, 13 vols. London: John Stockdale, 1807.

[JosG2000] Joseph, G. G. *The Crest of the Peacock*. Princeton: Princeton University Press, 2000.

[JosK1994] Jośī, Kedāradatta. *Graha-lāghava*. Delhi: Motilal Banarsidass, repr. 1994.

[Jus1996] Justus, Carol F. Numeracy and the German upper decades. *Journal of Indo-European Studies* 24, 1996, 45–80.

[Kad2007] Kadavny, John. Positional value and linguistic recursion. *Journal of Indian Philosophy* 35, 2007, 487–520.

[Kak1996] Kak, Subhash. Vena, Veda, Venus. *Brahmavidyā: The Adyar Library Bulletin* 60, 1996, 229–239.

[Kak2000a] Kak, Subhash. Birth and early development of Indian astronomy. In [Sel2000], pp. 303–340.

[Kak2000b] Kak, Subhash. *yamātārājabhānasalagām*: An interesting combinatoric *sūtra*. *Indian Journal of History of Science* 35, 2000, 123–127.

[Kak2005] Kak, Subhash. Vedic astronomy and early Indian chronology. In [BrPa2005], pp. 309–331.

[Kan1960] Kangle, R. P. *The Kauṭilīya Arthaśāstra*, 3 vols. Bombay: Univ. of Bombay, 1960–1965.

[Kap1937] Kāpadīā, H. R. *Gaṇita-tilaka by Śrīpati.* Baroda: Oriental Institute, 1937.

[Kat1995] Katz, Victor J. Ideas of calculus in Islam and India. *Mathematics Magazine* 68 (3), 1995, 163–174.

[Kat2007] Katz, Victor J., ed. *A Sourcebook in the Mathematics of Egypt, Mesopotamia, China, India, and Islam* Princeton NJ: Princeton University Press, 2007.

[Kel2005] Keller, Agathe. Making diagrams speak, in Bhāskara I's commentary on the *Āryabhaṭīya. Historia Mathematica* 32 (3), 2005, 275–302.

[Kel2006] Keller, Agathe. *Expounding the Mathematical Seed: A Translation of Bhāskara I on the Mathematical Chapter of the Āryabhaṭīya*, 2 vols. Basel: Birkhäuser, 2006.

[Kel2007] Keller, Agathe. Qu'est-ce que les mathématiques? les réponses taxinomiques de Bhāskara un commentateur, mathématicien et astronome indien du VIIème siècle. In Phillipe Hert and Marcel Paul-Cavalier, eds., *Sciences et Frontières: délimitations du savoir, objets et passages.* Fernelmont, Belgium: Éditions Modulaires Européennes, 2007, pp. 29-61.

[Ken1978] Kennedy, E. S. al-Bīrūnī. In [Gille1978], vol. 2, pp. 148–158.

[Ken1983a] Kennedy, E. S. An early method of successive approximations. Reprinted in [Ken1983d], pp. 541–543.

[Ken1983b] Kennedy, E. S. The history of trigonometry. Reprinted in [Ken1983d], pp. 3–29.

[Ken1983c] Kennedy, E. S. A medieval iterative algorism. Reprinted in [Ken1983d], pp. 513–516.

[Ken1983d] Kennedy, E. S., et al. *Studies in the Islamic Exact Sciences.* Beirut: American University, 1983.

[KiSa1987] King, David A., and Saliba, George, eds. *From Deferent to Equant: A Volume of Studies in the History of Science in the Ancient and Medieval Near East in Honor of E. S. Kennedy.* Annals of the New York Academy of Sciences 500. Special issue, 1987.

[Kli2005] Klintberg, Bo. Hipparchus's 3600′-based chord table and its place in the history of ancient Greek and Indian trigonometry. *Indian Journal of History of Science* 40, 2005, 169–203.

[Knu2005] Knudsen, Toke. Square roots in the *Śulba-sūtras. Indian Journal of History of Science* 40, 2005, 107–111.

[Knu2008] Knudsen, Toke. *The Siddhānta-sundara of Jñānarāja.* Ph.D. dissertation, Brown University, Providence, RI, 2008.

[Kun2003] Kunitzsch, Paul. The transmission of Hindu-Arabic numerals reconsidered. In [HoSa2003], pp. 3–21.

[Kup1957] Kuppanna Sastri, T. S. *Mahā-bhāskarīya of Bhāskarācārya with the Bhāṣya of Govindasvāmin and the Super-commentary Siddhānta-dīpikā of Parameśvara* (Madras Government Oriental Series 130). Madras: Government Oriental Manuscripts Library, 1957.

[Kus1981] Kusuba, Takanori. Brahmagupta's sutras on tri- and quadrilaterals. *Historia Scientiarum* 21, 1981, 43–55.

[Kus1993] Kusuba, Takanori. *Combinatorics and Magic Squares in India: A Study of Nārāyaṇa Paṇḍita's "Gaṇitakaumudī," chapters 13–14.* Ph.D. dissertation, Brown University, Providence, RI, 1993.

[Kus2004] Kusuba, Takanori. Indian rules for the decomposition of fractions. In [BurC2004], pp. 497–516.

[KusGeom] Kusuba, Takanori. Geometrical demonstrations in Sanskrit texts. In [PanaProc], forthcoming.

[Lal1693] de La Loubere, Simon. *A New Historical Relation of the Kingdom of Siam,* 2 vols. London, 1693.

[LamAn1992] Lam, Lay Yong, and Ang, Tian Se. *Fleeting Footsteps.* Singapore: World Scientific, 1992.

[Leg1776] Le Gentil, G. Mémoire sur l'Inde. *Mémoires de l'Académie Royale des Sciences* Part 2, 1776, pp. 169–214, 221–266.

[Lin2005] Linton, F. E. J. Shedding some localic and linguistic light on the tetralemma conundrums. In [Emc2005], pp. 63–73.

[Liu2002] Liu, Dun. A homecoming stranger: Transmission of the method of double false position and the story of Hiero's crown. In [Dol2002], pp. 157–166.

[MacKe1958] Macdonnell, A. A., and Keith, A. B. *Vedic Index of Names and Subjects*, 2 vols. Varanasi: Motilal Banarsidass, 1958.

[Mal1996] Malamoud, Charles. *Cooking the World: Ritual and Thought in Ancient India*, tr. David White. Delhi: Oxford University Press, 1996.

[Mar2000] Martzloff, Jean-Claude. Chinese mathematical astronomy. In Helaine Selin, ed., *Mathematics Across Cultures: The History of Non-Western Mathematics*. Dordrecht: Kluwer, 2000.

[Max1873] Max Müller, F. *Ṛgvedasaṃhitā*. London: Trübner & Co., 1873.

[Mer1984] Mercier, Raymond. The astronomical tables of Rājah Jai Singh Sawā'ī, *Indian Journal of History of Science* 19, 1984, 143–171.

[Min2001] Minkowski, Christopher Z. The paṇḍit as public intellectual: The controversy of *virodha* or inconsistency in the astronomical sciences. In Axel Michaels, ed., *The Paṇḍit. Proceedings of the Conference in Honour of Dr. K. P. Aithal*. Heidelberg: Sudasien Institute, 2001, pp. 79–96.

[Min2002] Minkowski, Christopher Z. Astronomers and their reasons: Working paper on jyotiḥśāstra. *Journal of Indian Philosophy* 30 (5), 2002, 495–514.

[Min2004a] Minkowski, Christopher Z. Competing cosmologies in early modern Indian astronomy. In [BurC2004], pp. 349–385.

[Min2004b] Minkowski, Christopher Z. On Sūryadāsa and the invention of bidirectional poetry. *Journal of the American Oriental Society* 124, 2004, 325–333.

[MinNil] Minkowski, Christopher Z. Nīlakaṇṭha Caturdhara and the genre of Mantrarahasyaprakāśika. In Y. Ikari, ed., *Proceedings of the Second International Vedic Workshop*. Kyoto: Institute for Research in Humanities, forthcoming.

[Mish1991] Mishra, Satyendra. *Karaṇa-kutūhala* (Krishnadas Sanskrit Series 129). Varanasi: Krishnadas Academy, 1991.

[Misr1932] Miśra, Babuāji. *The Siddhānta-śekhara of Śrīpati*, 2 vols. Calcutta: University of Calcutta, 1932, 1947.

[Muk2003] Mukherjee, B. N. Kharoshṭī numerals and the early use of decimal notation in Indian epigraphs. In [BagSa2003], pp. 87–100.

[Mur2005] Murthy, S. S. N. Number symbolism in the Vedas. *Electronic Journal of Vedic Studies* 12 (3), 2005, 87–99 (http://www.ejvs.laurasianacademy.com/issues.html).

[Nam2002] Namboodiri, V. Govindan. *Śrauta Sacrifices in Kerala* (Calicut University Sanskrit Series 13). Calicut: University of Calicut, 2002.

[NarR2007] Narasimha, Roddam. Epistemology and language in Indian astronomy and mathematics. *Journal of Indian Philosophy* 35, 2007, 521–541.

[NarP1949] Narayana Pillai, P. K. *Laghu-bhāskarīya of Bhāskara*. Trivandrum: Government Press, 1949.

[Neu1952] Neugebauer, O. Tamil astronomy. *Osiris* 10, 1952, 252–276.

[Neu1969] Neugebauer, O. *The Exact Sciences in Antiquity*. New York: Dover, repr. 1969.

[NeuPi1970] Neugebauer, O., and Pingree, D. *The Pañcasiddhāntikā of Varāhamihira*, 2 vols. Copenhagen: Danish Royal Academy, 1970–1971.

[Nor1974] North, John. The astrolabe. *Scientific American* 230 (1), 1974, 96–101.

[Oak2007] Oaks, Jeffrey A. Medieval Arabic algebra as an artificial language. *Journal of Indian Philosophy* 35, 2007, 543–575.

[Oha1985] Ōhashi, Yukio. A note on some Sanskrit manuscripts on astronomical instruments. In G. Swarup et al., eds., *History of Oriental Astronomy*, Cambridge: Cambridge University Press, pp. 191–195.

[Oha1986] Ōhashi, Yukio. Sanskrit texts on astronomical instruments during the Delhi Sultanate and Mughal periods. *Studies in History of Medicine and Science* X–XI, 1986–1987, 165–181.

[Oha2000] Ōhashi, Yukio. Remarks on the origin of Indo-Tibetan astronomy. In [Sel2000], pp. 341–369.

[Oha2002] Ōhashi, Yukio. The legends of Vasiṣṭha—a note on the Vedāṅga astronomy. In [Ans2002b], pp. 75–82.

[PanaProc] Panaino, Antonio, ed. Proceedings of a seminar on history of science in memory of David Pingree. Forthcoming.

[PandeC1981] Pāṇḍeya, Candrabhānu. *Śiṣya-dhī-vṛddhida-tantra* with commentary of Bhāskara. Varanasi: Sampurnanand Sanskrit University, 1981.

[PandeS1991] Pāṇḍeya, Śrīcandra. *Sūrya-siddhānta* (M.M. Sudhākara Dvivedi Granthamālā 2). Varanasi: Sampurnanand Sanskrit University, 1991.

[Pandi1993] Pandit, M. D. *Mathematics as Known to the Vedic Saṃhitās* (Sri Garib Das Oriental Series 169). Delhi: Sri Satguru Publications, 1993.

[Pandi2003] Pandit, M. D. Reflections on Pāṇinian zero. In [BagSa2003], pp. 101–125.

[Pani1960] Panikkar, K. M. *A History of Kerala, 1498–1801* (Annamalai University Historical Series 15). Annamalainagar (India): Annamalai University, 1960.

[PatF2004] Patte, François. *Le Siddhantasiromani: l'oeuvre mathématique et astronomique de Bhaskaracarya*. Geneva: Librarie Froz, 2004.

[PatF2005] Patte, François. The *karaṇī*: How to use integers to make accurate calculations on square roots. In [Emc2005], pp. 115–134.

[PatK2001] Patwardhan, K. S., et al., trans. *Lilavati of Bhaskaracarya: A Treatise of Mathematics of Vedic Tradition*. Delhi: Motilal Banarsidass, 2001.

[Per1999] Perrett, Roy W. History, time, and knowledge in ancient India. *History and Theory* 38 (3), 1999, 307–321.

[Pil1957] Pillai, S. K. *Āryabhaṭīya with the bhāṣya of Nīlakaṇṭha*, part 3. Trivandrum: Government Press, 1957.

[Pil1958] Pillai, S. K. *The Tantra-saṅgraha, a Work on Ganita by Gārgyakerala Nīlakaṇṭha Somasutvan* (Trivandrum Sanskrit Series 188). Quilon: Sree Rama Vilasam Press, 1958.

[Pin1967] Pingree, David. The *Paitāmaha-siddhānta* of the *Viṣṇudharmottarapurāṇa*. *Brahmavidyā* 31–32, 1967–1968, 472–510.

[Pin1968a] Pingree, David. The fragments of the works of Ya'qūb ibn Ṭāriq. *Journal of Near Eastern Studies* 26, 1968, 97–125.

[Pin1968b] Pingree, David. *Sanskrit Astronomical Tables in the United States*. Transactions of the American Philosophical Society n.s. 58 (3). Special issue, 1968.

[Pin1970a] Pingree, David. *Census of the Exact Sciences in Sanskrit*, series A, vols. 1–5. Philadelphia: American Philosophical Society, 1970–1994.

[Pin1970b] Pingree, David. The fragments of the works of al-Fazārī. *Journal of Near Eastern Studies* 28, 1970, 103–123.

[Pin1972a] Pingree, David. Precession and trepidation in Indian astronomy before A. D. 1200. *Journal for the History of Astronomy* 3, 1972, 27–35.

[Pin1972b] Pingree, David. *Sanskrit Astronomical Tables in England.* Madras: Kuppuswami Sastri Research Institute, 1972.

[Pin1973a] Pingree, David. The Greek influence on early Islamic mathematical astronomy. *Journal of the American Oriental Society* 93, 1973, 32–43.

[Pin1973b] Pingree, David. The Mesopotamian origin of early Indian mathematical astronomy. *Journal for the History of Astronomy* 4, 1973, 1–12.

[Pin1974] Pingree, David. Concentric with equant. *Archives internationales d'histoire des sciences* 24, 1974, 26–29.

[Pin1975] Pingree, David. Al-Bīrūnī's knowledge of Sanskrit astronomical texts. In Peter J. Chelkowski, ed., *The Scholar and the Saint: Studies in Commemoration of Abu'l-Rayhan al-Biruni and Jalal al-Din al-Rumi.* New York: NYU Press, 1975, pp. 67–81.

[Pin1976] Pingree, David. The recovery of early Greek astronomy from India. *Journal for the History of Astronomy* 8, 1976, 109–123.

[Pin1978a] Pingree, David. History of mathematical astronomy in India. In [Gille1978], vol. 15, pp. 533–633.

[Pin1978b] Pingree, David. Islamic astronomy in Sanskrit. *Journal for the History of Arabic Science* 2, 1978, 315–330.

[Pin1978c] Pingree, David. *The Yavanajātaka of Sphujidhvaja*, 2 vols. Cambridge, MA: Harvard University Press, 1978.

[Pin1980] Pingree, David. Reply to B. L. van der Waerden, Two treatises on Indian astronomy. *Journal for the History of Astronomy* 11, 1980, 58–61.

[Pin1981a] Pingree, David. *Jyotiḥśāstra.* Wiesbaden: Harrassowitz, 1981.

[Pin1981b] Pingree, David. A note on the calendars used in early Indian inscriptions. *Journal of the American Oriental Society* 102, 1982, 355–359.

[Pin1983] Pingree, David. Brahmagupta, Balabhadra, Pṛthūdakasvāmin and al-Bīrūnī. *Journal of the American Oriental Society* 103 (2), 1983, 353–360.

[Pin1985] Pingree, David. Power series in medieval Indian trigonometry. In Peter Gaeffke and David A. Utz, eds., *Science and Technology in India (Proceedings of the South Asia Seminar 1981/1982)*, vol. II. Philadelphia: Department of South Asia Regional Studies, University of Pennsylvania, 1985, pp. 25–30.

[Pin1987a] Pingree, David. Babylonian planetary theory in Sanskrit omen texts. In J. L. Berggren and B. R. Goldstein, eds., *From Ancient Omens to Statistical Mechanics: Essays on the Exact Sciences Presented to Asger Aaboe.* Copenhagen: Munksgaard, 1987, pp. 91–99.

[Pin1987b] Pingree, David. Indian and Islamic astronomy at Jayasiṃha's court. In [KiSa1987], pp. 313–328.

[Pin1987c] Pingree, David. Venus omens in India and Babylon. In Francesca Rochberg-Halton, ed., *Language, Literature, and History: Philological and Historical Studies Presented to Erica Reiner.* New Haven, CT: American Oriental Society, 1987, pp. 293–315.

[Pin1988] Pingree, David. *The Astronomical Works of Daśabala* (Aligarh Oriental Series 9). Aligarh: Viveka Publications, 1988.

[Pin1989] Pingree, David. MUL.APIN and Vedic astronomy. In Hermann Behrens et al., eds., *DUBU-E$_2$-DUB-BA-A. Studies in Honor of Ake W. Sjöberg.* Philadelphia: University Museum, 1989, pp. 439–445.

[Pin1990] Pingree, David. The Purāṇas and jyotiḥśāstra: Astronomy. *Journal of the American Oriental Society* 110, 1990, 274–280.

[Pin1992a] Pingree, David. Mesopotamian omens in Sanskrit. In Dominique Charpin and Francis Joannes, eds., *La circulation des biens, des personnes et des idées dans le Proche-Orient ancien, XXXVIIIe R.A.I.* Paris: Editions Recherche sur les Civilisations, 1992, pp. 375–379.

[Pin1992b] Pingree, David. On the date of the *Mahā-siddhānta* of the second Āryabhaṭa. *Gaṇita Bhāratī* 14, 1992, 55–56.

[Pin1993] Pingree, David. Āryabhaṭa, the *Paitāmaha-siddhānta*, and Greek astronomy. *Studies in History of Medicine and Science* n.s. 12 (1–2), 1993, 69–79.

[Pin1996a] Pingree, David. Bīja-corrections in Indian astronomy. *Journal for the History of Astronomy* 27, 1996, 161–172.

[Pin1996b] Pingree, David. Indian astronomy in medieval Spain. In Josep Casulleras and Julio Samsó, eds., *From Baghdad to Barcelona: Studies in the Islamic Exact Sciences in Honour of Prof. Juan Vernet*, 2 vols. Barcelona: Instituto Millás Vallicrosa, 1996, vol. 1, pp. 39–48.

[Pin1996c] Pingree, David. Indian reception of Muslim versions of Ptolemaic astronomy. In F. Jamil Ragep et al., eds., *Tradition, Transmission, Transformation.* Leiden: Brill, 1996, pp. 471–485.

[Pin1996d] Pingree, David. Translating scientific Sanskrit. In [Gar1996], pp. 105–110.

[Pin1997] Pingree, David. *From Astral Omens to Astrology: from Babylon to Bīkāner.* Rome: Istituto Italiano per l'Africa e l'Oriente, 1997.

[Pin1999] Pingree, David. An astronomer's progress. *Proceedings of the American Philosophical Society* 143, 1999, 73–85.

[Pin2000] Pingree, David. *Amṛtalaharī* of Nityānanda. *SCIAMVS* 1, 2000, 209–217.

[Pin2001] Pingree, David. Nīlakaṇṭha's planetary models. *Journal of Indian Philosophy* 29, 2001, 187–195.

[Pin2002a] Pingree, David. The *kalpa* system of time. Presented at exhibit "Destiny and design: perceptions and uses of time in South Asia," University of Pennsylvania, 15 January 2002.

[Pin2002b] Pingree, David. Philippe de La Hire at the court of Jayasiṃha. In [Ans2002b], pp. 123–131.

[Pin2002c] Pingree, David. Philippe de La Hire's planetary theories in Sanskrit. In [Dol2002], pp. 428–453.

[Pin2003a] Pingree, David. *A Descriptive Catalogue of the Sanskrit Astronomical Manuscripts Preserved at the Maharaja Man Singh II Museum in Jaipur, India.* Philadelphia: American Philosophical Society, 2003.

[Pin2003b] Pingree, David. The *Sarva-siddhānta-rāja* of Nityānanda. In [HoSa2003], pp. 269–284.

[Pin2003c] Pingree, David. Zero and the symbol for zero in early sexagesimal and decimal place-value systems. In [BagSa2003], pp. 137–141.

[Pin2004] Pingree, David. Kevalarāma's *Pañcāṅgasāraṇī.* In Vidyānivās Miśra, ed., *Padmabhūṣaṇa Professor Baladeva Upādhyāya Birth Centenary Volume* Varanasi: Sharda Niketan, 2004, pp. 725–729.

[PinAVEI] Pingree, David. Astronomy in the Vedas. *Enciclopedia Italiana*, forthcoming.

[PinEMEI] Pingree, David. Education in monasteries. *Enciclopedia Italiana*, forthcoming.

[PinGEI] Pingree, David. Gaṇita: Mathematics. *Enciclopedia Italiana*, forthcoming.

[PinILEI] Pingree, David. Indian libraries. *Enciclopedia Italiana*, forthcoming.

[PinSSEI] Pingree, David. Special "schools." *Enciclopedia Italiana*, forthcoming.

[PinVJEI] Pingree, David. Varṇas and jātis of Indian scientists and artisans. *Enciclopedia Italiana*, forthcoming.

[PinMo1989] Pingree, David, and Morrissey, Patrick. On the identification of the *yogatārās* of the Indian *nakṣatras*. *Journal for the History of Astronomy* 20, 1989, 99–119.

[Pla1790] Playfair, John. Remarks on the astronomy of the Brahmins. *Transactions of the Royal Society of Edinburgh* 2.1, 1790, 135–192. Reprinted in [Dha1971], pp. 9–69.

[Plo1996a] Plofker, Kim. An example of the secant method of iterative approximation in a fifteenth-century Sanskrit text. *Historia Mathematica* 23, 1996, 246–256.

[Plo1996b] Plofker, Kim. How to appreciate Indian techniques for deriving mathematical formulas? In Catherine Goldstein et al., eds., *L'Europe mathématique/Mathematical Europe*. Paris: Éditions de la Maison des sciences de l'homme, 1996, pp. 55–68.

[Plo2000] Plofker, Kim. The astrolabe and spherical trigonometry in medieval India. *Journal for the History of Astronomy* 31, 2000, 37–54.

[Plo2001] Plofker, Kim. The "error" in the Indian "Taylor series approximation" to the sine. *Historia Mathematica* 28, 2001, 283–295.

[Plo2002a] Plofker, Kim. Spherical trigonometry and the astronomy of the medieval Kerala school. In [Ans2002b], pp. 83–93.

[Plo2002b] Plofker, Kim. Use and transmission of iterative approximations in India and the Islamic world. In [Dol2002], pp. 167–186.

[Plo2004] Plofker, Kim. The problem of the sun's corner altitude and convergence of fixed-point iterations in medieval Indian astronomy. In [BurC2004], pp. 552–586.

[Plo2005a] Plofker, Kim. Derivation and revelation: The legitimacy of mathematical models in Indian cosmology. In T. Koetsier and L. Bergmans, eds., *Mathematics and the Divine*. Amsterdam: Elsevier, 2005, pp. 61–76.

[Plo2005b] Plofker, Kim. Relations between approximations to the Sine in Kerala mathematics. In [Emc2005], pp. 135–152.

[Plo2007a] Plofker, Kim. Euler and Indian astronomy. In Robert E. Bradley and Charles Edward Sandifer, eds., *Leonhard Euler: Life, Work and Legacy*. Amsterdam: Elsevier, 2007, pp. 147–166.

[Plo2007b] Plofker, Kim. Mathematics in India. In [Kat2007], pp. 385–514.

[Plo2008a] Plofker, Kim. Mesopotamian sexagesimal numbers in Indian arithmetic. In Micah Ross, ed., *From the Banks of the Euphrates: Studies in Honor of Alice Louise Slotsky*. Winona Lake, IN: Eisenbrauns, 2008, pp. 193–206.

[Plo2008b] Plofker, Kim. Sanskrit mathematical verse. In [Ste2008], pp. 519–536.

[Pol1996] Pollock, Sheldon. Philology, literature, translation. In [Gar1996], pp. 111–129.

[Pol1998] Pollock, Sheldon. The cosmopolitan vernacular. *Journal of Asian Studies* 57 (1), 1998, 6–37.

[Pol2000] Pollock, Sheldon. Cosmopolitan and vernacular in history. *Public Culture* 12.3, 2000, 591–625.

[Pol2001a] Pollock, Sheldon. The death of Sanskrit. *Comparative Studies in History and Society* 43.2, 2001, 392–426.

[Pol2001b] Pollock, Sheldon. The new intellectuals in seventeenth-century India. *The Indian Economic and Social History Review* 38.1, 2001, 3–31.

[Put1977] Puthenkalam, J. *Marriage and Family in Kerala*. New Delhi: Printaid, 1977.

[Rag1993] Ragep, F. J. *Naṣīr al-Dīn al-Ṭūsī's Memoir on Astronomy*, 2 vols. New York: Springer-Verlag, 1993.

[Rai2003] Raina, Dhruv. Betwixt Jesuit and Enlightenment historiography: Jean-Sylvain Bailly's History of Indian Astronomy. *Revue d'histoire des mathématiques* 9, 2003, 253–306.

[Raja1949] Rajagopal, C. T., and Venkataraman, A. Sine and cosine power series in Hindu mathematics. *Journal of the Royal Asiatic Society of Bengal* 15, 1949, 1–13.

[Raju2007] Raju, C. K. *Cultural Foundations of Mathematics: The Nature of Mathematical Proof and the Transmission of the Calculus from India to Europe in the 16th c. CE* (History of Science, Philosophy and Culture in Indian Civilization, vol. X, part 4). Delhi: Pearson Longman, 2007.

[Ram1994] Ramasubramanian, K., Srinivas, M. D., and Sriram, M. S. Modification of the earlier Indian planetary theory by the Kerala astronomers (c. 1500 AD) and the implied heliocentric picture of planetary motion. *Current Science* 66 (10), 1994, 784–790.

[Ran1912] Rangacarya, M. *Ganita-Sara-Sangraha.* Madras: Government Press, 1912.

[Rao2004] Rao, S. Balachandra. *Indian Mathematics and Astronomy: Some Landmarks*, 3rd ed. Bangalore: Bhavan's Gandhi Centre, 2004.

[RaoUm2006] Rao, S. Balachandra, and Uma, S. K. *Graha-lāghava of Gaṇeśa Daivajña: An English Exposition.* New Delhi: Indian National Science Academy, 2006.

[Rat1999] Ratnagar, Shereen. Does archaeology hold the answers? In [BroDe1999], pp. 207–238.

[Ric1997] Richards, John F. Early modern India and world history. *Journal of World History* 8 (2), 1997, 197–209.

[Rob1999] Robson, Eleanor. *Mesopotamian Mathematics, 2100–1600 BC: Technical Constants in Bureaucracy and Education.* Oxford: Clarendon Press, 1999.

[Row1992] Rowell, Lewis. *Music and Musical Thought in Early India.* Chicago: Univ. of Chicago Press, 1992.

[Sac1992] Sachau, Edward. *Alberuni's India*, 2 vols. New Delhi: Munshiram Manoharlal, repr. 1992.

[Sal1995] Salomon, Richard. On the origin of the early Indian scripts. *Journal of the American Oriental Society* 115, 1995, 271–279.

[Sal1998] Salomon, Richard. *Indian Epigraphy.* New York: Oxford University Press, 1998.

[Sara1979] Sarasvati Amma, T. A. *Geometry in Ancient and Medieval India.* Delhi: Motilal Banarsidass, 1979.

[SarE1999] Sarma, E. R. Sreekrishna. *Vedic Tradition in Kerala* (Calicut University Sanskrit Series 8). Calicut: University of Calicut, 1999.

[SarK1972] Sarma, K. V. *A History of the Kerala School of Hindu Astronomy (in Perspective)* (Vishveshvaranand Indological Series 55). Hoshiarpur: VVBIS & IS, Panjab University, 1972.

[SarK1975] Sarma, K. V. *Līlāvatī of Bhāskarācārya with Kriyā-kramakarī.* Hoshiarpur: VVBIS & IS, Panjab University, 1975.

[SarK1976a] Sarma, K. V. *Āryabhaṭīya of Āryabhaṭa with the Commentary of Sūryadeva Yajvan.* New Delhi: Indian National Science Academy, 1976.

[SarK1976b] Sarma, K. V. *Siddhāntadarpaṇa of Nīlakaṇṭha Somayāji.* Hoshiarpur: VVBIS & IS, Panjab University, 1976.

[SarK1977a] Sarma, K. V. *Jyotir-mīmāṃsā of Nīlakaṇṭha Somayāji* (Panjab University Indological Series 11). Hoshiarpur: VVBIS & IS, Panjab University, 1977.

[SarK1977b] Sarma, K. V. *Tantrasaṅgraha of Nīlakaṇṭha Somayāji.* Hoshiarpur: VVBIS & IS, Panjab University, 1977.

[SarK2002] Sarma, K. V. *Science Texts in Sanskrit in the Manuscripts Repositories of Kerala and Tamilnadu.* New Delhi: Rashtriya Sanskrit Sansthan, 2002.

[SarK2003] Sarma, K. V. Word and alphabetic numeral systems in India. In [BagSa2003], pp. 37–71.

[SarK2008] Sarma, K. V. *Gaṇita-Yukti-bhāṣa of Jyeṣṭhadeva*, 2 vols. With explanatory notes in English by K. Ramasubramanian, M.D. Srinivas, M.S. Sriram. Delhi: Hindustan Book Agency, 2008.

[SarKu1984] Sarma, K. V., and Kuppanna Sastri, T. S. *Vedāṅga-jyotiṣa* of Lagadha (in its Ṛk and Yajus recensions) with the translation and notes of T. S. Kuppanna Sastry. *Indian Journal of History of Science* 19, 1984, suppl.; repr. 1985.

[SarKu1993] Sarma, K. V., and Kuppanna Sastri, T. S. *The Pañcasiddhāntikā of Varāhamihira*, tr. T. S. Kuppanna Sastry. Madras: PPST Foundation, 1993.

[SarS1966] Sarma, Sreeramula Rajeswara. *The Pūrvagaṇita of Āryabhaṭa's <II> Mahā-siddhānta*, 2 vols. Ph.D. dissertation, Philipps Universität, Marburg, Germany, 1966.

[SarS1989] Sarma, Sreeramula Rajeswara. Vedic mathematics vs. mathematics in the Veda. *Vāṇījyotiḥ* 4, 1989, 53–65.

[SarS1992] Sarma, Sreeramula Rajeswara. Astronomical instruments in Mughal miniatures. *Studien zur Indologie und Iranistik* 16–17, 1992, pp. 235–276.

[SarS1997] Sarma, Sreeramula Rajeswara. Some medieval arithmetical tables. *Indian Journal of History of Science* 32, 1997, 191–198.

[SarS1999a] Sarma, Sreeramula Rajeswara. Kaṭapayādi notation on a Sanskrit astrolabe. *Indian Journal of History of Science* 34 (4), 1999, 273–287.

[SarS1999b] Sarma, Sreeramula Rajeswara. Yantrarāja: the astrolabe in Sanskrit. *Indian Journal of History of Science* 34 (2), 1999, 145–158.

[SarS2000] Sarma, Sreeramula Rajeswara. Sulṭān, Sūri and the astrolabe. *Indian Journal of History of Science* 35 (2), 2000, 129–147.

[SarS2002] Sarma, Sreeramula Rajeswara. Rule of Three and its variations in India. In [Dol2002], pp. 133–156.

[SarS2003] Sarma, Sreeramula Rajeswara. Śūnya in Piṅgala's Chandaḥsūtra. In [BagSa2003], pp. 126–136.

[SarS2004] Sarma, Sreeramula Rajeswara. Setting up the water clock for telling the time of marriage. In [BurC2004], pp. 302–330.

[SarS2008] Sarma, Sreeramula Rajeswara. *The Archaic and the Exotic: Studies in the History of Indian Astronomical Instruments.* New Delhi: Manohar, 2008.

[SarSBD] Sarma, Sreeramula Rajeswara. Bilingual dictionaries and the dissemination of Islamic astronomy and astrology in India. In [PanaProc], forthcoming.

[SarSAK1993] Sarma, S. R., Ansari, S. M. R., and Kulkarni, A. G. Two Mughal celestial globes. *Indian Journal of History of Science* 28 (1), 1993, 55–65.

[SasB1989] Śāstrī, Bāpū Deva. *Siddhānta-śiromaṇi* (Kashi Sanskrit Series 72), rev. ed. Varanasi: Chaukhambha Sanskrit Sansthan, 1989.

[SasK1930] Śāstrī, K. S. *Āryabhaṭīya with the Bhāṣya of Nīlakaṇṭha*, part 1 (Trivandrum Sanskrit Series 101). Trivandrum: Government Press, 1930.

[SasK1931] Śāstrī, K. S. *Āryabhaṭīya with the bhāṣya of Nīlakaṇṭha*, part 2 (Trivandrum Sanskrit Series 110). Trivandrum: Government Press, 1931.

[Sch2004] Schwartz, Randy K. Issues in the origin and development of *ḥisāb al-khaṭ'ayn* (calculation by double false position). Presented at the Eighth North African Meeting on the History of Arab Mathematics [COMHISMA 8], 2004, (http://facstaff.uindy.edu/~oaks/Biblio/COMHISMA8paper.doc).

[Sei1978] Seidenberg, A. The origin of mathematics. *Archive for History of Exact Sciences* 18, 1978, 301–342.

[Sel1997] Selin, Helaine, ed. *Encyclopaedia of the History of Science, Technology, and Medicine in Non-Western Cultures.* Boston: Kluwer Academic, 1997.

[Sel2000] Selin, Helaine, ed. *Astronomy Across Cultures: The History of Non-Western Astronomy.* Dordrecht: Kluwer, 2000.

[SenBa1983] Sen, S. N., and Bag, A. K. *The Śulbasūtras.* New Delhi: Indian National Science Academy, 1983.

[Ses2003] Sesiano, Jacques. Quadratus mirabilis. In [HoSa2003], pp. 199–233.

[Ses2004] Sesiano, Jacques. Magic squares for daily life. In [BurC2004], pp. 715–734.

[SewDi1896] Sewell, Robert, and Dikshit, S. B. *The Indian Calendar.* London: Swan Sonnenschein, 1896.

[ShaLi1999] Shaffer, Jim G., and Lichtenstein, Diane A. Migration, philology, and South Asian archaeology. In [BroDe1999], pp. 239–260.

[Sham1960] Shamasastry, R. *Kauṭilya's Arthaśāstra.* Mysore: Mysore Printing and Pub. House, 1960.

[Shar1995] Sharma, Virendra Nath. *Sawai Jai Singh and His Astronomy.* Delhi: Motilal Banarsidass, 1995.

[Shu1954] Shukla, K. S. Ācārya Jayadeva, the mathematician. *Gaṇita* 5, 1954, 1–20.

[Shu1959] Shukla, K. S. *The Pāṭī-gaṇita of Śrīdharācārya with an ancient Sanskrit commentary.* Lucknow: Lucknow University, 1959.

[Shu1970] Shukla, K. S. *Bīja-gaṇitāvataṃsa*, part 1. Lucknow: Ṛtam, 1970.

[Shu1976] Shukla, K. S. *Āryabhaṭīya of Āryabhaṭa with the Commentary of Bhāskara I and Someśvara.* New Delhi: Indian National Science Academy, 1976.

[Shu1986] Shukla, K. S. *Vaṭeśvara-siddhānta and Gola of Vaṭeśvara*, 2 vols. New Delhi: Indian National Science Academy, 1986.

[Shu1990] Shukla, K. S. *A Critical Study of the Laghu-mānasa of Mañjula.* New Delhi: Indian National Science Academy, 1990.

[ShuSa1976] Shukla, K. S., and Sarma, K. V. *Āryabhaṭīya of Āryabhaṭa.* New Delhi: Indian National Science Academy, 1976.

[Srid2005] Sridharan, R. Piṅgala and binary arithmetic. In [Emc2005], pp. 34–62.

[Srin2004a] Srinivas, M. D. Indian approach to science: The case of jyotiśśāstra. In [Rao2004], pp. 283–291.

[Srin2004b] Srinivas, M. D. Methodology of Indian mathematics. Preprint, 2004.

[Srin2005] Srinivas, M. D. Proofs in Indian mathematics. In [Emc2005], pp. 209–250.

[Sta1983] Staal, Frits, et al. *AGNI: The Vedic Ritual of the Fire-Altar*, 2 vols. Berkeley: Asian Humanities Press, 1983.

[Sta1999] Staal, Frits. Greek and Vedic geometry. *Journal of Indian Philosophy* 27, 1999, 105–127.

[Sta2006] Staal, Frits. Artificial languages across sciences and civilizations. *Journal of Indian Philosophy* 34, 2006, 89–141.

[Ste2008] Stedall, Jacqueline, et al., eds. *The Oxford Handbook of the History of Mathematics.* Oxford: Oxford University Press, 2008.

[Swa1922] Swamikannu Pillai, L. D. *An Indian Ephemeris, A.D. 700 to A.D. 1799*, 7 vols. in 8 parts. Madras: Government Press, 1922–23.

[Swa1985] Swamikannu Pillai, L. D. *Panchang and Horoscope.* Repr. New Delhi: Asian Educational Services, 1985.

[ThDv1968] Thibaut, G., and Dvivedī, Sudhākara. *The Pañchasiddhāntikā of Varāha Mihira* (Chowkhamba Sanskrit Studies 68), 2nd ed. Varanasi: Chowkhamba Sanskrit Series, 1968.

[Tie2001] Tiefenauer, Marc. *Les enfers purāṇiques d'après des morceaux choisis: description linéaire des textes sanscrits.* Lausanne: Mémoir de licence, University of Lausanne, 2001.

[Tir1992] Tīrthajī, Bhāratī Kṛṣṇa. *Vedic Mathematics*, rev. ed. Delhi: Motilal Banarsidass, 1992.

[Too1994] Toomer, G. J. *Ptolemy's Almagest.* London: Duckworth, 1984.

[TuBo2006] Tubb, Gary A., and Boose, Emery R. *Scholastic Sanskrit: A Handbook for Students.* New York: American Institute of Buddhist Studies, 2006.

[VanB2008] Van Brummelen, Glen. *Measuring the Stars: The History of Trigonometry from the Beginning to the Eve of Symbolic Algebra.* Princeton: Princeton University Press, forthcoming.

[VanN1993] Van Nooten, B. Binary numbers in Indian antiquity. *Journal of Indian Philosophy* 21, 1993, 31–50.

[Vas1962] Vasu, S. C. *The Aṣṭādhyāyī*, 2 vols. Delhi: Motilal Banarsidass, 1962.

[Wae1980] van der Waerden, B. L. Two treatises on Indian astronomy. *Journal for the History of Astronomy* 11, 1980, 50–58.

[Wilk1834] Wilkinson, Lancelot. On the use of the Siddhántas in the work of native education. *Journal of the Asiatic Society of Bengal* 3, 1834, 504–519.

[Will2005] Williams, Clemency. *Eclipse Theory in the Ancient World.* Ph. D. dissertation, Brown University, Providence, RI, 2005.

[Wit1999] Witzel, Michael. Aryan and non-Aryan names in Vedic India: Data for the linguistic situation, c. 1900–500 B.C. In [BroDe1999], pp. 337–404.

[Wit2001] Witzel, Michael. Autochthonous Aryans? The evidence from old Indian and Iranian texts. *Electronic Journal of Vedic Studies* 7 (3), 2001, 1–93 (http://www.ejvs.laurasianacademy.com/issues.html).

[Wuj1993] Wujastyk, Dominik. *Metarules of Pāṇinian Grammar* (Groningen Oriental Studies 5), 2 vols. Groningen: Egbert Forsten, 1993.

[Yan1986] Yano, Michio. Knowledge of astronomy in Sanskrit texts of architecture. *Indo-Iranian Journal* 29, 1986, 17–29.

[Yan1997] Yano, Michio. Distance of planets in Indian astronomy. In Il-Seong Nha and F. Richard Stephenson, eds., *Oriental Astronomy from Guo Shoujing to King Sejong: Proceedings of the First International Conference on Oriental Astronomy.* Seoul: Yonsei University Press, 1997, pp. 113–120.

[Yan2006] Yano, Michio. Oral and written transmission of the exact sciences in Sanskrit. *Journal of Indian Philosophy* 34, 2006, 143–160.

Index